THE ENGINEERING OF SPORT

PROCEEDINGS OF THE 1ST INTERNATIONAL CONFERENCE ON THE ENGINEERING
OF SPORT/SHEFFIELD/UNITED KINGDOM/2-4 JULY 1996

The Engineering of Sport

Edited by

STEVE HAAKE

The University of Sheffield, United Kingdom

Taylor & Francis

Taylor & Francis Group

LONDON AND NEW YORK

The texts of the various papers in this volume were set individually by typists under the supervision of either each of the authors concerned or the editor.

Published by Taylor & Francis
2 Park Square, Milton Park, Abingdon, Oxon, OX14 4RN
270 Madison Ave, New York NY 10016

Transferred to Digital Printing 2006

ISBN 90 5410 822 3
© 1996 A.A. Balkema, Rotterdam

Publisher's Note
The publisher has gone to great lengths to ensure the quality of this reprint but points out that some imperfections in the original may be apparent

Printed and bound by CPI Antony Rowe, Eastbourne

The Engineering of Sport, Haake (ed.) © 1996 Balkema, Rotterdam. ISBN 90 5410 822 3

Table of contents

Preface IX

1 *Aerodynamics*

Going faster, higher and longer in sport with CFD 3
R. K. Hanna

Cricket ball swing: A preliminary analysis using computational fluid dynamics 11
J.M.T. Penrose, D.R. Hose & E.A. Trowbridge

Effect of cyclist's posture and vicinity of another cyclist on aerodynamic drag 21
M.M. Zdravkovich, M.W. Ashcroft, S.J. Chisholm & N. Hicks

2 *Biomechanics*

The effect of an exercise regime on Lumbar Spine Curve 31
M. Hossein Alizadeh & J. Standring

Reliability of indices of anterior tibio-femoral ligamentous function in the normal 37
and anterior cruciate ligament-deficient knee
N. Gleeson, D. Rees & S. Rakowski

Musical expertise: The dynamic movement of the trombone slide 43
M. Kruger, M. Lammers, L.J. Stoner, D. Allyn & R. Fuller

Finite element analyses of additional stressing during (light) sporting events following 51
total hip arthoplasty
S. P.G. Madabhushi, A.S. Usmani, D.R. Fairbairn & M. Rajalakshmi

The influence of body weight on ski jumping performance 63
W. Müller & T.T.J. DeVaney

Comparison of the powers at the lower limb joints during walking at different velocities 71
and their significance for a possible optimal walking velocity
W. Wang, R.H. Crompton, Y. Li & M.M. Gunther

3 Design

The modern tennis racket
H.Brody
79

The design optimisation of poles for pole vaulting
S.C.Burgess
83

Shuttlecock design and development
A.J.Cooke
91

Design and prototype manufacture of a composite bicycle frame
D.Katsanis & S.M.Grove
97

The design and development of a shuttlecock hitting machine for training badminton players
at all levels of the game
J.E.Morgan
103

How to win the America's Cup: Optimum control of a yacht having an optimally designed
sail
T.Sugimoto
111

4 Instrumentation

Accuracy of kinematic data collection, filtering and numerical differentiation
P.Dabnichki, S.Aritan, M.Lauder & D.Tsirakos
119

Measuring the longitudinal force during javelin release
S.Iwnicki, P.Dabnichki & S.Aritan
125

Experimental study of the perception of body position in space
I.Kobenz, T.T.J.DeVaney, W.Müller, W.Habermann & M.Samastur
131

Instrumentation of the Concept II ergometer for optimization of the gesture of the rower
P.Pudlo, F.Barbier & J.C.Angue
137

Three-dimensional kinematic analysis of upper extremity in the soft-tennis forehand drive
L.-H.Wang, H.-W.Wu, Y.-W.Chang, F.-C.Su & K.-C.Lo
141

The mechanical analysis of kicking during physical effort
Z.Waśkiewicz
147

5 Materials in sport

Developments of manufacturing of metal matrix composites for applications in the sports
and leisure industries
M.B.Barker, A.M.Davidson & D.Regener
153

Materials in mountaineering equipment: A look at how processing and heat treatment
influences the structure and properties of aluminium alloy karibiners
J.R.Blackford
161

The effect of microstructure on the impact dynamics of a cricket bat
C.Grant & S.A.Nixon
169

Materials selection for sports equipment   175
U.G.K.Wegst & M.F.Ashby

6 Mechanics

Mechanics and design of a windsurfer mast   187
A.J.Barker & J.L.Wearing

Friction coefficient of golf balls   193
W.Gobush

A motion based virtual reality training simulator for bobsled drivers   195
R.K.Huffman & M.Hubbard

Mechanics of the modern target archery bow and arrow   205
S.H.Joseph & S.Stewart

The fatigue life of nylon monofilament as fishing gear material   211
H.Kanehiro, S.Takeda, T.Kakihara & K.Satoh

Bicycle chain efficiency   217
M.D.Kidd, N.E.Loch & R.L.Reuben

The dynamic response of a golf club head   221
J.S.B.Mather & J.Immohr

The development of protection systems for rock climbing   229
R.A.Smith

Experimental mechanics and artificial turf   239
C.A.Walker

7 Modelling of sport

Parametric modelling of the dynamic performance of a cricket bat   245
C.Grant & S.A.Nixon

Normal impact models for golf balls   251
S.H.Johnson & B.B.Lieberman

A proposed mechanical model for measuring propulsive forces in front crawl swimming   257
M.A.Lauder & P.Dabnichki

Analyzing championship squash match-play: In search of a system description   263
T.McGarry, M.A.Khan & I.M.Franks

Derivation of a rope behaviour model for the analysis of forces developed during a rock   271
climbing leader fall
M.Pavier

Symbolic dynamics for motion analysis in sports   281
A.J.Subic & S.B.Preston

8 *Motion analysis*

Static and dynamic accuracy determination of a three-dimensional motion analysis system   289
L.W.Alaways, M.Hubbard, T.M.Conlan & J.A.Miles

3-D kinematic analysis of the forward stroke of white-water paddlers using   297
a paddle-ergometer
A.Kranzl, J.Kollmitzer & E.B.Zwick

The accuracy of kinematic data collected from underwater three-dimensional analysis   303
M.A.Lauder, P.Dabnichki, R.M.Bartlett & S.Aritan

9 *Vibration analysis*

Vibrations on the golf course   315
E.A.Ekstrom

The validation and updating of dynamic models of golf clubs   323
M.I.Friswell, G.Horwood, M.G.Smart & S.M.Hamblyn

Engineering 'feel' in the design of golf clubs   333
A.Hocknell, R.Jones & S.Rothberg

Cricket bat design and analysis through impact vibration modelling   339
S.Knowles, J.S.B.Mather & R.Brooks

Author index   347

The Engineering of Sport, Haake (ed.) © 1996 Balkema, Rotterdam. ISBN 90 5410 822 3

Preface

Sports technology encompasses a wide range of disciplines from biomechanics to engineering. As a consequence, researchers working in sports technology find that their work falls neither in the field of traditional sports science nor in mainstream engineering. Finding a suitable forum for publishing this kind of work becomes very difficult. The Engineering of Sport fulfils this role by bringing together workers from sports science, engineering physics, and indeed all scientific disciplines, looking at the use of technology in sports and leisure.

The work found in this book represents papers that were presented at the 1st International Conference on the Engineering of Sport held in Sheffield, UK in July 1996. The papers are divided into themes; aerodynamics, biomechanics, design, instrumentation, materials, mechanics, modelling, motion analysis and vibration analysis. The variety of sports is wide and shows the extent to which technology and science is used to improve and enhance many of our leisure activities. You will see that the authors of the papers come from around the world. Sport, as with science, crosses all international boundaries.

1 Aerodynamics

The Engineering of Sport, Haake (ed.)© 1996 Balkema, Rotterdam. ISBN 90 5410 822 3

Going faster, higher and longer in sport with CFD

R.K. Hanna
Fluent Europe Ltd, Sheffield, UK

ABSTRACT

Digital computers have increased dramatically in both speed and capacity over the last 5-10 years. More and more powerful machines are being used routinely on the desks of design engineers. In parallel with this growth of computer platform power and usage, more sophisticated software is being developed and commercially marketed to exploit these advances. One such area is Computational Fluid Dynamics (CFD) where the fundamental non-linear differential equations that describe fluid flow, heat transfer and turbulence phenomena are now routinely solved numerically for a range of industrial applications from hypersonic flows around rockets to low speed cooling of electronic components in computers.

Some sports and sporting teams have been keeping abreast of the rapid rise of CFD technology over recent years. In the vanguard has been motor racing teams such as the Benetton Formula 1 team who started to use CFD as a design tool to improve the aerodynamics of their championship-winning cars four years ago. Very soon other teams followed suit and a new era in F1 design was ushered in. Now, a number of teams use CFD to model their cars computationally prior to wind tunnel testing to both optimise their proposed designs and understand the flow fields around new chassis configurations. CFD is proving to be more cost effective than conventional approaches. It also provides a wealth of data for aerodynamicists to analyse and produces rapid project turnaround times in this highly competitive field of sport. Now ten configurations can be studied in the same time as two would have been looked at before.

Areas of sport that are showing the benefits of computational methods for improving equipment design and techniques are yachts (both the hull and sail designs), ski-jumpers, golf and even the humble frisbee. With reference to the above examples, this paper aims to highlight the approach to these problems designers have adopted using software supplied by Fluent Europe Ltd, the market leader in commercial CFD software. Nowadays, many designers and aerodynamicists can ask the "what if …?" question and get answers before a ball is struck, a yacht built or a racing car model constructed.

1 INTRODUCTION

This paper seeks to provide an overview of the current state of Computational Fluid Dynamics (CFD) as applied to the field of sports engineering. It is by no means exhaustive and since CFD is still developing as a technology no claim is made that the complete spectrum of possible sporting applications for CFD is covered. However, an attempt is made to map out the territory of CFD in sport by way of examples related to the commercial code FLUENT™. The current limitations of CFD technology will be discussed together with suggestions for possible future areas of application for CFD in sport.

As the title of this paper suggests, the old Olympian ideal of going faster, jumping longer and leaping higher than the opposition is embedded within competitive sports. The difference between winning and losing in sport may be fractions of a second which in turn may be related to a number of factors including the physique (even posture) of the athlete, the skill of the athlete and the equipment being used. With global media interest in all types of sport and the multi-million pound industries leading sports support it was inevitable that science and engineering technologies would be applied systematically to a variety of sports to give the leading competitors that extra winning edge.

Computational Fluid Dynamics deals with the computer simulation of aerodynamics and hydrodynamics of bodies in the presence of moving fluids. Historically, wind tunnel and water modelling techniques have dominated the field of aircraft, ship and chemical industry equipment research and development. These techniques have had the drawbacks however of being expensive to run; there are difficulties with generating good experimental measurements; they involve time consuming tasks and they have usually been limited to the more sophisticated government, university and industrial laboratories around the world.

The basic mathematical equations that describe fluid flow, the transfer of heat and some turbulence phenomena in fluids have been known in the scientific community for a long time (Versteeg et al, 1995). It has only been with the rapid advances in digital computers over the last 30 years that a numerical solution of these fundamental non-linear differential equations has been attempted (Patankar, 1980). The general class of equations commonly referred to as the "Navier-Stokes equations" describe the behaviour of a Newtonian fluid. These equations can, however, be written in many different forms; differential, integral, 2D, 3D, axisymmetric, stationary-grid, moving grid, laminar, turbulent, etc. A representative selection of the formulations for these governing equations are presented below in integral form.

Navier-Stokes Equation

Conservation of Mass

$$\frac{\partial}{\partial t}\iiint \rho \, dV + \iint \rho \upsilon_i dA_i = 0$$

Conservation of Momentum

$$\frac{\partial}{\partial t}\iiint \rho \upsilon_i dV + \iint \rho \upsilon_j \upsilon_i dA_i + \iint \left(p\delta_{ij} + \tau_{ij}\right) dA_i = \iiint \rho g_i \, dV$$

Conservation of Energy

$$\frac{\partial}{\partial t}\iiint \rho E \, dV + \iint \rho E \upsilon_i dA_i + \iint \left(\left(p\delta_{ij} + \tau_{ij}\right)\upsilon_j + q_i\right) dA_i = 0$$

Conservation of Turbulent Kinetic Energy

$$\frac{\partial}{\partial t}\iiint \rho k \, dV + \iint \rho k \upsilon_i \, dA_i - \iint \frac{\mu_e}{\sigma_k}\frac{\partial k}{\partial \chi_i} dA_i = \iiint (P - \rho\tilde{\varepsilon}) dV$$

Conservation of Turbulent Dissipation

$$\frac{\partial}{\partial t}\iiint \rho \varepsilon \, dV + \iint \rho \varepsilon \upsilon_i dA_i - \iint \frac{\mu_e}{\sigma_\varepsilon}\frac{\partial \varepsilon}{\partial \chi_i} dA_i = \iiint \left[\frac{\varepsilon}{k}(C_1 P - C_2 \rho \varepsilon) - R\right] dV$$

where, ρ = fluid density, t = time, V = cell volume, A = cell interface area, p = local static pressure, υ = velocity, τ = fluid shear stress, g = gravitational constant, E = total energy of the fluid, δ_{ij} = Kronecker Delta, q = heat flux source, k = turbulent kinetic energy, μ_e = fluid effective viscosity, P = production of turbulence, ε = dissipation of turbulent kinetic energy, σ_k, σ_ε, C_1, and C_2 are empirical constants, R = source term for turbulent dissipation and i, j, are vector notations.

It is not the purpose of this paper to deal with the detailed mathematics associated with CFD. The reader is referred to Weiss et al, 1994, for more information on the techniques used in some of the applications reported here.

Modern computers have the ability to solve millions of mathematical expressions per second and this has been one of the keys to the emergence of CFD and commercial codes over the last ten years. Another driving force for CFD has been the rapid development of Computer Aided Design (CAD) software, structural stress analysis codes and even automotive crash dummy computer simulations. Both stress analysis and crash dummy modelling rely on computational grids as does CFD. Grids or meshes are made up of finite blocks of physical space (topologically usually quadrilateral, triangular, hexahedral, tetrahedral or even mixed elements) that break up real fluid space around an object into smaller imaginary computational elements (see Figure 1).

Figure 1. Unstructured Triangular Grid around
Formula 1 Car Rear Wing Aerofoils.

Since numerical solution of the equations that
describe flow and heat transfer in a fluid can be
carried out computationally on these grids, CFD
thus provides grid cell by grid cell predictions of
local fluid velocities, pressures, temperatures etc in
the presence of an aerodynamic body simulated
within the grid as shown below.

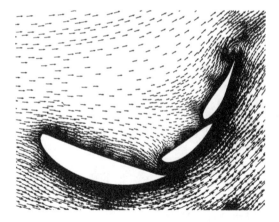

Figure 2. Predicted CFD Flow Field in the Grid
shown in Figure 1.

The CFD user therefore effectively has a
computational wind tunnel or water model that will
simulate fluid flow around any body shape that the
design engineer can conceive of. As a
consequence, CFD has taken the chemical,
automotive, aerospace and power generation
industries by storm over the last ten years. The
market for CFD is growing rapidly worldwide and
it is projected to grow for the next ten years at
least. Many large and small scale industrial

companies around the world have benefited from CFD
as a design tool in tandem with traditional laboratory
and empirical methods. CFD has also led to an
increase in designers' understanding of the detailed
flow phenomena occurring in complex industrial
processes. This in turn has resulted in huge cost
savings in terms of more efficient modifications to
existing equipment as well as a need for fewer
prototypes in the critical early stages of design.
Eliminating bad designs at the research stage will
remove the need for costly modifications and
corrections towards the end of a product's
development cycle.

2 EXAMPLES OF CFD IN SPORT

Figure 3 shows a picture of Johnny Herbert celebrating
victory in the 1995 British Grand Prix in a Benetton
Formula 1 racing car. Motor racing is a global big
business and Formula 1 is perhaps the pinnacle of the
sport. With a worldwide audience of 500,000,000
each race and a contribution of 3 billion pounds to the
British economy alone each year, it is a significant
industry in this country. Many of the top teams are
based in the UK and over the years Formula 1 has
proven to be a high-tech crucible for automotive
research and development. The fastest cars reach
speeds of 200 mph and since the governing Technical
Regulations for the sport include tight guidelines on
vehicle size and width, aerodynamics plays a major
role in the competitiveness of these vehicles. The
competing teams are allowed to modify the basic
shape of the car by using wings, flaps and boards to
keep the optimum balance between downforce (needed
for cornering and keeping the car from lifting off the
ground) and drag (which slows the car down on the
straights). This balancing act is the preserve of the
aerodynamicists in the Formula 1 teams. Nearly all
teams have access to wind tunnels with simulated
moving ground planes (in the form of conveyor belts)
under small-scale models of the actual racing car. By
measuring drag, lift and pitching forces in the model
for a range of wing heights, angles of attack and multi-
wing configurations, the optimum arrangement for a
given racetrack can be found. Formula 1 teams have
to retune the wing arrangements for each Grand Prix
throughout the season depending on the particular
track they are racing at that weekend. Timescales are
therefore very tight and it is essential to try as many
permutations of wing and body shapes as possible
between races during the season in the search for the
elusive fractions of a second per lap a good car chassis
will bring. In the close season (and towards the end of

5

the current season) next year's car has to be designed, built and tested. More radical season-to-season design modifications can be experimented with such as new internal radiator cooling duct systems, air-box shapes and even engine designs.

The Benetton Formula 1 team was the first racing team to approach commercial CFD vendors in 1992 to see if significant competitive advantage could be gained by this technique. An unstructured CFD code supplied by FLUENT was chosen and after an extensive validation period it was used to improve the rear wing design of the race car as well as the radiator side pod channels. Very quickly they found they could analyse ten wing configurations in the same time that they would have looked at two in the past. They would then choose the best two or three designs from the trends given by the CFD analysis and test them in the wind tunnel (which they were still running at the same level of testing as previously). In effect the aerodynamic design was being optimised *a priori* of the wind tunnel in a cost effective way since wind tunnel models are quite expensive. The

best wind tunnel configurations still have to be tested on the track by the drivers before their race debut.

Figure 4 helps to illustrate the sequence of events in a typical CFD simulation of a Benetton F1 car. First, a CAD definition of the car's geometry is created in terms of points, lines, curves and surfaces relative to some spatial origin as shown by the nose region of the car in the figure. Next, a computational grid is created in the fluid phase around the car surface. A typical triangular surface mesh is illustrated by the cockpit region of Figure 4. Suitable boundary conditions (the velocity of the car, the wheels and the ground plane that effectively dictate the flow around the vehicle) are given to the CFD code and the flow in the grid solved.

The rear of the Benetton car in Figure 4 illustrates a CFD prediction for airflow streamlines over, under and around the car body. Contours of static pressure on the walls of the car, the wheel and the rear wing are also shown in the figure.

These surface pressures will obviously influence the balance of forces that the car experiences moving at

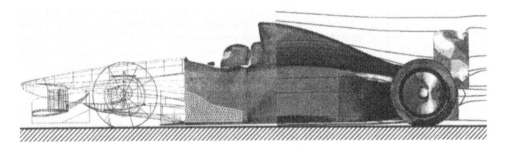

Figure 3. Johnny Herbert Celebrating Victory in the 1995 British Grand Prix.

Figure 4. The Simulation Process From CAD-GRID-CFD.

the simulated speed of 180 mph. A number of motor racing teams are currently using CFD technology routinely to develop more aerodynamic shapes for their race cars. Turnaround times for simulations are high (of the order of one day for a quarter of a million computational cells) especially for teams like Benetton who run their Hewlett Packard computers in parallel on their network overnight.

A different type of sporting application is shown in Figure 5 which summarises some CFD work carried out in 1992 (Spragle). At that time the aerodynamics of ski jumping was being explored when some skiers adopted the so-called "V" style jumping technique (Figure 5(a)) rather than the classical parallel ski position (Figure 5(b)). The "V" style jumpers were averaging about 10 metres more distance (10% increase) than jumpers using the classical style. Although the CFD model geometries were crude and the calculations inviscid the comparison showed the proper trends. The "V" style configuration generated approximately 8% more lift than the classical style and had slightly lower drag. The majority of the extra lift contribution was seen to be due to the position of the skis (after analysis of individual forces on the jumper and the skis in the CFD predictions) since

the lift and drag forces on the jumper were approximately the same in each case. Note also the larger wake region behind the jumper using the classical ski arrangement as opposed to the "V" style jumper.

Another area of sport where CFD is increasingly playing a significant role is that of yachting. The America's Cup race is one of the most prestigious sporting events in the world and has been dominated by high-tech (and highly secretive) hull designs over the last 20 years. Considerable expense goes into developing these ocean-going yachts and the recent cup-winning yacht from New Zealand was known to have used CFD techniques in the development of its hull. Commercial and military submarine builders already use CFD codes extensively in the preliminary design of hulls. It is certainly well within the realms of existing CFD technology to model multi-hull yachts and stabilisers. Figure 6 illustrates a surface grid from a study going on in the U.S. at present looking at various sail configurations for ocean racing yachts experiencing winds coming from a range of directions. Again, CFD is proving to be invaluable at the early stage of trend analysis prior to prototype testing. It has permitted the study of sail/sail flow interactions which can cause serious loss of forward momentum on a yacht.

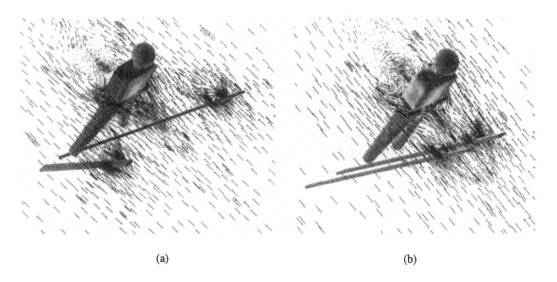

(a)   (b)

Figure 5. CFD Simulation of (a), "V" Style and (b), "Classical" Style Ski Jumpers.

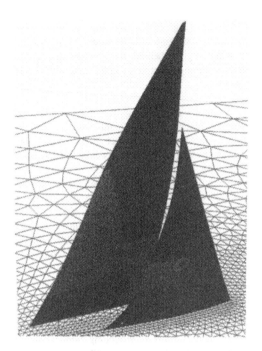

Figure 6. Surface Grid of a Multi-Sail Yacht Configuration used in a Parallel CFD Simulation.

Figure 7. CFD simulation of a Golf Driver Moving at 100 mph.

In a different vein, Figure 7 shows a CFD prediction of flow over a left-handed driver during the early part of a golfer's swing. Here, a club speed of 100 mph has been modelled. This particular design shows some flow separation at the rear of the top surface of the club head. This would lead to greater drag as the club head moves through the air and slower momentum of the club on impact with the ball. Consequently, the ball would not be propelled as far as it could be with a more aerodynamic club head. It is also likely that this detached flow region would lead to the phenomenon of vortex shedding during the swing. This in turn might lead to "flutter" or a slight lateral movement of the club head just prior to impact with the ball leading to a less-than-sweet strike. This analysis took one day to complete.

Finally, a rather innocuous example of a neo-sporting CFD application. A rotating frisbee is shown in Figure 8. Although it appears to be a simple body to model; in reality a rotating object produces asymmetric forces on the surface of the projectile causing a curved trajectory to occur dependent on the forward speed of the frisbee and the prevailing gravitational and wind forces. In the example shown, it is more correct to say that the picture describes a simulation of a stationary frisbee rotating about its axis in a uniform cross-flow similar to that it might experience in a wind tunnel for instance. This is not a "real world" simulation but nevertheless Figure 8 highlights the asymmetric pressure distribution on the surface of the frisbee and the rolling vortice on the underside of the leading edge of the rotating body. Clearly, this sort of work could have implications in terms of the discus throwing event in the Olympics and might lead to suggestions on what is the best angle to tilt the discus on release.

3   DISCUSSION, CONCLUSIONS AND FUTURE DIRECTIONS FOR CFD IN SPORT

CFD is already being used in a large-scale way to understand sporting aerodynamics in some sports as shown above. Other sports are relatively ignorant of its existence and the potential advantages to be gained from its use. Certainly, the leading edge of CFD lies with unstructured codes that have allowed realistic geometries to be modelled in timescales of days and weeks rather than months and years. This has led to parametric design studies on computers of both racing cars

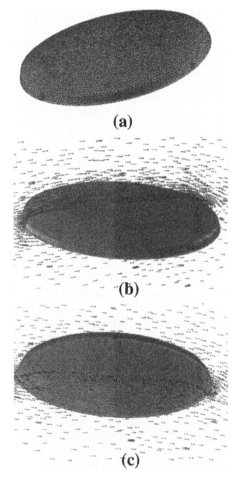

(a)

(b)

(c)

Figure 8. Surface Grid (a), Top Side (b), and
Under Side (c), Velocity Predictions for
a Spinning Frisbee in a Cross-Flow.

notoriously difficult to predict accurately in CFD (Launder, 1991). The best we can expect at present is to achieve results within 90% of experimental measurements. However, there is a wide body of evidence to show that CFD at present is good at picking up trends correctly in parametric aerodynamic studies. For a number of applications this can be good enough to exploit for commercial (or sporting) gain or at least as a precursor to follow-up wind tunnel and/or water modelling tests and eventually real life testing.

One area of the application of CFD to industrial (and sporting) applications that needs further research is turbulence modelling (Launder, 1991, Hinze, 1975). Science still does not have a complete understanding of turbulence phenomena which are related to chaotic events and Chaos Theory. We can for instance pick up macro effects and trends quite accurately as witnessed by national weather reporting, but the finer details or micro effects can be predicted very poorly. Sometimes, these micro effects can be the starting points for major flow changes in a system and to ignore them (or to miss them in existing models) is to inaccurately predict the physical flow processes from a CFD point of view. Commercial CFD has relied on empirical turbulence models such as the k-ε, RNG k-ε, and RSM models for a number of years. These have been adopted industrially for pragmatic reasons in terms of their ease of use and ability to glean some information and insights into industrial processes rather than none at all. However, they are inadequate in terms of predicting the onset of transitional turbulent instabilities when flow moves over bodies as undoubtedly occurs in the movement of sporting balls through the air for instance. In the author's view, CFD is not mature enough to deal with sports ball modelling in anything but a rudimentary way not least because of allied spinning effects during flight and the influence of atmospheric humidity in the fluid that the ball is moving through.

Perhaps we will see massive advances in turbulence modelling in CFD over the next 10 years. It is difficult to predict what way the technology will be driven. Some advocate Large Eddy Simulation methods, some even envisage modelling of the movement of individual molecules in fluids with huge computational resources (Direct Numerical Simulation). The real difficulty will be predicting random flow perturbations that set in motion complex turbulent phenomena. These will be the "Holy Grail" of CFD for the foreseeable future.

and America's Cup yachts as a matter of course. To do realistic aerodynamic CFD modelling still requires a significant allocation of computational resources beyond that available from PC s for instance. But rapid advances in computational hardware is making the technology accessible to a wider audience. Despite this, in the author's view only a computer resource with sufficient memory to run over a quarter of a million cells is acceptable for detailed viscous CFD simulations of aerodynamic bodies. This is due largely to the fact that boundary layers which are usually physically thin around moving bodies have to be resolved accurately in CFD predictions as indeed do wake effects, shear layers and vortex shedding. These physical phenomena are

In terms of sports that could benefit from CFD at present apart from motor-racing, ski-jumping, yachting and golf, the author believes the following would be interesting applications to consider. Much depends on rules and regulations and design envelopes for equipment in the particular sport:

1. The discus and javelin events in the Olympics could be analysed with CFD. A study on the respective projectiles for various angles of attack of the oncoming flow and rotational speeds might reveal the optimum throwing conditions.
2. Bob-sleigh and Luge designs could be studied with CFD for these Winter Olympic events. It would be interesting to look at different nose shapes for bob-sleighs in a similar vein to Formula 1 cars. Is there also an optimum posture/height/position of athlete for each sport?
3. Cyclists have already shown the benefit of several aerodynamic modifications to helmets and bikes in terms of their improved race times. Some more exotic modifications could be tested with CFD prior to wind tunnel and real life testing. Similarly, motorbike design modifications could be tested computationally.
4. Surf boards are more of a challenge for existing CFD technology because of free surface and multi-phase fluid effects. Nevertheless, various fin/board configurations could be tested with CFD.
5. Is there an optimum physique/running posture for a track athlete in a given event? The world 400m champion, Michael Johnson, has a very upright running style. Does this contribute (from an aerodynamic point of view) to his success?
6. The shape of racquet heads used in tennis, squash and badminton could benefit from a CFD analysis. If the frames were more aerofoil - like in cross-section would they move faster through the air and strike their respective balls harder? Similarly, could their handles be redesigned for better movement through the air to minimise drag? Does the head have to be flat? Would there be advantages to a curved profile?

There are probably many other applications that CFD could be applied to in sport with a bit of lateral thinking. The surface has barely been scratched in terms of useful applications of CFD in the arena of sport. What is certain in the future is that CFD will become a tool used increasingly within sport to achieve that extra edge needed for winning.

ACKNOWLEDGEMENTS

The author is indebted to Ross Brawn of Benetton Formula Ltd for permission to publish the Formula 1 information in this paper and the assistance of Dipankar Choudhury, Bart Patel and Ren Liu from the Fluent group of Companies. The patience of Robert Palmer who loaned the author the golf club used in one of the simulations is also very much appreciated. The diligence of Claire Jones and Diane Amos in the preparation of this manuscript is also acknowledged with gratitude.

REFERENCES

Hinze, J.O, "Turbulence", McGraw-Hill, N.Y., 1975.

Launder, B.E., "Current Capabilities for Modelling Turbulence in Industrial Flows", J. of Appl. Sci.Res., Vol 48, pp247-269, 1991.

Pantankar, S.V. "Numerical Heat Transfer and Fluid Flow", Hemisphere, Washington D.C., 1980.

Spragle, G.S., "RAMPANT Ski Jumper", Internet location http://www.fluent.com/ Applications/RAMPANT/ski/skiier.html.

Versteeg, H.K and Malaskera, W.," An Introduction to Computational Fluid Dynamics, The Finite Volume Method", Longman, 1995.

Weiss, J.M. and Smith, W.A., "Preconditioning Applied to Variable and Constant Density Time-Accurate Flows on Unstructured Meshes". AIAA Report No. 94-2209.

The Engineering of Sport, Haake (ed.)© 1996 Balkema, Rotterdam. ISBN 90 5410 822 3

Cricket ball swing: A preliminary analysis using computational fluid dynamics

J.M.T. Penrose, D.R. Hose & E.A. Trowbridge

Department of Medical Physics and Clinical Engineering, The University of Sheffield, UK

ABSTRACT : A cricket ball is said to swing when it moves in a plane that is perpendicular to the vertical plane of projection. Cricketers have observed this phenomenon for at least 100 years yet even today, the conditions and techniques required to exploit swing effectively are still hotly debated by cricket practitioners. Studies have been performed in wind tunnels[1,2,4], where force imbalances leading to swing have been demonstrated. Such experiments, however, are expensive, have large numbers of different variable parameters, and reveal little about the actual mechanisms of swing.

In this study the air flow patterns around a cricket ball were analysed by using Computational Fluid Dynamics (CFD) software (FLOTRAN). An isothermal, incompressible, Newtonian fluid model was used throughout. Techniques based on aerofoil analysis were used to investigate the effect of Reynolds number and the angle of incidence of the seam at delivery. Initially an idealised 2D model (infinite cylinder representation) was used. Both steady-state (turbulent) and transient (laminar) flow solutions were obtained. Then using this experience, 3D spherical models were developed.

Visualisation of the flow patterns was achieved by colour coded contour maps. Pressure differences between the two sides of the ball were identified and were accompanied by asymmetric boundary layer separation. Transient solutions of velocity vectors and speed contours were animated to show asymmetric flow development and vortex shedding. The calculated lift and drag coefficients compared favourably with experimental data[2,4,7]. The topical phenomenon of 'reverse swing' was also observed.

CFD provides a detailed view of the airflow and turbulence around a cricket ball in flight. Despite the fact that these models need to be validated externally, they do shed light on the mechanisms of swing. Indeed this approach has the potential for analysing 'form drag' in other sports where speed and drag are intimately connected.

1. INTRODUCTION

The phenomenon of cricket ball swing has been prominent in the game, probably ever since the main seam was introduced on the ball. In the modern game, the combination of swing and its control have become arguably the most potent weapon in the bowler's armoury[8]: any batsman will tell you that trying to strike a ball which is curving in mid-flight either towards or away from him is much harder than trying to strike a straight delivery.

Precisely how to produce this combination, however, has been a source of much debate amongst bowlers. In order to make the ball swing, that is make it move in a horizontal plane perpendicular to the vertical plane of projection, the bowler tries to deliver the ball so that the seam remains in a vertical plane throughout its flight. The bowler may also try to release the ball with the seam at an angle to a line drawn between himself and the batsman (See Figure 1). In practice, some degree of backspin is also imparted to the ball.

It is agreed that whilst atmospheric conditions undoubtedly play a part in the swing phenomenon[2], it is the precise condition of the seam and surface of the ball which chiefly governs the amount of swing a bowler can create. Bowlers will often be seen trying to shine one side of the ball on their clothing, with sweat as polish, whilst leaving the other to get dull and roughened. This helps to create an imbalance in the airflow over the two sides of the ball, which in turn leads to an asymmetry in pressure on the two sides of the ball, and thus a resultant 'swing' force. When the bowler delivers the ball with the seam pointing to, say, the left of the batsman and the shiny side on the right, the seam itself 'exaggerates' the flow imbalance, and the result is the ball swinging in the direction of the seam to the left (see Figure 1). This is known as creating orthodox swing[1,3] - swing in the direction the seam is pointing - and is perfectly legal under the laws of the game.

However, in recent years the quest for more swing as led to accusations that players are breaking the rules by deliberately roughing and scarring one side of the ball, as well as 'lifting' the seam to make it more prominent. This ball tampering has also led to the 'discovery' of reverse swing[1,3,8]. This happens when the altered ball is delivered at high speed, with the seam pointing in the same direction as before, but with the shiny side on the opposite side to that found in orthodox swing (see Figure 1). The result is the ball swinging unexpectedly in the opposite direction to that observed in the orthodox motion.

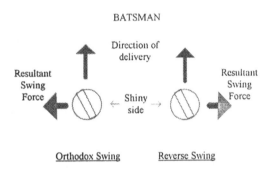

Figure 1.

Experiments have been carried out in wind tunnels for many years now and have successfully recreated swing[1,2,4]. However, with such setups it is difficult to control precisely the atmospheric, seam angle and ball condition parameters, and thus they are limited in their use for the evaluation of exact swing mechanisms.

Over the last few years, advances in software and computing power have allowed aerodynamic experiments to move away from the wind tunnel and into the world of computer simulation. Computational finite element analysis of a dynamic fluid like air utilises software which numerically solves the Navier-Stokes and continuity equations across an elemental mesh[5,7]. This is done using an iterative process similar to the Newton-Raphson method. Provided that the problem is approached

properly and that the mesh is fine enough, accurate representations of the velocity and pressure distributions in the fluid can be calculated for a steady-state flow situation or as part of a transient situation. It is also now possible to simulate turbulence as part of the computational process using the k-epsilon model[12], and so facilitate accurate simulation of high Reynolds number systems.

The aim of the present study is to use this computational approach, evaluate the precise airflow over the surface of a cricket ball and perhaps give greater insight into the mechanisms of cricket ball swing.

2. MATERIALS AND METHODS

2.1 Software and Hardware

The software used was a finite element analysis (FEA) system called ANSYS 5.8a which is packaged with a computational fluid dynamics (CFD) module called FLOTRAN. The ANSYS system contains modules for designing the finite element mesh model and for analysing and post-processing the results. The actual solution calculations are performed by the FLOTRAN module using data files created by ANSYS.

The software was run on two separate platforms: firstly a 486DX2 66Mhz PC with 512Mb hard disk and 8Mb RAM, and then latterly a Silicon Graphics Indigo workstation with 1.8Gb hard disk and 48Mb RAM. Most of the pre and solution processing was performed on the PC whilst the SGI machine was used predominantly for post-processing, and graphical output.

2.2 Dimensional Analysis

In order to reduce the number and complexity of the variables in this problem, a technique called Dimensional Analysis[7] was utilised. The force F on a particular body immersed in a fluid is a function of the fluid density ρ, the free stream velocity V, the fluid viscosity μ and the characteristic body length L.
ie. $F = f(L, V, \mu, \rho)$

With dimensional analysis, the equation can be reduced to an equivalent form where F is simply a function of the dimensionless Reynolds Number Re, where Re $= (\rho VL)/\mu$. That is, the dimensionless force coefficient $C_F = g(Re)$. Thus, the system can be evaluated by controlling only one variable, Re.

Another benefit of the dimensional analysis is that is has dynamic similarity and scaling properties. The force data evaluated from one model at a particular Reynolds Number can be used to directly produce force data on another similar model of different scale, provided that the Reynolds Number is the same. Thus the dimensions of the model can be 'chosen' to assist the computational control in the problem.

In this study, it was convenient to choose the density of the fluid ρ, the free stream velocity V and the characteristic length L (the ball diameter) to all equal unity. Thus the Reynolds Number of the system could then be controlled by simply varying the fluid viscosity directly, where $Re=1/\mu$.

2.3 Finite Element Mesh Design

The design of the elemental mesh is fundamental to the analysis. A poorly designed mesh, will inevitably produce poor results. In this study, it was advantageous to begin by designing a 2-dimensional cylinder based model. We based the design upon one used to analyse airfoils, known as an 'O-mesh'. It is an efficient mesh that uses solely quadrilateral shaped elements and is as homogeneous as possible about the ball. The mesh density is increased towards the ball surface and seam areas as this is where there is the greatest change in the local flow parameters (see Figure 2). The seam itself was modelled to be the equivalent of 2cm wide, protruding a maximum of 0.9mm from the ball surface, to conform to the laws of the game.

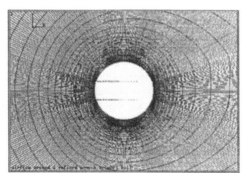

Figure 2. The finite element 'O-mesh' in the region of the ball. The seam lies in the direction of the x-axis and is marked for clarity.

The area of fluid modelled should be large enough so that the flow effects produced by the ball cannot be interfered with by the specified outer boundary conditions. This would lead to software convergence problems, and less accurate results. In this study, the model extended from 15 ball diameters in front of the

ball to 50 diameters behind it, with 30 diameters fluid area either side (see Figure 3).

The inlet of the model was considered to be all of the outer boundary lines that lay in front of the ball, and also those which lay up to 25 ball diameters behind it. The outlet of the model was considered to be all of the outer boundary lines that lay beyond 25 ball diameters behind the ball (see Figure 3).

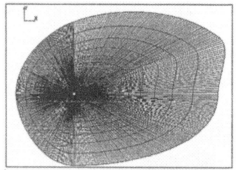

Figure 3. The finite element mesh. To the left along the x-axis is upstream, to the right is downstream. The ball lies in the centre-left of the model.

The completed 2-D finite element mesh consisted of 17,300 quadrilateral elements made from 17,724 nodal points. There were 444 nodes around the circumference surface of the ball.

2.4 Boundary Conditions

In order for the FEA software to begin its solution calculations, the boundary conditions of the mesh model were specified as follows: All elemental nodes that lay on the surface of the ball were considered to have non-slip conditions. That is, all components of velocity at those nodes were set to zero. It would be possible to model the back spin of the ball, but in this study the ball was considered as static for simplicity.

All elemental nodes that lay on the inlet of the model were constrained to have the free stream velocity components, the magnitude of those components being unity in our dimensional analysis model. The effective angle of incidence of the ball seam could be controlled by varying the velocity components of these inlet nodes, without changing the FE model. For example, for an angle of incidence of 0° to the x-axis (seam axis), the x-component of the free stream velocity would be one, whilst the y-component would be zero. For an angle of incidence of θ°, the x-component would be Cosθ, and the y-component would be Sinθ.

Elemental nodes that lay on the outlet of the model where constrained to be at a free-stream pressure of zero. This allows the software to create a flow directly from the inlet to the outlet. All measured pressures are then relative to this 'atmospheric' pressure.

The fluid itself was considered to be isothermal, incompressible and Newtonian, ie. of constant viscosity. The fluid properties are defined by the dimensional analysis method, with the fluid viscosity chosen for a corresponding Reynolds Number system. For clarity, Table 1. shows the approximate delivery speeds corresponding to given Reynolds Number systems for a cricket ball in air at room temperature. ($\nu = \mu/\rho = 1.5e^{-5}$, $L = 0.0716$m).

Table 1. Reynolds Number with its approximate equivalent in delivery speed, for a cricket ball in air.

Reynolds No.	Delivery Speed
1.90×10^5	40 ms^{-1} or 90 mph
1.27×10^5	26.6 ms^{-1} or 60 mph
0.64×10^5	13.3 ms^{-1} or 30 mph
1.00×10^4	2.1 ms^{-1} or 4.7 mph

2.5 Solution Conditions

FLOTRAN allows solutions to be computed under two different temporal regimes: steady-state and transient. It also has the facility to use the k-epsilon turbulence model in each of these regimes although the transient-turbulent combination really has no practical meaning.

In this study, both transient and steady-state solutions were obtained. The steady-state process involves the software iterating until a stable flow is established. However, because of the relatively high Reynolds Numbers and consequent chaotic flow behaviour in our system, the turbulence model was also required in order for the software to converge on any kind of solution at all. This usually meant around 600 global iterations (around 8hrs processing). The drawback with this combination is that local flow effects such as vortices are unlikely to be seen as there is a time averaging effect, and the turbulence is assumed to be isotropic. However, the k-epsilon model also produces turbulent kinetic energy and turbulent energy dissipation data in addition to the usual velocity and pressure distributions.

The transient process is slightly different. It involves the software iterating to find the change in flow over a small time step. The time step itself has to be small enough so that the Advection Limit[12] is not violated, ie. the step must be small enough so that a flow particle does not travel further than the width of one element in the chosen mesh. Usually 0.002-0.005 dimensional analysis time units is sufficient. This means that the software requires few global iterations to converge at each time step (5-10 perhaps), but that many time steps, perhaps one or two thousand, are required to produce a developed flow. The turbulence model is not required in this regimen, and so local flow effects can be seen, and the results are usually more representative of a snapshot of a 'real' flow.

FLOTRAN also has the facility to integrate the pressure results over the surface of the ball model. This allows the lift and drag forces, which arise from differences in pressure between two sides of the ball, and the front and back of the ball respectively, to be directly computed. For a cylindrical model, the drag coefficient is given by:

C_D = drag force D / dynamic force = $D/(^1/_2\rho U^2 Ld)$

In the dimensional analysis regimen this reduces simply to $C_D = 2D$ (d is the cylinder length which can also be considered as unity), and similarly for the lift coefficient (resultant perpendicular force Y / dynamic force), $C_L = 2Y$. Also, the pressure coefficient C_P can be expressed in this manner, using the fact that $P_\infty = 0$ as dictated by the outlet boundary conditions:

$$C_P = (P - P_\infty)/(^1/_2\rho U^2) = 2P$$

2.6 Mesh refinement

With the experience gained from producing and analysing a 2-D mesh, it was possible to develop a 3-dimensional model (see Figure 4) of similar size. However, there was difficulty in designing a homogeneous mesh as only hexahedrals, and not wedges, are allowed.

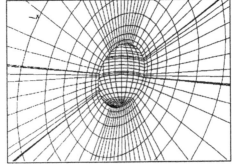

Figure 4. A y/z plane section of the finite element 'O-mesh' of the 3-D model, in the region of the ball.

Also, The number of nodes and the calculation times are greatly increased with a 3-D system, hence the fine density of the mesh could not be maintained. The completed 3-D finite element mesh consisted of 16,320 hexahedral elements and 17,440 nodes with 1090 nodes on the surface of the ball. Similar boundary conditions and solution controls were then applied.

3. RESULTS

3.1 Steady-State 2-D Solutions

Figure 5 show a typical fluid speed distribution, derived from a steady-state solution with turbulence modelling. Upstream is to the left of the figure, with the flow running left to right horizontally across the figure. The ball seam has been drawn onto the figure to clarify its orientation.

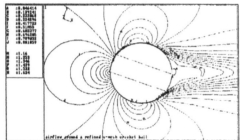

Figure 5. Fluid speed distribution around a 2D cricket ball, steady-state turbulent solution. Re = 1.27x10^5, seam angle = 20°.

Figure 6 shows a typical fluid pressure distribution, derived from a steady state solution with turbulence modelling, and Figure 7 shows the turbulent kinetic energy distribution obtained from the same solution. Again, the seam has been drawn onto these figures.

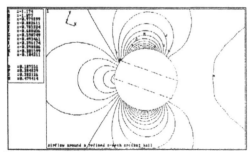

Figure 6. Fluid pressure distribution around a 2D cricket ball, steady-state turbulent solution. Re = 1.27x10^5, seam angle = 20°.

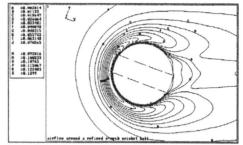

Figure 7. Fluid turbulent kinetic energy distribution around a 2D cricket ball, steady-state turbulent solution. Re = 1.27x10^5, seam angle = 20°.

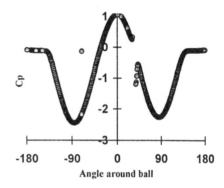

Figure 8. Variation of pressure coefficient around a 2-D cricket ball. Re = 1.9x10^5, seam angle = 20°.

Figure 8 shows the variation of the pressure coefficient with angle around the circumference of the ball model. The gap in the curve is due to a problem with the data conversion program caused by the ball seam area.

Table 2 shows how the lift and drag coefficients varied with Reynolds Number and seam incidence angle, for turbulent steady-state solutions. All the results quoted were obtained from solutions which had completed the same number of global iterations.

Table 2. Variation of Lift and Drag coefficients with Reynolds Number and seam angle, for a steady-state turbulent 2-D solution. Negative values for C$_{Lift}$ indicate reverse swing.

Re (x10^5)	0.64	1.00	1.27	1.90	
Seam θ	20°	20°	20°	20°	
C$_{Drag}$	0.276	0.264	0.258	0.250	
C$_{Lift}$	-0.055	-0.078	-0.088	-0.094	
Re (x10^5)	1.27	1.27	1.27	1.27	1.27
Seam θ	0°	10°	20°	30°	40°
C$_{Drag}$	0.236	0.246	0.258	0.290	0.364
C$_{Lift}$	0.016	-0.014	-0.088	-0.082	-0.029

To give some idea of the state of the convergence of these solutions, we can state the normalised change in each of the parameters over the last iteration. The normalised change, for say Pressure P, is given by $\Delta P = (\Sigma_1^N |P_{new} - P_{old}|) / (\Sigma_1^N |P_{new}|)$, where N is the total number of nodes, P_{new} is the current calculated value, and P_{old} is the calculated value for the previous iteration. In each case, the normalised change for the velocity components was of the order 3×10^{-5}, for pressure of the order 2×10^{-3}, and for the turbulence parameters k and ε of the order 1×10^{-3}.

3.2 Transient 2-D Solutions

Transient solutions were obtained using time steps of 2.5 'milliseconds', with 6 global iterations at each.

Figure 9 show a fluid speed distribution, derived from a transient solution. It is a 'snapshot' of the flow at an instant in time. Upstream is to the left of the figure, with the flow running left to right horizontally across the figure. The ball seam has been drawn onto the figure to clarify its orientation

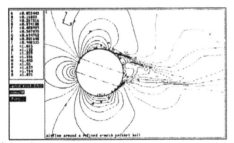

Figure 9. Fluid speed distribution around a 2D cricket ball, transient solution, at time = 7. Re = 1.27×10^5, seam angle = 20°.

Figure 10 shows a fluid pressure distribution, derived from the same transient solution as Figure 9.

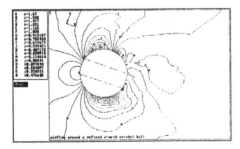

Figure 10. Fluid pressure distribution around a 2D cricket ball, transient solution, at time = 7. Re = 1.27×10^5, seam angle = 20°.

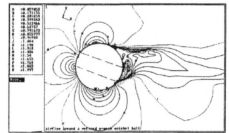

Figure 11. Fluid speed distribution around a 2D cricket ball, transient solution, at time = 8. Re = 1.27×10^5, seam angle = 20°.

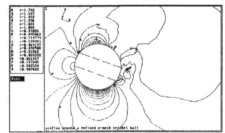

Figure 12. Fluid pressure distribution around a 2D cricket ball, transient solution, at time = 8. Re = 1.27×10^5, seam angle = 20°.

Fluid velocity and pressure distributions a short time later are shown in Figures 11 and 12.

3.3 Steady-State 3-D Solutions

Figure 13 shows a top view of the fluid speed distribution derived from a steady-state solution with turbulence modelling. Upstream is to the left of the figure, with the flow running left to right horizontally across the figure.

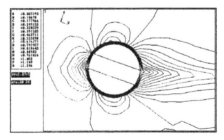

Figure 13. Fluid speed distribution around a 3D cricket ball, steady-state turbulent solution. Re = 1.27×10^5, seam angle = 20°.

Figure 14 shows an isometric slice view of the fluid pressure distribution derived from a steady-state solution with turbulence modelling. Upstream is to the upper left of the figure, with the flow running towards the bottom right of the figure. Pressure contours are also shown on the surface of the ball.

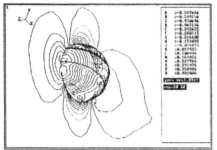

Figure 14. Fluid pressure distribution around a 3D cricket ball, steady-state turbulent solution. Re = 1.27×10^5, seam angle = 20°.

4. DISCUSSION

Theories about the possible mechanism of orthodox swing have been around for some years[11], and it is widely agreed that the state of the boundary layer around the surface of the ball plays the major role[1,3,4]. The boundary layer can be in one of two distinct states: laminar, where the layers of air flow smoothly over one another, or turbulent where the air eddies chaotically. Experiments show that a turbulent boundary layer actually separates away from the surface of the ball and into the wake much further towards the back of the ball, than a laminar boundary layer[7].

The theory of orthodox swing suggests then that when there is an angle of incidence between the ball seam and the direction of projection, at medium to fast delivery speeds, the seam actually 'trips' the otherwise laminar boundary layer on that side of the ball into turbulence. On the seam side then, the turbulent boundary layer separates from the surface further towards the back of the ball than on the other side, which is still laminar. The speed of the air in the turbulent boundary layer is greater than that in the laminar boundary layer, and so its pressure is relatively less. Thus there is a pressure imbalance between the two sides of the ball and a resultant swing force towards the turbulent (seam) side is created.

The theory put forward for the mechanism of reverse swing, however, is less certain. It is thought that the combination of high delivery speed and the forward facing rough side of the ball causes a turbulent boundary layer to form over both sides of the ball. On one side, the seam 'thickens' or 'lifts off' this turbulent boundary layer causing separation to occur nearer the *front* of the ball on the seam side, compared with the other side[1,10]. Thus a pressure imbalance is created and the ball swings *away* from the seam side.

This study is a preliminary parametric study to evaluate the effects of Reynolds Number and seam angle. As such, the discussion of results will be restricted to trends and qualitative observations, rather than quantative results.

4.1 Steady-State Solutions

Figure 5 is typical of the steady-state results and shows the speed of the fluid flow over one side of the ball being greater than over the other side, due to the influence of the seam. This is reflected in a difference in pressure exerted on the two sides of the ball (shown in Figure 6), and this leads then to a resultant swing force. Figure 8 showing the asymmetric pressure coefficient around the ball is in excellent agreement with both theory[7], and experimental[1] results. Table 2 shows that the angle of incidence of the seam effects C_L, which can be considered a non-dimensional measure of the amount of swing created. It shows that there is a maximum swing with a seam angle of between 20° and 30°, which is in broad agreement with previous studies[1,4], whilst as you might expect there is little discernible swing at a seam angle of 0°.

Table 2 also shows that the amount of swing increases as the Reynolds Number increases. However, this may not translate to the ball deviating through a greater distance during its flight down the pitch because of the corresponding increase in delivery speed[1]. Whilst direct comparison between a 2D model and a real 3D cricket ball is obviously difficult, it should be noted that the values in Table 2 for lift and drag coefficients are of the order of those quoted in previous studies[1,2,4,7], and exhibit similar trends. The trend whereby the drag decreases with increasing Reynolds Number is unsurprising as it is at this order of Reynolds Number that boundary layer state transitions occur (sometimes known as the 'drag crisis').

Almost all of the steady-state turbulent results produced flows and lift coefficients of sign consistent with reverse swing, and this was somewhat surprising. Figure 7 shows there is some degree of turbulence in front of the ball, and this turbulence does seem crucial in instigating the reverse swing. As it moves around the ball, on one side it appears it is effectively 'thickened' by the influence of the seam, the result being it separates from the ball earlier on the seam side than on the other (see Figure 7). This appears consistent with existing reverse swing theory. However, the authors are not confident as to the source of this turbulence and it warrants further investigation. There is some evidence to suggest that the k-epsilon model is prone to producing this type of result where bluff bodies are concerned[13].

The 3-D model results produced similar pressure and velocity distributions to the 2-D model results. Differences in pressure between the two sides of the ball were observed, and a small lift coefficient calculated with the seam at an angle of 20°. Very little lift was detected with a seam angle of 0°. Drag coefficients calculated were of similar order to those reported in the literature However, these results should be treated with caution as the mesh used had many fewer nodes in the regions close to the ball surface, and certain areas where the mesh density transitions were not smooth.

4.2 Transient Solutions

The transient solutions show 'snapshots' of the flow over the ball at an instant in time. As such, they can be misleading as the pressure and velocity profiles will change, oscillating as vortices are shed alternately from each side of the ball. It is difficult then to be precise about the magnitude and direction of any swing, and lift and drag coefficients from these solutions should only be quoted if calculated from many results integrated over time.

Figure 11 shows the influence of the seam on the flow over one side of the body, distinct changes occurring in the flow speed near the surface of the ball. Vortices also can be seen being shed from the surface of the ball, but are quickly forced back into the wake by the surrounding flow. It can also be seen how this vortex shedding affects the boundary layer separation point on each side of the ball, and this would then affect any lift/swing created.

5. CONCLUSIONS

In this study, qualitative results have successfully demonstrated how a pressure imbalance between the two sides of cricket ball, which leads directly to ball swing, can be re-created by computer simulation. It has also been shown how that resultant swing force can vary according to the angle of incidence of the

seam, and the Reynolds Number of the system. The fluid flow patterns over the surface of the ball have been shown to confirm some of the postulated mechanisms of cricket ball swing[1].

However, this is only a preliminary study to assess the feasibility of this approach, and there is still much work that can be done to explore the validity and reliability of these results. Ideally, these techniques should be applied more rigorously, over a wider range of Reynolds Number to obtain quantitative results, and the exact shape and angle of incidence of the seam should be more closely studied. This would certainly require better models with finer meshes, in three dimensions, and thus more computing power. It could then provide for a closer insight into the mechanisms of cricket ball swing and give a better understanding to bowlers and ball manufacturers alike.

Indeed, these techniques could be applied very cheaply to analyse drag and increase performance[6] in any number of sports where speed and concomitant drag is critical.

REFERENCES

1) Mehta, RD 1985. Aerodynamics of Sports Balls. *Annual Review of Fluid Mechanics*, **17**: pp151-189

2) Bowen, LO 1995. Torque and Force measurements on a Cricket Ball and the Influence of Atmospheric Conditions. *Transactions of Mechanical Engineering*, IE Aust **ME20(1)**: pp15-20

3) Bown, W & Mehta, RD 1993. The seamy side of swing bowling. *New Scientist*, Aug 21st 1993: pp21-24

4) Mehta, RD Bently, K et al. 1983. Factors affecting cricket ball swing. *Nature*, 303: pp787-788

5) Rice, JG Schnipke, RJ et al. Navier-Stokes computation of a typical high-lift airfoil system. Compuflo Inc. Charlottesville, Virginia 22901 USA.

6) Macleish, MS Cooper, RA et al. 1993. Design of a composite monocoque frame racing wheelchair. *Journal of Rehabilitation Research & Development*, 30(2): pp233-249

7) White, FM 1994. *Fluid Mechanics, 3rd Ed.* McGraw-Hill Inc. ISBN 0-07-113765-3

8) Simon Hughes. Balance swings seamer's way in reversal of fortune. *Daily Telegraph*. Telegraph Publishing Company, London. 9th Aug '95: pp34

9) FLOTRAN Theoretical Manual, 1992
    Compuflo, Inc. Houston, PA 15342, USA

10) Barrett, RS & Wood, DH 1993. Aerodynamics
    of the cricket ball: Understanding the reverse
    swing phenomena. *Abstracts - International
    Society of Biomechanics XIV Congress* 1: pp654

11) Lyttleton, RA 1957. The swing of a cricket ball.
    Discovery, 18: pp186-191

12) Abbott, MB & Basco, DR 1989. *Computational
    Fluid Dynamics*. Longman Scientific & Technical.
    ISBN 0-582-01365-8

13) Murakami, S & Mochida, A 1988. 3-D numerical
    simulation of airflow around a cubic model by
    means of the k-ε Model. *J. Wind Eng. Ind.
    Aerodyn.*, 31: pp283-303

The Engineering of Sport, Haake (ed.) © 1996 Balkema, Rotterdam. ISBN 90 5410 822 3

Effect of cyclist's posture and vicinity of another cyclist on aerodynamic drag

M. M. Zdravkovich
Telford Institute, University of Salford, UK

M. W. Ashcroft, S. J. Chisholm & N. Hicks
Department of Aeronautical, Mechanical, and Manufacturing Engineering, University of Salford, UK

ABSTRACT: Two long-established and well-known methods of drag reduction are the posture of the rider and aerodynamic interference when two riders are in close vicinity. The aim of the tests on the model and full-scale cyclists is to quantify the drag reduction. The measurements are carried out for four postures: hands on brakes, straight hands on drops, bent hands on drops (body crouched), and hands on 'aero' handlebars. Drag coefficients are calculated for each of the four postures based on riders' projected area. The interference drag is measured for twenty tandem and staggered positions of two cyclists. Significant drag reduction up to *49%* is found.

1. INTRODUCTION

Cycling has entered a new era when aerodynamic aids such as tri-spoke and disc wheels, aero handlebars, low-drag frame, teardrop shaped helmet, skin suit *(Kyle 1981, Kim 1990, and Zdravkovich,1992)*, to mention a few, are allowed to be used in races. The Union Cyclist Internationale (UCI) forbade, in 1938, any alteration of the traditional racing bicycle which would lead to an improvement of its aerodynamics. The UCI first revised its ruling in 1976 when disc wheels were allowed in competition. Since then, many further revisions have taken place allowing clipless pedals, bladed spokes, teardrop helmets, skin suits, and most recently, 'aero' handlebars, developed at the Texas A and M University.

The most important contribution to the overall drag force exerted on the cyclist is the aerodynamic force which may exceed *90%* of the overall drag at top speed. The rest being made up of rolling resistance between the tyres and road, mechanical friction of gears, chain, hubs, etc.

The advantage of aerodynamic shaping of the bicycle has limited overall effect because most of the aerodynamic drag is caused by the rider. The governing factor in drag reduction is the posture of the rider. The measurements of drag forces with the rider at various postures have been undoubtedly carried out in the past but the results are not published in open literature. This is one aim of the present paper.

Another important aerodynamic factor in races is the interference between riders in close proximity. The well known and experienced drag reduction, when two riders race one behind the other, seems not to have been

quantified in the open literature. This is the second aim of the present paper.

2. EXPERIMENTAL ARRANGEMENT

Two wind tunnels have been used in tests of a model and full-scale bicycles with dummy and rider on them, respectively. The first is a closed-circuit wind tunnel having a test section of *0.92 m* high, *1.22 m* wide, and *1.83 m* long. To ensure uniformity of flow in the test section, air is passed through a honeycombe followed by a *3.5:1* contraction section. In order to reduce the blockage effect, the ceiling of the test section is perforated (test section is at atmospheric pressure). The maximum blockage ratio with the bicycle model and dummy rider on it is *11.7%*. The high precision six-component balance is used to measure drag force component only. The speed can be varied up to *40 m/s*.

The second, an open circuit wind tunnel, has the test section of *1.6 m* high, *2.2 m* wide, and *10 m* long, capable of speeds up to *8.2 m/s (29.5 km/h)* with bicycle and rider in it (Figure 1). The blockage ratio varies from *12.4%* to *16.2%* depending on rider's posture. The wind tunnel has gauze at the entrance but no contraction section. The uniformity of the velocity field in the test section is ±2%.

The bicycle model used in the first wind tunnel is shown in Figure 2 and the dummy rider (known as Jill) is seen in Figure 3. The scale of the bicycle model is *1:2.5* and main dimensions are given in Table 1. It is made of wood, copper tubing (handlebars and aerobars), and wire (*32 spokes per wheel*).

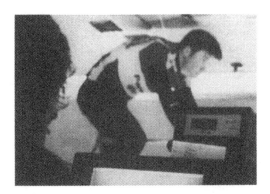

Fig. 1. Cyclist on the bicycle suspended in test section.

Fig. 2. 1:2.5 bicycle model.

Table 1. Dimensions of bicycle model and full-scale in cm.

Bicycle		Model	Full-Scale
Tube	Top	22	56
	Seat	23	58
	Down	25	63.5
	Head	1	15
Triangle	Chain	16.5	41
	Seat Length	10.6	
Forks	Width	4.6	10
	Rake Length	15.2	40
Handlebar	Width	16	42
	Drop	7.6	13.5
Spokes		32	32

The dummy is also made in scale *1:2.5* of the rider. Metal rods are inserted to form a 'skeleton' to support the dummy limbs in various postures during tests. The bicycle model and dummy on it are held near the bottom of the test section by a streamwise rod attached to two vertical streamlined struts protruding from the ceiling and attached to the balance outside and above the test section.

Table 2. Riders' dimensions in cm.

Size		Dummy	Rider #1	Rider #2
			SJC	NH
Height:	Head	13	6.5	25
	Neck	4.5	55	6.5
	Back	3.1		58
	Upper Leg	24		46
	Lower Leg	25.5		48.5
Width:	Head	15	19	20
	Shoulder	29	46	48
	Waist	22.5	36	31
	Hips	24	39	35
Arms:	Upper	14	29.5	32
Length:	Lower	13	29.5	27
	Heads	3	18.5	19
Height		68	170	174
Weight		N/A	69	69

The main dimensions of the full-scale bicycle and rider are given in Tables 1 and 2, respectively.

The rider on the bicycle, shown in Figure 4, is suspended on two *2 mm* steel wires attached to the top corners of the test section. The wires form two letters, *V*. The bicycle is suspended *5.6 cm* above the test section floor outside the boundary layer. The strain gauged cantilever, *1cm × 2cm × 2.5cm*, is clamped to the floor of the test section and the top end gently rests against the rear bicycle bracket. Any deflection of the rider and bicycle in the drag direction does not exceed *0.1 mm*. This is essential to prevent uplift of the suspended rider.

3. EXPERIMENTAL RESULTS AND DISCUSSION

3.1 *Classification of Rider's Postures*

Six different riding postures are initially considered:

(a) Hands placed on the tops of handlebars, torso upright, Figures 3a and 4a.

(b) Hands placed on brake hoods, arms straight, Figures 3b and 4b.

(c) Hands placed on handlebar drops, arms straight, Figures 3c and 4c.

(d) Hands placed on handlebar drops, arms bent, body crouched, Figures 3d and 4d.

(e) Time-trial posture, hands on aerobars, Figures 3e and 4e.

(f) Descending posture, hands on middle of handlebars, Figures 3f and 4f.

The tests are carried out for all six postures but only results for postures (b), (c), (d), and (e) will be presented. There are three reasons for this omission:

(1) The measured drag for postures (a) and (f) are very close to that for postures (b) and (e), respectively.

(2) For posture (a), the head of the rider touches the ceiling of the test section when straight, and that is unacceptable.

(3) Posture (f) is extremely uncomfortable and can hardly be maintained by the rider for sufficiently long periods of time for measurements.

The reference area for the drag coefficient calculation is taken to be the frontal area of the rider and bicycle. The frontal area is evaluated from photographs by making a grid based on *1 m* rule seen in Figures 3 and 4. The values of frontal areas for the dummy and rider in various postures are given in Table 3.

$$C_D = \frac{D_t - D_s}{\frac{1}{2}\rho V^2 S} \qquad \qquad \dots (1)$$

where D_t is the total drag, D_s is the support drag, ρ is the density of air, V is the free-stream velocity, and S is the rider frontal area (Table 3).

Figure 5 shows the calculated drag coefficient for four postures (b) to (e) in terms of Reynolds number, based on waist width. The drag coefficient for all postures does not vary when Reynolds number is greater than 3.1×10^5. The variation for a Reynolds number less than 3.1×10^5 is probably due to an error caused by a small force rather than due to a change in flow around the dummy and bicycle model.

The average values of drag coefficient are given in Table 4. They are similar to that measured on the short circular cylinders with two free ends *(Zdravkovich, 1989)* of the same aspect ratio. The trend of a decreasing drag coefficient from postures (b) to (e) reflects closely the decrease in the respective frontal areas. The lowest drag coefficient is found for aerobars (e). It demonstrates that the effectiveness of aerobars is based on the riding posture which appears more aerodynamic than the others.

Table 3. Frontal areas of dummy and riders in m².

	Dummy		Rider #1		Rider #2	
Posture	Area	Height	Area	Height	Area	Height
(a) Tops	0.094	0.81	0.52	1.57	0.43	1.57
(b) Brakes	0.105	0.77	0.51	1.57	0.43	1.57
(c) Drops	0.079	0.72	0.49	1.54	0.42	1.54
(d) Crouch	0.021	0.68	0.48	1.50	0.38	1.50
(e) Aerobar	0.062	0.68	0.39	1.50	0.35	1.50
(f) Descending	0.058	0.52	0.38	1.43	0.33	1.43

Table 4. Average drag coefficient.

Posture	Model	Full-Scale	
		NH	SJC
(b) Brakes	0.61	0.75	0.6
(c) Drops	0.63	0.69	0.54
(d) Crouch	0.6	0.67	0.52
(e) Aerobars	0.58	0.6	0.49

It is evident that there is a considerable difference in the frontal areas of two riders (SJC and NH) which will effect the value of drag coefficient.

3.2 *Drag of Dummy on Bicycle Model*

The measured drag force consists of that exerted by wind on the dummy on the bicycle model and on the supporting rod attached to the balance struts. The latter is measured subsequently by detaching dummy and bicycle model from the rod and struts. The dummy and bicycle model are attached to the floor in order to simulate the same aerodynamic interference as in previous tests. The drag of the supporting rod and struts is subtracted from the total measured drag force and expressed through the drag coefficient:

3.3 *Drag of Full-Scale Bicycle and Rider*

The large wind tunnel has not got the aerodynamic balance. The home-made strain-gauge cantilever used shows a wide variation of initial zeroing. To avoid this error the tests are carried out, starting with the highest velocity (*8.2 m/s*) and decrementally going down to *6.2 m/s*. The effect of suspension wires on the measured drag is small. For example, the drag of two *2 mm* wires, *2.8 m* long, and assuming drag coefficient of *1.2* yields the drag of *0.54 N* at *8.2 m/s*. This is less than 5% of the total measured drag for posture (b).

Another important effect on accuracy is the un-steadiness of the rider during the run. This effect is partly counter-balanced by taking ten readings and then averaging the drag value. Typical variation of readings is up to *+1.7%* and *-6%*. However, if the top and bottom extreme values are ignored, the scatter is almost halved.

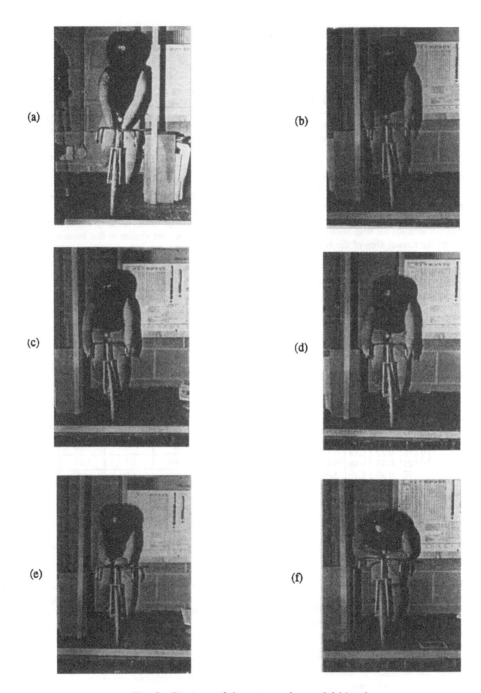

Fig. 3. Postures of dummy on the model bicycle;
hands on: (a) tops of handlebars, (b) brake hoods,
(c) handlebar drops, (d) same but body crouched,
(e) aerobars (time-trial), (f) descending.

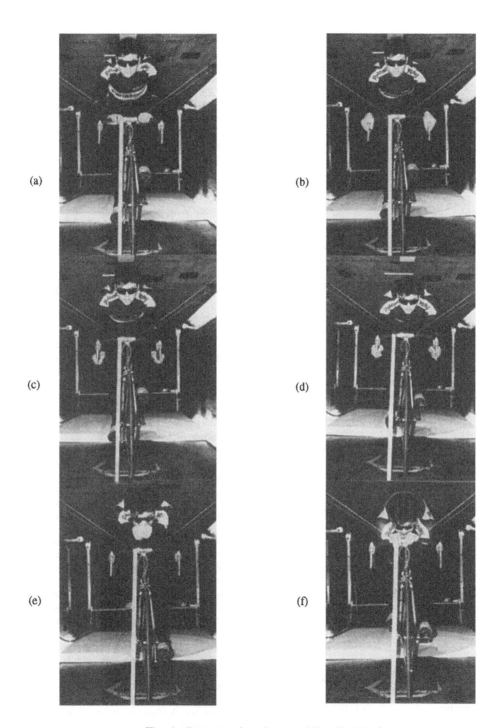

Fig. 4. Postures of cyclists on full-scale bicycle;
hands on: (a) tops of handlebars, (b) brake hoods,
(c) handlebar drops, (d) same but body crouched,
(e) aerobars (time-trial), (f) descending

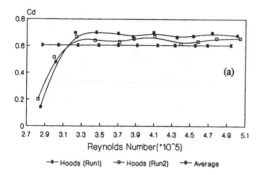

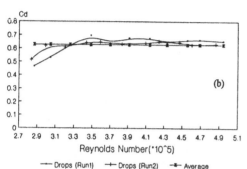

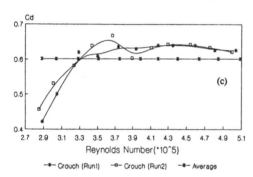

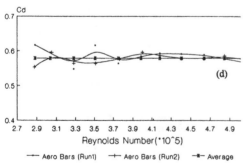

Fig. 5. Drag coefficient in terms of reynolds number; hands on: (a) brake hoods, (b) handlebar drops, (c) same but body crouched, (d) aerobars

Note that blockage effects, and stationary legs and wheels may increase and decrease the drag, respectively.

Figure 6 shows the drag coefficient variation in terms of Reynolds number for four postures. The drag coefficient does not change much for a Reynolds number greater than 3.8×10^5. The aerobar posture (e) is the most effective in reducing drag coefficient, *17%*. This improvement is considerably higher than for the dummy, presumably due to the inadequate posture of the latter. The average values of drag coefficient are given in Table 4. The different values obtained with two riders of different body size and shape suggests that not a single value, but rather a range of C_D values, seem to be more appropriate to be given. Note that *Hoerner (1962)*, in his book, expressed his data through the drag factor *f[m²]*, being C_DA:

$$f = \frac{D}{\frac{1}{2}\rho V^2} \qquad\qquad , \qquad \dots (2)$$

in order to avoid cumbersome frontal area altogether.

3.4 *Interference Drag of Two Cyclists*

It has been known to cyclists for a long time that riding closely behind another rider requires less power. In road racing tactics, riders are being sheltered by team mates in order to conserve energy for the finish. The purpose of present tests is to quantify that effect.

Figure 7 shows the position of two cyclists in the test section of the large wind tunnel. Note that the monitored downstream cyclist is always kept suspended in the same place, while the lead cyclist assumes various positions in tandem and staggered arrangements. Figure 8 displays twenty positions of the lead cyclist being *20 cm* apart in the streamwise direction and *10 cm* apart in the transverse direction. The tests are carried out only at the maximum speed of *8.2 m/s*. Also, only posture (b), as seen in Figure 7, is tested.

Figure 9 shows the measured drag force on the downstream cyclist *(a)* SHC and *(b)* NH in tandem, *1-5*, and staggered, *6-20*, arrangements. The maximum reduction in drag of *49%* is attained in the closest tandem arrangement, position *1*. However, such a close proximity to the lead cyclist may be dangerous. Safer positions are at *10 cm* stagger, *6* and *7*, and tandem, *2*. Note that in staggered arrangements there is not only less chance of catching the rear wheel of the lead cyclist, but also the view of the road ahead is less obstructed. Therefore, positions *6* and *7* may be more appropriate and these yield a reduction in drag of *37%* and *35%*, respectively. The curves for tandem and *10 cm* stagger are closer to each other than *20 cm* and *30 cm* stagger. It seems that the 'shielding' effect is rather abruptly reduced for *20 cm* and *30 cm* stagger. This discontinuous feature is also found for two circular

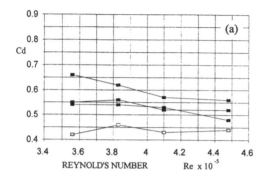

Fig. 6. Drag coefficient in terms of Reynolds number for four postures: (a) SJC, (b) NH.

Fig. 7. Two cyclists in slightly (10cm) staggered arrangement.

cylinders in tandem and staggered arrangements *(Zdravkovich 1977).*

4. CONCLUSIONS

The systematic measurement of the rider-on-the-bicycle drag confirmed what is generally expected. However,

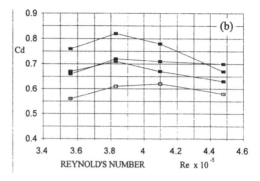

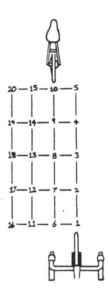

Fig. 8. Tandem and staggered arrangements tested.

there are some new revelations about the nature of drag coefficient as follows:

1. It is found that a single value of drag coefficient cannot be specified for any of the four postures.

2. The drag coefficient shows a strong dependence on the size and shape of riders. It seems that the height-to-waist width ratio strongly influences the drag coefficient. The difference in drag coefficient for the two riders and dummy consistently exceeds the scatter of data and shows a definitive strong effect of the height-to-waist width ratio.

3. The lowest value of drag coefficient is found for the aerobar rider's posture. The aerobars streamline the

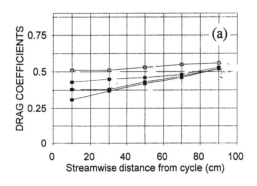

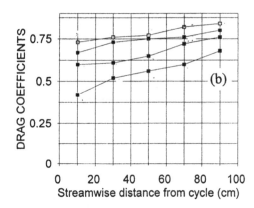

Fig. 9. Drag force in terms of streamwise and transverse postures behind the lead cyclist: (a) SJC, (b) NH.

rider's posture and also reduces the frontal area. The low drag coefficient is the result of both effects.

4. There is a considerable drag reduction when two cyclists are in close proximity. The lowest drag coefficient is measured when cyclists are arranged in the closest tandem position. This position may not be safe and the neighbouring *10 cm* stagger positions are slightly less effective at *37%* reduction compared with *49%* in the neighbouring tandem position.

REFERENCES

Hoerner, S. 1962. *Fluid Dynamic Drag*, published by the author, 2nd edition.

Kim, I. 1990. Racers, Rough Riders, and Recumbants, *Mechanical Engineering*, 112: pp 52-59.

Kyle, C. R. 1984. Improving the racing bicycle, *Mechanical Engineering*, 105: 34-35.

Zdravkovich, M. M. 1977. Review of flow interference between two circular cylinders at various arrangements, *J Fluids Engineering*, 99: 33-47.

Zdravkovich, M. M. 1992. Aerodynamics of bicycle wheel and frame, *J Wind Engg Ind Aerodyn*, 55-70.

Zdravkovich, M. M., Brand, P., Weston, A., and Mathew, G. 1989. Flow past short circular cylinders with two free ends, *J Fluid Mechanics*, 203: 557-575.

Note: There is a continuous flow of popular articles written in *Bicycling, Cycling Weekly, BCCS Coaching News, Winning Treathlets, etc.*

2 Biomechanics

The Engineering of Sport, Haake (ed.) © 1996 Balkema, Rotterdam. ISBN 90 5410 822 3

The effect of an exercise regime on Lumbar Spine Curve

M. Hossein Alizadeh
Physical Education, University of Manchester, UK

Janette Standring
School of Physiotherapy, University of Manchester, UK

Abstract: The purpose of this study was to determine the effects of a combination exercise regime on the Lumbar Spine curve (LSC). Measurements of the LSC used a flexible ruler (FR) were taken of 32 healthy male subjects aged 30-44 (mean 36.3) years. A universal goniometer was used for measuring the hip range of motion (HROM), and a tape measure for measuring lumbar flexion (LF). Following analysing of the preliminary results 2 sub-groups were formed. Groups contained subjects who had an increased LSC (the experimental group, 9 men) and subjects who their LSC being about the mean were selected as the control group (9 men). The control group did not participate in the exercise regime. The experimental group underwent active stretching exercises of their ilio psoas, lumbar erector spine muscles, and strength exercises for their abdominal and hip extensor muscles. The exercise program was performed three sessions per week for 4 weeks. A one-way analysis of variance (ANOVA) for repeated measurer ($P<0.05$) was used to examine the subjects' differences before and after the exercise regime. The results indicated that there was a significant difference in the experimental group's LSC which could be the result of the effects of exercise regime on the LSC (the mean of LSC decrease from 46.26 degree to 40.89 degree). Non significant differences were observed in the HROM and the LF in the experimental group. ANOVA for the HROM, LSC and the LF demonstrated no significant differences in the control group. The findings indicated that the exercise regime may be an effective combination of exercises for influencing LSC. We thus suggest further re-examination of combination exercise on the lordotic subjects.

Key words: exercise, posture, pelvic, corrective physical education, lumbar spine, stretching, muscle, strength, flexible ruler.

1. INTRODUCTION

In recent years there has been an increasing interest in posture. Posture is defined as the position of different segments of the body. Correct posture is the position in which minimum stress is applied to each joint of the body. The appropriate muscle groups are carrying a balance workload to maintain the body upright in correct posture (Russell and Highland 1990). Conversely, muscular strain and inadequate muscle tone cause poor posture. It seems that in poor posture muscles are working more than usual and the amount of energy expenditure is increased. This is because excess energy expenditure indicates hypertonicity or poor neuromuscular coordination, or both (Luttagens, and Well, 1982). Kendall et al (1983) suggest that

muscle weakness or shortness may cause faulty alignment and faulty alignment may give rise to stretch-weakness or adaptive shortness of muscles. Poor posture or faulty posture may also lead to increase stress on joint or body segments. In fact normal muscle tone is essential for good posture. The exceeded anterior pelvic tilt is an example of poor posture when attached muscles on the pelvis have a tendency to shorten or to lengthen and the lumbar spine adopts an increased curvature. The pelvis is an important part of the body because it is the base for the trunk. It also is located in the middle of the body and joins the upper extremities to the lower extremities. As the pelvis tilts forward the lumbar vertebrae are forced anteriorly and as a result the LSC (lumbar convexity) is increased. Some of the studies show that the anterior pelvic tilt is related to the function of the abdominal muscles. For example, Kendall et al (1983) state that the weakness of abdominal muscles may result in the forward pelvic tilt and a low back lordotic (Hollow-back) position. Walker et al (1987) examined lumbar lordosis, pelvic tilt, and abdominal muscle performance during normal standing and found no relationship between them. Since different muscles affect the pelvis and tilt it forward, backward or sidewards, a comprehensive assessment of these muscles in this region is essential in determining their effects function on the pelvic alignment. An important question concerning these studies is whether only the weakness of abdominal muscle should be considered in the LSC while a couple of forces are working together to establish the pelvis position. It seems that not only the improvement of pelvic alignment needs the abdominal muscle strength but also there must be an adequate muscle strength between opposing groups of muscles for the pelvic balancing.

The purpose of this study was to examine the effects of an exercise regime on the LSC. The exercise regime consisted of two stretching exercises to increase the flexibility of iliopsoas and lumbar erector spine muscles and two strength exercises to increase the strength of abdominal and hip extensor muscles. The hypotheses was that the exercise regime would lead to a decrease in the LSC in the experiment group.

2. METHOD

2.1 *Subjects*
Thirty two male university students aged 30-44 (mean 36.3) years, with no complaint of back dysfunction, were invited to take part in the preliminary phase of the study called posture screening. The posture screening consisted of an interview (the subjects were asked to give information about their age, weight and history of injury in the low back ...) and a physical examination. All subjects signed an informed consent form.

2.2 *Procedure*
After the interview, the subjects were given an explanation about the physical measurement procedure. Physical measurements were taken immediately after the interview. This part consisted of measurements of the LSC, the HROM and the LF. All measurements were repeated three times and the mean of the values was calculated.

2.3 *Lumbar Spine Measurement*
The subjects stood barefoot and assumed a comfortable standing posture with body weight evenly distributed between both feet while their trunk and arms were relaxed. We measured the LSC from L2 to S2 with the subject in the comfortable standing position. The FR was moulded to the skin in the mid line contour of the lumbosacral spine. The site of the FR that intersect L2 and S2 were marked. then the FR was lifted carefully from the spine and placed on a piece of paper without altering its configuration. the line of the FR was traced onto the paper. This procedure was repeated three times with one minutes rest between each measurement. Mathematical characterisation of the curve was made and expressed as an angle. The mean of three times of measurements was selected as the Lumbosacral angle. The angle of the curve was calculated using the following formula:

$$\Theta = 4\,\text{Arc Tan}\,(2H/L)$$

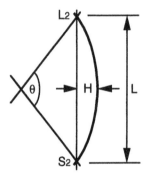

Fig. 1. Schematic representation of formula used for the LSC computation. Where L represents the length of the line drawn between the points L_2 and S_2 and H represents the length of the perpendicular from the mid point of this line to the moulded curve.

2.4 *The Hip Assessment*
A universal goniometer was employed for measuring HROM in the prone position. The method suggested by Gilewich and Clarkson (1989) was used for measuring both hips ROM. The axis of the goniometer was placed on the greater trochanter of the hips with the stationary arm parallel to the midaxillary line of the trunk and the moveable arm parallel to the longitudinal axis of the femur. The angle between the midaxillary line of the trunk and femur represents the degree of hip flexion. Once the

first leg returned back to its start position the other leg was flexed.

2.5 *The Lumbar Flexion Assessment*
the subjects were instructed to be in the standing position with feet and shoulders wide apart. A tape measure was used to measure a distance and mark a point 10 cm above the spinous process of S2. We first marked a point on the S2, then, using a tape measure, we marked another point 10 cm above it. The subjects were asked to bend forward as far as possible. The new distance between the two points was measured again. This method of measurement is referred to as the modified Schober test (Clarkson and Gilewich 1989). (Figure 2)

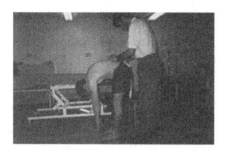

Fig. 2. The lumbar spine flexibility measurement.

2.6 *Pilot Study*
the pilot study was carried out before the commencement of the main part of the exercise regime so that the intensity, duration and arrangement of the exercise was determined according to the subjects' abilities.

2.7 *Exercise Program Procedure*
This study was designed to investigate the effects of exercises on the LSC, the results obtained by the FR measurements were the main criterion for dividing the subjects. The subjects were split into two groups, the experimental group and the control group, on the basis of the above criterion. Each group consisted of 9 subjects. The experimental group performed the following exercise program for four weeks, three sessions a week.

2.8 *Warm-up*
the subjects were asked to warm-up for five minutes. The warm-up supposed to have the following effects: increasing the muscle temperature, reducing the muscles's viscosity, decreasing muscular tension and raising the total body temperature (Astrance and Rodahl 1986). A five minute warm-up exercise was implemented as a pre-exercise regime. Astrance and Rodahl (1986) state that for ordinary exercise a five minute warm-up consisting of light to moderate exercise is usually adequate.

2.9 *Stretching Exercises*
To increase the flexibility in the experimental group, two muscle stretching exercises were performed. The first exercise was aimed at stretching lumbar erector spine muscles (Fig. 3) and the second at stretching iliopsoas muscles (Kisner and Colby 1990). Each stretching exercise was performed for 30 seconds (Alter, M 1988) and repeated three times. (Fig. 4).

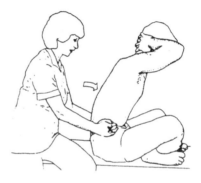

Fig. 3. The lumbar spine muscle stretching (Kisner and Colby 1990).

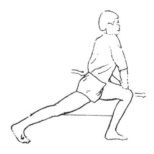

Fig. 4. The hip flexor muscle stretching (Alter, M 1988).

2.10 *Strength Exercise*

To increase the capacity of the muscles to produce the adequate tension necessary for maintaining the pelvis the De Lorme and Watkines approach was followed to improve abdominal and hip extensor muscle strength (Galley and Forset 1987). This is because, in lordosis people the abdominal muscles, especially external oblique and hip extensor are in lengthened position (Kendall et al 1983). This study was concerned with weakness of those muscles which are attached to the pelvis and their weakness may cause exceeded anterior pelvic tilt. The heavy progressive resisted exercise was performed to increase the strength of the program was used to strengthen weakened muscles. Fleck and Kramen (1987), cited by Brown (1991), state "that novices with little or no pre-training have a greater significant strength gain at higher repetitions than an athlete".

The subjects were asked to take part in a test before commencing the exercise regime to determine their maximal strength by the trial and error technique. The curl-up exercise was used to strengthen the abdominal muscles while determined resistance load was placed on the chest (Fig. 5).

The hip extensor muscle strength exercises were performed whilst the subjects flexed their trunks over on the table. They bent each knee and lifted the legs up as far as possible with a resistance load on the thighs (Fig. 6). This was repeated with the other leg.

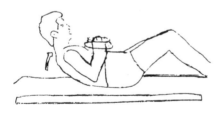

Fig. 5. The abdominal muscle strength exercise.

Fig. 6. The hip extensores muscle strength exercise.

3. DATA ANALYSIS

Analysis of variance with repeated measures was used because the dependent variable measure more than once. Measurement of the same dependent variable more than once provides the experimenter with the opportunity to control individual differences among subjects (Thomas and Nelson 1990).

We calculated the ICC intratester reliability by comparing paired measurements. The ICC was used to examine the degree of measurement between three sets of measures (Rothstein, 1985).

4. RESULTS

The descriptive statistics for the mean and the SD of the three measurements for the LSC, the LF and the left and right HROM in the two groups are reported in Table 1. The analysis of variance indicated significant differences of the LSC in the experiment group (F= 9.57). A similar analysis for the LSC in the control group did not show a significant difference (F= 1.57). The level of significance was set at P<0.05.

Apart from the LSC analysis, the ANOVA in the experimental group for the LF, the HROM (both legs), was not of significant difference. The value of the ICC for intratester reliability for the LSC in the experiment group and the control group was .72 and 87 respectively.

5. DISCUSSION

Proper lumbar spine curve requires normal muscle tone. Some of the studies proposed that lumbar lordosis be related to the function of abdominal muscles (Kendall et al 1983, Cuchemme. G.B. 1966. Williams, P.C. 1965 and Caillet, R. 1962 cited by Waler et al 1987). They supposed that the abdominal muscles pull up the pubic area and as a result tilt the pelvis backward. Thus the lumbar lordosis is maintained in a normal position. Since various muscles are attached to the pelvis and tilt it forward. backward and sidewards, pelvis maintenance requires adequate balance of these muscles. The purpose of this study was to examine the effects of a combined exercise of stretching and strength approaching on the anterior pelvic tilt, for reducing the LSC.

Subjects receiving exercises demonstrated reduction in the LSC in the standing position. The significant difference was in the LSC of the experiment group (F= 9.57). The amount of the LSC in the control group was not significantly different. These findings may support the idea that the frequent repetition of careful exercise with control and with constant orientation changes the habitual postural pattern (Goldthwait et al, 1952, Hawely, 1949, Kelly, 1965; Mensedick, 1937; Rathbone and Hunt, 1965; cited by Luttages and Wells 1982) Flexibility is an important component of good posture. In fact maintenance of flexibility is an essential part of lumbar spine curve re-education We used an active self-stretching technique (Kisner and Colby 1990) to improve the flexibility for the lumbar erector spine and iliopsoas muscles. The subjects carried out exercise himself by his body weight. The results indicated that there was no significant difference in the experiment group's LF however, they showed some improvement in the LF flexibility (from 6.34 cm increased to 6.64 cm).

Despite the attempts of the subjects to carry out iliopsoas muscle stretching exercises, no improvement was observed in the flexibility of iliopsoas muscle. A possible explanation of this finding is that the subjects were not prevented from doing 'sit ups' when they were carrying out the trunk curl-up exercise (Kendall et al 1983) Therefore it seems that iliopsoas flexibility and strength were trained simultaneously. We suggest further examination of combination exercise on the LSC,

Table 1. Details of pre, mid, and post measurements of the two groups

| | Experimental group | | Control group | |
Measurements	Mean	SD	Mean	SD
Lumbosacral curve				
1	46 26	3.46	39	9.95
2	44.83	3.05	41 71	9 54
3	40.89	4.05	39 67	9 05
The right HROM				
1	119	10.27	122.37	8.07
2	121.11	13.33	121.11	7 73
3	118.55	7.21	119.22	6.07
The left HROM				
1	120.88	9.18	121.88	5 56
2	117.33	7.46	119 33	5.02
3	118.22	6 76	118.33	3.87
Lumbar flexion				
1	6.34	1.11	6 24	1 11
2	6.72	.96	5 77	.6
3	6.64	1.16	6.03	1 05

The measurements for the LSC and the HROM were according to degree. The LF flexibility was by centimetre.

REFERENCES

Alter, M. (1988) *Science of Stretching* Human kinetics fools. Champaign Il. USA

Astrand, P O Rodahl, K (1986) *Text Book of Work Physiology*. physiological bases of exercise 3rd ed McGraw-Hill, Inc.

Bloomfield, J. Ackland, T R. and Elliott, B C. (1949) *Applied Anatomy and Biomechanics in Sport*. Blackwell Scientific publication

Brown, R A. (1991) *A study to determine the optimum percentage of the one repetition maximum at which to train using a ten repetition maximum programme*.. Unpublished dissertation School of Physiotherapy, Manchester University

Galley, P M. Forster, A L (1987) *Human Movement* 5th ed. UK Limited, Churchill Livingstone

Kendall, F.R. McCreay, E K and Provance, P G (1983) *Muscles: testing and foundation* 3rd ed Baltimore. William and Willcins

Kisner, C. and Colby, L A (1990) *Therapeutic Exercise*: foundation and techniques 2nd ed Philadelphia Davis Company

Luttagens, K. and Wells, K F (1982) *Kinesiology* 7th ed. Philadelphia College Publishing.

Rothstein, J.M. (1985) *Measurement in Physical Therapy*. Churchill Livingstone Inc

Russell G.S. Highland, T R (1990) *Care of the Low Back* Spine Publications USA

Rose, S.J. Jart, D.L. (1982) *Reliability of a Non invasive Method for Measuring the Lumbar Curve*. Abstract Phys ther. Vol 62, p 642

Thomas, J.R. and Nelson J K (1990) *Research Methods in Physical Activity* 2nd ed Human kinetics books. Champaign. Illinois USA

Walker, M.L. Rothstein, J M Finucane, S D. and Lamb, R.L. (1987) *Relationship Between Lumbar Lordosis, Pelvic Tilt, and Abdominal Muscle Performance*. Vol. 67, No 4. p 512-516.

The Engineering of Sport, Haake (ed.) © 1996 Balkema, Rotterdam. ISBN 90 5410 822 3

Reliability of indices of anterior tibio-femoral ligamentous function in the normal and anterior cruciate ligament-deficient knee

N. Gleeson
Division of Sport, Health & Exercise, School of Sciences, Staffordshire University, Stoke-on-Trent, UK

D. Rees
N.H.S. Centre for Sports Injury Surgery, Leighton Hospital, Mid-Cheshire Hospitals Trust, Crewe, UK

S. Rakowski
Division of Physics, School of Sciences, Staffordshire University, Stoke-on-Trent, UK

ABSTRACT: Anterior tibio-femoral displacement (ATFD) is regulated essentially by the anterior cruciate ligament (ACL). The aim of this study was to document the intra-subject variability and reliability associated with the intra-day non-invasive instrumented measurement of ATFD and knee stiffness in normal and ACL-deficient knees. Such markers of function may aid the identification and prophyactic management of lesser sport-related ligamentous trauma which, without appropriate rehabilitative intervention, may form a precursor to subsequent and more orthopaedically serious musculoskeletal trauma.

Nineteen active adult male recreational team-game players (age 27.5 ± 4.9 years; height 1.78 ± 0.04 m; body mass 79.2 ± 4.0 kg [mean $\pm$ SD]) with unilateral complete ACL rupture (arthroscopically verified) gave their informed consent to participate in this study. Following habituation to procedures, each subject completed a single evaluation session. All subjects were tested as near to the same time of day as possible (± 1 hour). The same test administrator performed all measurements. Within the evaluation session, the subjects completed a series of three assessments of the normal contralateral (NORM) and then ACL-deficient (ACL) knee using a computer-interfaced instrument constructed to measure ATFD and knee stiffness. Indices of ATFD were measured at anterior tibial displacement forces of 120 N and 200 N during incremental loading applied in the sagittal plane with a knee flexed to 0.44 rad (25°). An index of knee stiffness was calculated as the gradient of the resultant applied force-displacement curve between 120 N and 200 N. To localise the action of the tibial displacement forces, subjects were seated and securely strapped at the distal femur and tibia.

Repeated measures ANOVA of V% scores revealed that ATFD measured at increased levels of applied force (200 N: 5.8 ± 4.9 % [NORM]; 2.5 ± 1.8 % [ACL]) demonstrate greater reproducibility than measurements at lower applied forces (120 N: 8.7 ± 7.1 % [NORM]; 3.6 ± 3.2 % [ACL]) ($p<0.005$). Assessment of ATFD in the ACL knee demonstrated greater measurement reproducibility compared to the NORM knee ($p<0.005$). Similarly, the measurement of stiffness in the ACL knee (10.4 ± 10.2 %) demonstrates greater reproducibility compared to the NORM knee (17.3 ± 12.2 %) ($p<0.05$). Consideration of intra-class correlation coefficient (R_I) scores for ATFD measurements suggests that ACL and NORM knees demonstrated similar levels of single measurement reliability across applied force conditions, ranging between 0.98 and 0.99. By contrast, the measurement of stiffness in the NORM knee (0.84) reveals reduced single measurement reliability compared to the ACL knee (0.97). The SEM% scores (95% confidence intervals) for ATFD demonstrated greater error in the NORM knee (± 10.6 % [200 N]; ± 15.8 % [120 N]) compared to the the ACL knee (± 7.0 % [200 N]; ± 7.5 % [120 N]). A similar trend was observed for the measurement of knee stiffness.

Measurements of ATFD undertaken at higher levels of applied force offer greater reproducibility and should be preferred on this basis, assuming subject tolerance. Results suggest that intra-day measurement of ligamentous function in the NORM knee requires greater kinanthropometric rigour compared to the ACL-deficient knee. Since the contralateral uninjured limb is used as a reference to appraise levels of dysfunction in contemporary clinical practice, then it should serve also to identify necessary multiple-trial rigour in the measurement protocol. Results suggest that the mean score of 5 trials is required to achieve an ATFD (200 N) measurement error of better than ± 5% in ACL-deficient and contralateral knee comparisons. By comparison, the mean score of 15 trials is required to achieve equivalent measurement sensitivity in the assessment of knee stiffness.

1 INTRODUCTION

The knee joint is one of the most complex and most frequently injured joints in the body (DeHaven and Lintner, 1986; Zarins and Adams, 1988). A model which defines the limits of normal knee movement comprises primary ligamentous restraints interacting with the other static stabilisers (osseous geometry, capsular structures, and menisci) and with the dynamic muscle stabilisers (Fu, 1993). The anterior cruciate ligament (ACL) is the principal ligamentous restraint to anterior tibio-femoral displacement (ATFD) (Butler, Noyes, and Grood, 1980).

An increasing orthopaedic interest has focussed on ACL insufficiency, reflecting in part, a general increase in the number of orthopaedically serious ACL injuries, and clinical interest in the relationship between injury, mode of surgical treatment intervention, and rehabilitation on short- and long-term knee joint function (Rees, 1994).

The assessment of pathologic anterior laxity of the knee has focussed on tests such as Anterior Drawer, Lachman's and Pivot Shift tests carried out clinically or under anaesthesia. While positive Lachman and Pivot Shift tests show excellent qualitative correlation with ACL dysfunction by comparison to arthroscopic findings, such tests are subject to inter-clinician execution and perception biases (Katz and Fingeroth, 1986). Consequently, instrumented knee testing, which uses mechanical devices attached to the leg to measure applied anterior force-ATFD response curves indirectly via the change in relative position of bony landmarks about the knee, is becoming more frequently evaluated as an objective and quantitative method of assessing ACL insufficiency. The indices of ATFD at specific applied anterior tibial forces and knee stiffness have become popular as potential discriminators of ACL function (Daniel, Malcolm, Losse, et al., 1985; Sherman, Markolf and Ferkel, 1987).

A primary purpose of an index of ligamentous function is to provide a reliable estimate of performance capacity. The fundamental question of how many trials to establish a true measure of an individual's ATFD or knee stiffness characteristics has been the subject of relatively few and methodologically diverse investigations. Markolf, Kochan and Amstutz (1984) investigated the reproducibility of ATFD using an instrumented research tool. Multiple intra-subject trials on separate occasions in two subjects revealed coefficient of variation scores of ±29% to ±33% for ATFD at 100N of applied anterior tibial force.

The aim of this study was to document the intra-subject variability and reliability associated with the intra-day non-invasive instrumented measurement of ATFD and knee stiffness in normal and ACL-deficient knees. Such markers of function may aid the identification and prophyactic management of lesser sport-related ligamentous trauma which, without appropriate rehabilitative intervention, may form a precursor to subsequent and more orthopaedically serious musculoskeletal trauma.

2 METHODS

Nineteen active adult male recreational team-game players (age 27.5 ± 4.9 years; height 1.78 ± 0.04 m; body mass 79.2 ± 4.0 kg [mean ± SD]) with unilateral complete ACL rupture (arthroscopically verified) gave their informed consent to participate in this study. Following familiarisation, each subject returned to the laboratory to complete a sequence of three test sessions each separated by a period of 7 days. All subjects were tested as near to the same time of day as possible (± 1 hour). The same test administrator performed all measurements.

Within the evaluation session, the subjects completed a series of three assessments of the normal contralateral (NORM) and then ACL-deficient (ACL) knee using a computer-interfaced instrument constructed to measure ATFD and knee stiffness. The apparatus and subject orientation during assessment is shown schematically in Figure 1. This was constructed to provide an adjustable rigid chair-like framework and was designed to maintain the knee in a standardised position during the measurement of ATFD. The involved leg was secured by Velcro™ straps and a clamping device at the distal femur and tibia, respectively. The knee joint was maintained at 0.44 rad of flexion (0 rad = full extension) and foot position at 0.26 rad of external rotation (Markolf, Kochan and Amstutz, 1984) and 0.35 rad of plantar flexion. The subject was seated in an upright position with a 0.26 rad angle between the back and seat supports.

Instrumentation to measure ATFD and knee stiffness consisted of two linear inductive displacement transducers (DCT500C, RDP Electronics Ltd., Wolverhampton, U.K.: 0.025 m range). The latter incorporate spring-loaded plungers which were adjusted accurately in three planes to provide perpendicular attachment to the patella and tibial tubercle. During measurements, both transducers were secured to the skin surface using

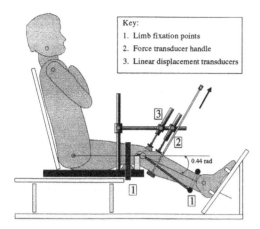

Key:
1. Limb fixation points
2. Force transducer handle
3. Linear displacement transducers

Fig. 1. Subject and ATFD measurement
apparatus orientation.

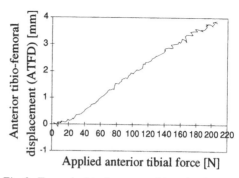

Fig. 2. Example data from one subject showing the
instrumented anterior tibio-femoral displacement
(mm) in response to applied anterior force (N).

tape and able to move freely only in the anterior-posterior plane relative to the supporting framework. The instrument monitored only the relative motion between the patella and tibial sensors and so facilitated the exclusion of measurement artifacts caused by extraneous movements of the leg during the application of anterior displacement forces.

Anterior force was applied in the sagittal plane, at a rate of 66 ($\pm$8) N·s⁻¹, and in a perpendicular direction relative to the tibia by an instrumented force-handle incorporating a load cell (Model 31E500N0, RDP Electronics Ltd., Wolverhampton, U.K.: range 500 N). This device was positioned behind the leg at a level 0.02 m distal to the tibial tubercle. The transducers were interfaced to an IBM compatible microcomputer via a 16 channel A/D 12 bit converter (Model PC-28A, Amplicon Liveline Ltd., Brighton, U.K.). Data from all transducers was sampled at 50 Hz.

Measurements on each knee was preceded by two practice trials. During each measurement, subjects were instructed to relax the musculature of the involved limb. Rapid but gentle manual anterior-posterior drawer oscillations were used to facilitate relaxation and to establish a neutral tibio-femoral position from which all measurements were initiated.

Indices of ATFD were calculated as the mean of three intra-session replicates of the net displacement of the patella and tibial tubercle transducers at anterior tibial displacement forces of 120 N and 200 N applied in the sagittal plane during incremental loading with the knee flexed to 0.44 rad (25°). An index of knee stiffness was calculated as the gradient

of the resultant applied force-displacement curve between 120 N and 200 N.

The displacement transducers were calibrated against known lengths throughout the range of linear operation specified by the manufacturer (0.025 m). The standard error of the estimate associated with the recording of displacement was $\pm$1.6 x 10⁻⁵ m. The force transducer was calibrated against known masses through a physiologically valid range (0 N - 220 N) with correction for the mass of the apparatus and angle of force application. The standard error of the estimate associated with the application and recording of the applied force was $\pm$0.003 N. Throughout the period of testing, the calibration of the force and displacement transducers was verified against objects of known mass and length, respectively.

The selected indices of ATFD and knee stiffness were described using ordinary statistical procedures (mean $\pm$ SD). Coefficient of variation (V%) corrected for small sample bias was used to assess variability of indices across 3 trials. The index of V% was calculated according to the expression (SD/mean).(1 + [1/4N]) where N is the number of trials. Intra-class correlation coefficients (R_I) were computed to describe single-measurement reliability (Winer, 1981). Standard error of a single measurement (SEM%) (95% confidence limits, computed as a percentage of the group mean score) was calculated for each index (Thomas and Nelson, 1990). Variability across 3 trials and absolute scores were compared using a 2 (leg [normal and ACL-deficient]) x 2 (level of applied anterior force [120 N, 200 N]) factorial repeated measures ANOVA. An a priori alpha level of 0.05 was applied in all statistical procedures. All statistical analyses were programmed using SPSS/PC+ (V3.1) software (SPSS Inc., 1989).

Table 1. Group mean scores for three intra-session assessments of indices of ATFD and knee stiffness for NORM and ACL knees. Data are mean [SD].

	NORM		ACL	
Index:	120 N	200 N	120 N	200 N
ATFD [mm]:				
trial 1	1.8 (0.8)	2.5 (0.9)	5.9 (2.2)	8.1 (2.8)
trial 2	1.8 (0.7)	2.4 (0.8)	5.9 (2.2)	8.1 (2.7)
trial 3	1.8 (0.7)	2.5 (0.8)	5.9 (2.3)	8.2 (2.9)
Knee stiffness [x 10^{-3} mm•N^{-1}]:				
trial 1	9.0 (4.3)		27.0 (14.1)	
trial 2	8.5 (3.6)		28.3 (13.7)	
trial 3	8.9 (4.2)		29.5 (13.6)	

Table 2. Group mean coefficient of variation (V%) [SD] for intra-session assessments of indices of ATFD and knee stiffness for NORM and ACL knees.

	V% [%]			
	NORM		ACL	
Index:	120 N	200 N	120 N	200 N
ATFD	8.7 (7.1)	5.8 (4.9)	3.6 (3.2)	2.5 (1.8)
Knee Stiffness	17.3 (12.2)		10.4 (10.2)	

Table 3. Intra-class correlation coefficient (R_I) for intra-day assessments of indices of ATFD and knee stiffness for NORM and ACL knees.

	R_I			
	NORM		ACL	
Index:	120 N	200 N	120 N	200 N
ATFD	0.98	0.98	0.99	0.99
Knee Stiffness	0.84		0.97	

3 RESULTS

Table 1 shows group mean ATFD and knee stiffness scores (±S.D.) in three intra-session trials under the selected applied experimental conditions. Repeated measures ANOVA revealed no significant differences in absolute ATFD and knee stiffness scores across intra-session trials under the experimental conditions.

Table 2 shows group mean coefficient of variation (V%) scores for intra-session assessments of indices of ATFD for NORM and ACL knees. Tables 3 and 4 show intra-class correlation coefficient (R_I) and standard error of the measurement (SEM%) (95% confidence intervals, expressed as a percentage of the group mean score) for intra-session assessments of indices of ATFD for NORM and ACL knees.

Repeated measures ANOVA of V% scores revealed that ATFD measured at increased levels of applied force (200 N: 5.8 ± 4.9 % [NORM]; 2.5 ± 1.8 % [ACL]) demonstrate greater reproducibility than measurements at lower applied forces (120 N: 8.7 ± 7.1 % [NORM]; 3.6 ± 3.2 % [ACL]) ($F_{[1,18]}$ = 11.8; p<0.005). Assessment of ATFD in the ACL knee demonstrated greater measurement reproducibility compared to the NORM knee (($F_{[1,18]}$ = 13.0; p<0.005). Similarly, the measurement of stiffness in the ACL knee (10.4 ± 10.2 %) demonstrates greater reproducibility compared to the NORM knee (17.3 ± 12.2 %) ($F_{[1,18]}$ = 43.4; p<0.05). Consideration of intra-class correlation coefficient (R_I) scores for ATFD measurements suggests that ACL and NORM knees demonstrated similar levels of single measurement reliability across applied force conditions, ranging between 0.98 and 0.99. By contrast, the measurement of stiffness in the NORM knee (0.84) revealed reduced single measurement reliability compared to the ACL knee (0.97). SEM% scores (95% confidence intervals) for ATFD demonstrated greater error in the NORM knee (± 10.6 % [200 N]; ± 15.8 % [120 N]) compared to the the ACL knee (± 7.0 % [200 N]; ± 7.5 % [120 N]). A similar trend was observed for the measurement of knee stiffness.

4 DISCUSSION

Results suggest that intra-session changes in indices of ATFD and knee stiffness can be attributed to biological and technological error sources rather than to systematic learning effects. This finding suggests further that the assessment of the number of trials to achieve appropriate measurement sensitivity can be made on reproducibility and reliability criteria alone.

40

Table 4. Standard error of the measurement (SEM%) (95% confidence intervals, expressed as a percentage of the group mean score) for intra-day assessments of indices of ATFD and knee stiffness for NORM and ACL knees.

Index:	SEM% [%]			
	NORM		ACL	
	120 N	200 N	120 N	200 N
ATFD	15.8	10.6	7.5	7.0
Knee Stiffness	37.8		18.1	

The V% data suggest that the application of higher levels of applied force during the intra-leg estimation of ATFD will provide greater levels of measurement reproducibility and sensitivity for both normal and ACL-deficient knees. It may be that the latter observations are associated with a more consistent interaction of those factors limiting normal knee movement in the preferred leg under conditions of increased anterior force (Fu, 1993). Assessments of ATFD using low levels of applied force would appear to offer an unfavourable threat to the establishment of appropriate measurement sensitivity and utility. Measurements of ATFD undertaken at higher levels of applied force offer greater reproducibility and should be preferred on this basis, assuming subject tolerance.

Generally, R_I for the indices of ATFD and knee stiffness exceeds the clinically acceptable reliability coefficient threshold of greater than 0.80 (Currier, 1984). However, overall group mean V% and SEM% scores across the applied force and leg conditions for these indices, which range between ±2.5% (ATFD, ACL leg, 200 N) and ±17.3% (knee stiffness, NORM leg), and between ±7.0% (ATFD, ACL leg, 200 N) and ±37.8% (knee stiffness, NORM leg) [95% confidence limits], respectively, indicate a limited capacity to discriminate physiological change in ACL functional capacity based on single-trial assessments for both intra- and inter-leg comparisons. Furthermore, they suggest that in many applications which require high levels of measurement sensitivity, it is imperative to use a mean score of multiple trials as the basis for estimating ATFD and knee stiffness in order to reduce measurement error.

Results suggest that intra-day measurement of ligamentous function in the NORM knee requires greater kinanthropometric rigour compared to the ACL knee. Since the contralateral uninjured limb is used as a reference to appraise levels of dysfunction in contemporary clinical practice, it should serve also to identify necessary multiple-trial rigour in the measurement protocol.

Estimated error of the mean score of multiple trials would be expected to vary inversely with the square root of the number of intra-leg replicates, assuming a normal distribution of the replicates (Winer, 1981; Gleeson and Mercer, 1996). Using this criterion, the results suggest that the mean score of 5 trials is required to achieve an ATFD (200 N) measurement error of better than ± 5% in ACL-deficient and contralateral knee comparisons. By comparison, the mean score of 25 trials is required to achieve equivalent measurement sensitivity in the assessment of knee stiffness (95% confidence limits).

It should be noted that these estimates relate to the average group response which does not reflect fully the ATFD and knee stiffness heterogeneity of some subjects within this sample.

These findings present a logistical threat to measurement utility, particularly in relation to the detection and prophyactic management of lesser sport-related ligamentous trauma which, without appropriate rehabilitative intervention, may form a precursor to subsequent and more orthopaedically serious musculoskeletal trauma. For example, following ACL reconstructive surgery, a subject may demonstrate substantial reductions in ATFD of the involved knee, from 11.2 mm (4 months post-surgery) to 4.3 mm (18 months post-surgery) during aggressive physiological rehabilitation (Gleeson and Rees, unpublished observations). Thus by comparison to this relatively large change in capacity, discrimination of subtle but important ligamentous dysfunction would be expected to demand far superior levels of measurement sensitivity.

The measurement sensitivity concerns which have been identified in this paper provide a basis from which to question the rationale for the application of instrumented knee testing for the indirect measurement of ATFD and knee stiffness. The trade-off between measurement sensitivity and utility means that in applied clinical settings, the detection of potentially subtle and important differences in ATFD and knee stiffness may not be possible. The data suggests that the routine use of instrumented knee testing is justifiable in situations such as in adaptive recovery following injury or surgical interventions, where relatively large changes in capacity can be expected.

ACKNOWLEDGEMENTS: The authors acknowledge gratefully the contributions of Graham Barlow and Arthur Smallman, Senior Technicians Staffordshire University, who constructed the measurement apparatus, and thank Jane Kirkham and Sue Dugdale, Senior Physiotherapists M.C.H.T., for their assistance during preparatory work for this study.

5 REFERENCES

Butler, D.L., Noyes, F.R. and Grood, E.S. (1980). Ligamentous restraints to anterior-posterior drawer in the human knee; A biomechanical study. *Journal of Bone and Joint Surgery*, 62A: 259-270.

Currier, D.P. (1984). *Elements of Research in Physical Therapy* (2nd ed). Baltimore: Williams and Wilkins.

Daniel, D., Malcolm, L., Losse, G., Stone, M., Sachs, R. and Burs, R. (1985). Instrumented measurement of anterior laxity of the knee. *Journal of Bone and Joint Surgery*, 67A: 720-726.

DeHaven, K.E. and Lintner, D.M. (1986). Athletic injuries: Comparison by age, sport, and gender. *American Journal of Sports Medicine*, 14: 218-224.

Fu, F.H. (1993). Biomechanics of knee ligaments. *Journal of Bone and Joint Surgery*, 75A: 1716-1727.

Gleeson, N.P. and Mercer, T.H. (1996). The utility of isokinetic dynamometry in the assessment of human muscle function. *Sports Medicine*, 21: 18-34.

Katz, J.W., and Fingeroth, R.J. (1986). The diagnostic accuracy of rupture of the anterior cruciate ligament, comparing the Lachman Test, the anterior draw sign, and the pivot shift test in acute and chronic knee injuries. *American Journal of Sports Medicine*, 14, 1: 88-91.

Markolf, K., Kochan, A. and Amstutz, H. (1984). Measurement of knee stiffness and laxity in patients with documented absence of the anterior cruciate ligament. *Journal of Bone and Joint Surgery*, 66A: 242-253.

Rees, D. (1994). ACL reconstructions: Possible modes of failure. *Proceedings of the Royal College of Surgeons, Edinburgh - Football Association, Sixth Joint Conference on Sports injury*, Lilleshall, Shropshire, U.K., 2nd-3rd July.

Thomas, J.R. and Nelson, J.K. (1990). *Introduction to Research in Health, Physical Education, Recreation and Dance* (2nd ed.), Champaign, Illinois: Human Kinetics.

Winer, B.J. (1981). *Statistical principles in experimental design*. 2nd Edition. New York: McGraw-Hill.

Zarins, B. and Adams, M. (1988). Knee injuries in sports. *The New England Journal of Medicine*, 318: 950-961.

The Engineering of Sport, Haake (ed.)© 1996 Balkema, Rotterdam. ISBN 90 5410 822 3

Musical expertise: The dynamic movement of the trombone slide

Mark Kruger, Mark Lammers, Lela June Stoner & Richard Fuller
Gustavus Adolphus College, St. Peter, Minn., USA

Debra Allyn
Iowa State University, Ames, Iowa, USA

Abstract: Movement of the right arm of trombone players who differed in level of experience was studied. Forty-two professional, student, and beginning players participated. All performers played a series of musical exercises and excerpts at fast and slow tempos. An ultrasonic ranging system was used to test the hypothesis that professionals would move the slide faster than student performers. Electrogoniometers were used to assess movement of the wrist during performance. Cinematography was used to assess integration of shoulder, elbow, and wrist movement. Data taken from the trombone slide, wrist and elbow indicate that there are similarities of activity among players, but professional performers move the slide more efficiently and use less elbow angle. Professionals also accelerate the slide faster than amateurs. Individual variability in motion was greatest in the wrist. The researchers believe that the biomechanical study of musical performance will contribute to increased understanding of expertise, improved pedagogy, and potentially an understanding of movement related injuries.

Research on the biomechanics of performance in the fine arts is becoming an established area of research (Stoner, 1995). As in the biomechanics of sport, research on the biomechanics of musical performance has been motivated by an interest in preventing and treating performance related injuries (Morasky, Reynolds & Clark, 1981; Caldron et. al. 1986; Elbaum, 1986; Van Horn, 1987; An & Bejanni, 1990), improving individual performance (Basmajian & Newton, 1974; Sloboda, 1982; Cutietta, 1986), and refining pedagogical techniques (Lammers, 1983; Lammers & Kruger, 1991). This paper will provide a brief review of research on musical performance, identify the relevance of trombone performance to problems in the study of expertise and motor skill, and present our comparative data on the differences between expert and less expert trombone performers.

Several studies indicate that one strategy for improving musical performance is to provide feedback to performers based on biomechanical and physiological measures (Basmajian & Newton, 1974; Sloboda, 1982; Cutietta, 1986; Koehler, 1996). Sloboda (1982), in an early review of work on the psychology of musical performance, describes a study conducted by Tucker, Bates, Frykberg, Howarth, Kennedy, Lamb, and Vaughn (1977) in which a pianist was able to improve the quality of a trill by seeing visual feedback on the duration of the individual notes struck in the trill itself. Basmajian and Newton (1974) were able to improve the performance of wind players by providing biofeedback on muscles in the cheek. Koehler (1996) reports improved cello performance for students given feedback on the level of muscular tension present in the bow arms.

A second approach to improving musical performance relies on examination of the movements of successful performers. The goal is to identify the motor skills which are either most efficient or most typical of successful performers. In this tradition, Bejanni and Halpern (1989) examined the effect of body posture on wind support in trumpet players, White and Basmajian (1974) used electromyography (EMG) to examine the embouchures of trumpet players during performance, and Henderson (1979) used EMG to examine tension in the throat muscles during trumpet performance.

Careful description of the differences between successful and less successful performers also allows one to assess current pedagogical assumptions about the motor behaviors which lead to successful performance. For example, it has been assumed that trombone players, like tennis players, should avoid using their wrists while moving the trombone slide. However more successful performers use more rather than less wrist while playing a simple etude (Lammers & Kruger, 1991).

The study of successful trombone players has a number of advantages for the study of the acquisition of motor skills Trombone performance is a natural task in which individuals differ in ability. It is a complex task which requires rapid sequences of accurate movements from one slide position to the next, involving movement in several joints - the shoulder, elbow, and wrist. Winold, Thelen, & Ulrich (1994) provide an related example of the study of expertise using biomechanical techniques to study successful cello players.

Theoretical models concerned with the acquisition of motor skill predict that motion should become more and more programmed as expertise is developed (Schmidt, 1975a; Adams, 1977). Examination of successful hitters in baseball and successful table tennis players indicate that both types of performers have developed consistent strokes (Bootsma & van Wieringen, 1990). Consistency is presumed to mean that control of an action largely comes in its initiation. Once initiated each stroke appears to proceed automatically. Given this research, one would expect to find that expertise in trombone performance would involve, as one component, developing a set of motor programs for moving the trombone slide from one position to another. This expertise should be evident in faster, more consistent movements.

Movement during trombone performance has been most extensively studied by Lammers (1983). Seven professional trombonists and seven college students studying the trombone were observed while playing a short etude which required subjects to move between the first arm position and each of the other six positions of the slide used to produce different pitches on the instrument. Surface electromyography was used to observe activity in triceps and biceps of the upper arm and the extensor and flexor carpi radials of the lower arm. Electrogoniometers were attached to the elbow and wrist. Pen recordings of the measures were used to (1) calculate the angle of the elbow used for each of the seven positions on the trombone slide, (2) calculate the time used to move from position to position, 3) calculate the amount of time the slide was held in each position during the exercise, (4) calculate the angle of flexion-hyperextension of the wrist, and finally (5) the amount of time the wrist was held at a particular angle. Consistency was found across and within subjects in elbow angle as a function of position and relative change in muscle activity as a function of position. However, differences as a function of expertise did emerge. In comparison to student performers, professional trombone players used less elbow angle and less muscle activity while moving the slide more quickly from position to position. Observations on the wrist were inconclusive.

In a second study, Lammers and Kruger (1991) used a triaxial goniometer to observe abduction and adduction of the wrist as well as flexion and hyperextension. Ten college level performers participated in this study. Subjects performed the same etude as in the earlier study. Pen recordings provided data on changes in wrist angle from position to position. All subjects showed changes in both flexion-hyperextension and abduction-adduction of the wrist during performance. The more accomplished performers used the wrist more rather than less. This was most notable in the longest movements to the sixth and seventh positions. Differences between the inconclusive findings of the first study and those in the second study were attributed to differences in the goniometers used to assess performance. The research reported here compares the performance of professional performers, college level student performers, and beginning trombone students. All performers played a series of musical exercises and excerpts at fast and slow tempos. An ultrasonic ranging system was used to test the hypothesis that professionals would move the slide faster than student performers. Electrogoniometers were used to assess movement of the wrist during performance. Cinematography was used to assess integration of shoulder, elbow, and wrist movement.

METHOD

Subjects

Forty-two trombone players participated in this study. Participants were recruited from three groups - beginning players who had played less than three years, college students, and adults performing in either professionally or at an advanced amateur level. The subjects ranged in ability and age from sixth graders who had played for one year to senior players who are full-time professional musicians. Each subject was fully informed and consented to participate in this study. Subjects played their own instruments and were given a short period to warm up before their performance was recorded. All subjects were paid for their participation

Musical Exercises

The performance tasks involved three study exercises and two short musical excerpts. The first study exercise was the b flat scale played ascending and descending without repeating the top note and in common time in half notes with the quarter note equal to sixty beats per minute. The subject was asked to repeat the b flat scale with the half note equal to sixty beats per minute (twice as fast). Then the b flat scale was played again with the half note set at 120 beats per minute (again, twice as fast). Beginning students repeated these three exercises using the fastest tempo first and the slowest tempo last.

The second study exercise was constructed to make use of the entire slide. Starting in first-position "F" the exercise consisted of two notes played on each note descending chromatically to each position from one through seven. The subject was asked to perform this exercise with the quarter note set at ninety-two beats per minute and then twice that fast (half note at 92 beats per minute). Finally, the subject was asked to play this exercise with the whole note set at 92 beats per minute.). Beginning students repeated these three exercises using the fastest tempo first and the slowest tempo last.

The third study exercise was composed to use each of the forty-two possible motions of the trombone slide. Through a series of random numbers assigned to each motion an exercise was developed to be played with the metronome set at 72 quarter notes per minute. Because of the difficulty of this piece some beginning performers did not complete this or the remaining exercises.

The fourth performance task was an excerpt from "Sento nel core" by Stefano Donaudy published by G. Schirmer, New York, 1972. For this musical excerpt the metronome was set at 69 beats per minute. The subject performed that musical excerpt once.

The fifth task was another musical excerpt titled "Minuet alternat" from Sonata VI by Johann Ernst Galliard, published by G. Schirmer, New York, 1963. The performer played this excerpt once with the quarter note set at 116 beats per minute on the metronome.

Demographic Measures

Before beginning each subject responded to a brief questionnaire which requested the following information: age, number of years playing the trombone, number of hours per week spent playing the trombone. Each player's upper (shoulder to elbow) and lower (elbow to wrist) right arm were measured. The means for age, hours of performance per week, and length of the upper and lower arm in centimeters are found in Table 1.

Table I
Study Participants

group	age (years)	hours/week of performance	length in cm upper arm	lower arm
Pros	38	18.5	37	27.5
Amateurs	40	9.4	36	27.5
College Students	19.5	9.6	35	26
Beginners	12.8	5	32	24.9

Measures of Movement

Two measurement systems were used to assess the movement of performers. A Vernier Ultrasonic Motion Detector interfaced to a Hewlett Packard QS/165 microcomputer was used to record distance, velocity, and acceleration of the trombone slide. A 30.5 cm circular target made of posterboard was attached to the end of each performer's trombone slide. The motion detector was attached to a small stand placed in front of the performer and aimed at the modal position of the target on the end of the slide. Penny & Giles electrogoniometers were used to measure the relative change in the position of the elbow and the wrist in two planes. Data were recorded using a Colburn LabLinc A/D module attached to an IBM AT microcomputer. A Peak-5 system was used to digitize motion using reflective dots attached at the torso, back of the neck, the shoulder, the elbow, the wrist, and two locations on the trombone slide.

RESULTS AND DISCUSSION

Quality of Performance and Arm Length

Performances on each of the musical exercises were evaluated for musical quality and number of audible errors by an undergraduate music student with performance experience who was blind to the background of each performer. No correlation was found between either quality of performance or number of audible errors and length of the upper and lower arm. Professional performers, not surprisingly, made fewer audible errors summed across all of the exercises and they performed more musically than either adult amateurs or college student performers.

Movement of the Slide

Figure 1 shows a representative record of the movement of the slide by a professional performer playing the second study exercise at 92 beats per minute (two beats in each position). In this exercise the performer moves the slide from first position to second, second to first, first to third, third to first,

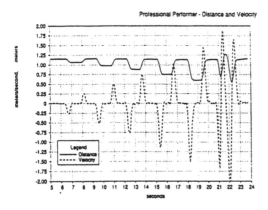

Fig 1. Movement of the trombone slide by a professional performer recorded in distance (meters) and velocity (meters/second).

first to fourth, and so on. As can be seen the velocity of the slide is dependent upon the distance to be traveled. This was true across all performers. The velocity of the slide also differed consistently as a function of expertise. Professional performers moved the slide faster than college aged student performers. This was most noticeable in the long motions from the first to the sixth position and from the first to the seventh position. Trombone students are often told that they should move the slide as fast as possible all of the time. Our data suggest that students and professionals alike move the slide as fast as the tempo of the music and distance to be traveled dictates. These findings replicate the earlier observation that professionals move more quickly from position to position while producing less overall muscle activity in the biceps and triceps (Lammers & Kruger, 1991). Comparison with beginning students on this exercise were not possible. Beginning students moved their slide from

side to side (rotating their trunk) considerably more than either the college students or the professional performers. Consequently, the reflective cone placed on the end of the slide moved out of position for accurate readings. However, the increased rotation side to side in beginning students is an interesting observation in itself. Some professional performers consistently rotated their trunk in the move to seventh position far enough to move the reflective cone outside of the ranger's view. Unlike the beginners, however, these motions only occurred in the movement to seventh position.

Motion of the Wrist

Figure 2 shows the relative change in position of the elbow and wrist in another representative performance. As can be seen this professional moved his wrist systematically during movement to each position. For this performer, motion of the wrist was most pronounced in the flexion-extension plane, but was also present in the abduction-adduction plane. All performers showed some systematic

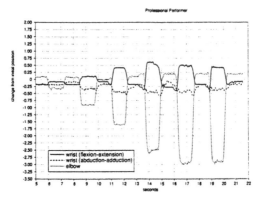

Fig. 2. Change in relative position of the elbow and wrist. Negative voltage changes in the elbow and flexion-extension plane of the wrist represent extension. Negative voltage changes in the abduction-adduction plane of the wrist represent adduction.

motion of the wrist in one or the other plane. However, the use of the wrist differed from performer to performer, and was not systematically associated with skill or experience. These findings are consistent with our earlier observation that highly skilled student performers produced more rather than less wrist motion when compared to less

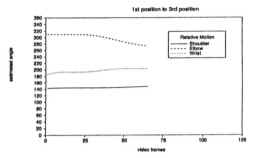

1st position to 3rd position

Figure 3: Motion of the shoulder, elbow, and wrist during the move from first to third position as estimated from digitized motion of markers in two dimensions with the Peak-5 system.

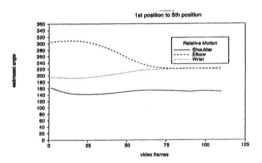

1st position to 5th position

Figure 4: Motion of the shoulder, elbow, and wrist during the move from first to fifth position as estimated from digitized motion of markers in two dimensions with the Peak-5 system.

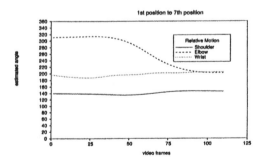

1st position to 7th position

Figure 5: Motion of the shoulder, elbow, and wrist during the move from first to seventh position as estimated from digitized motion of markers in two dimensions with the Peak-5 system.

skilled performers. This is more pronounced in the current study, most likely because of the shift from mechanical triaxal goniometers to the light weight electrogoniometers produced by Penny and Giles.

Motion of the Shoulder

Subjects used their shoulders most prominently in the longer motions. The relative use of the shoulder, elbow, and wrist can be seen in the three figures below. They represent change in the relative angle of each joint during the performance of the second study exercise. The motion shown is for the same professional performer whose movements are depicted in the earlier figures. We observed that professionals utilized their shoulder more effectively than students. This was partly due to the way in which many of them held their instruments. Rather than holding the instrument straight away from their face the tended to point the bell off to the right and down. This reduces the distance one needs to reach in order to accurately place the slide in the far positions. The figures below show the relative motion of the shoulder, elbow, and wrist during motions from the first position to the third, fifth, and seventh positions respectively.

Control Mechanisms

Finally, we present three phase plots which show the change in slide velocity as a function of slide position. Saltzman and Kelso (1987) and Abraham and Shaw (1992) argue that this type of graphical representation of task performance can be used to differentiate the control processes which underly a skilled motion. Winold, Thelen, and Ulrich (1994) use a similar analysis to examine cello performance. The three phase plots below show that change in velocity is smooth throughout each performance. Again, the figures represent the performance of a professional performer playing our second study exercise, this time at each of three tempos. Notice how velocity increases in the far motions (those that reduce the distance in meters from the ultrasonic ranger to the cone on the end of the trombone slide).

This series of studies has shown that the use of biomechanical techniques in studying music performance is a useful and important way to enhance the oral tradition of teaching instruments and voice. The work of Tulchinsky and Riolo (1994), Winold, Thelen, & Ulrich (1994) and others provides additional support for the utility of examining performance with biomechanical

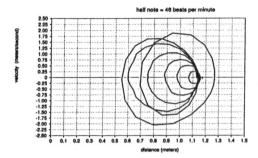

Figure 6: Phase-plane trajectories of the trombone slide position (distance in meters from the ultrasonic range finder) and velocity (meters/second) during the second study exercise played at 46 beats per minute by a professional performer. Each circle represents a complete movement from the first position out to another position and back to first position.

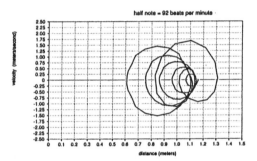

Figure 7: Phase-plane trajectories of the trombone slide position (distance in meters from the ultrasonic range finder) and velocity (meters/second) during the second study exercise played at 92 beats per minute by a professional performer.

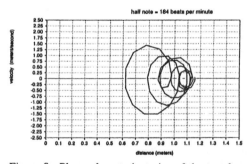

Figure 8: Phase-plane trajectories of the trombone slide position (distance in meters from the ultrasonic range finder) and velocity (meters/second) during the second study exercise played at 184 beats per minute by a professional performer.

techniques. The findings reported above should have an influence upon the trombone teacher in the way s/he approaches the student's right arm which, heretofore, was largely ignored in the literature. What do we know now about successful trombone performance that we did not know before? Professional trombonist use less muscle activity than do students and amateurs. Professional trombonists use less elbow angle in positions two through seven than do their student counter parts, and all players use as much speed as necessary to get to the next position, but the professionals get there faster.

Our future work on this problem will focus on increasing our understanding the effect of the visual, kinesthetic, and auditory feedback processes which influence movement of the slide arm, the comparison of exercise performance and musical performance, and impact of making random (unusual) sequences of motions rather than musically sensible sequences of motions.

REFERENCES

Abraham, R.H. & Shaw, C.D. (1982) *Dynamics - The Geometry of behavior*. Santa Cruz, CA: Ariel Press.

Adams, J.A. (1977) Feedback theory of how joint receptors regulate the timing and positioning of a limb. *Psychological Review*, 84, 504-523.

An, K. & Bejjani, F.J. (1990) Analysis of upper-extremity performance in atheletes and muscians. *Hand Injuries in Sports and Performing Arts*, 6(3), 393-403.

Basmajian, J.V. & Newton, W.J. (1974) Feedback training of parts of buccinator muscle in arm. *Psychophysiology*, 11, 92.

Basmajian, J.B. and White, E.R. (1973) Electromyography of muscles and their role in trumpet playing, *Journal of Applied Physiology*, 35(6), 892-897.

Bootsma, R.J. and van Wieringen, P.C.W. (1990) Timing an attacking forehand drive in table tennis. *Journal Experimental Psychology: Human Perception and Performance*, 16(1), 21-29.

Caldron, P.H., Calbrese, L.H., Clough, J.D., Lederman, R.J., Williams, G. & Leatherman, J. (1986) A survey of musculoskeletal problems encountered by high-level musicians. *Medical Problems of Performing Artists*, 1(4), 136-139.

Cutietta, R. (1986) Biofeedback training in music: From experimental to clinical applications. *Council for Research in Music Education Bulletin*, Spring, 35-42.

Elbaum, L. (1986) Musculatory problems of instrumental musicians. *The Journal of Orthapaedic and Sports Physical Therapy*, 8(6), 285-287.

Koehler, W. (1996) Physiological mapping of bowing technique. *Proceedings of the Third International Conference on Technology in Music Education*, Institute for Music Research, San Antonio, Texas.

Lammers, M. (1983) An electromyographical examination of selected muscles in the right arm during trombone performance. Unpublished doctoral dissertation, University of Minnesota.

Lammers, M. and Kruger, M. (1991) The right arm of trombonists. *International Trombone Association Journal*, Summer, 14-17.

Morasky, R.L., Reynolds, C. & Clark, G. (1981) Using biofeedback to reduce left arm extensor EMG of string players during musical performance. *Biofeedback and Self-Regulation*, 6, 565-572.

Saltzman, E. & Kelso, J.A. (1987) Skilled actions: A Task-Dynamic approach. *Psychological Review*, 94, 84-106.

Schmidt, R. A. (1975) A schema theory of discrete motor skill learning. *Psychological Review*, 82, 251-255.

Sloboda, J. (1982) Music performance. In Deutsch, D. (Ed.) *The Psychology of Music*. New York, N.Y: Academic Press.

Stoner, L. J. & Brink, C. (1995) Biomechanics in the Arts. 2nd ed. In Adrian, M. & Cooper, J. *Biomechanics*. Indianapolis, IN: Benchmark Press.

Tulchinsky, E., Riolo, L. (1994) A biomedical analysis of the violinist's bow arm. *Medical Problems of Performing Artists*, December, 119-114.

Van Horn, R. (1987) Medicine: The Miller Health Care Institute. *Modern Drummer*, May, 29-105.

Winold, H., Thelen, E. & Ulrich, B.D. (1994) Coordination and control in the bow arm movements of highly skilled cellists. *Ecological Psychology*, 6(1), 1-31.

The Engineering of Sport, Haake (ed.) © 1996 Balkema, Rotterdam. ISBN 90 5410 822 3

Finite element analyses of additional stressing during (light) sporting events following total hip arthoplasty

S. P. G. Madabhushi, A. S. Usmani & D. R. Fairbairn
Department of Civil & Environmental Engineering, University of Edinburgh, UK

M. Rajalakshmi
Liberton, Edinburgh, UK

ABSTRACT: Over 800,000 total hip replacement (THR) operations are performed each year in the UK. A third of these operations fail due to aseptic loosening of the prosthesis. The revision surgeries are carried out using the 'Impaction grafting technique' introduced in Exeter in 1987. This technique largely depends on the mechanical strength of the impacted bone graft into which the prosthesis is fixed. As the patient mobilises the bone graft must withstand forces of 2.5 to 3.5 times the body weight during normal walking. As the operation is essentially long-term, it is very likely that the patient will engage in light sporting events like jogging and swimming several months after the surgery. The light sporting activities will clearly induce larger forces on the bone graft and are cyclic in nature. In this paper we consider the effects of these cyclic forces on the bone graft-prosthesis system. The stresses and deformations in different regions of the bone graft under these forces will be analysed.

1 INTRODUCTION

Primary hip replacement operations are carried out on over 800,000 patients within the UK. A stainless steel prosthesis is press-fitted into the femur bone in a cementless surgery. In Fig.1 the prosthesis inserted into the femur bone is shown schematically. Finite element analysis was used successfully in estimating the stresses in the femur bone following these surgeries by Huiskes (1990). In this paper Huiskes identifies the load transfer mechanism in which the shear stress induced in the femur bone balances the axial force imposed by the prosthesis. Earlier, Crowninshield et al (1980) used an axisymmetric beam model in their finite element analysis which simulated the press-fit prosthesis and carried out a parametric study on different kinds of prothesis (Charnley, Aufranc-Turner etc.). Von Mises yield criterion was used in predicting the octahedral stresses in different regions of the femur bone. More recently Kang et al (1993) have carried out a three dimensional finite element analyses based on Computerised Tomography (CT) scan images of the femur and the prosthesis. The normal and shear stresses in the femur bone were predicted based on the these

analyses in which they use simple elastic constitutive models for the femur bone.

All the above analyses led to
a) better understanding of the load transfer mechanism from the prosthesis into the femur bone,
b) improvements in the design of the prosthesis for more efficient transfer of axial load by generating larger shear stresses and
c) establishment of finite element analyses as an effective tool in studying the prosthesis-femur bone system.

It is estimated, based on clinical studies, that a third of all the primary total hip replacements fail due to aseptic loosening of the prosthesis. The longevity of the revision surgeries is much less than that of the primary THR surgeries. One of the main reasons for this is ascribed to poor bone stock. Kershaw et al (1991) identified that the risk of failure in patients with poor bone stock is six times as much as in the patients with good bone stock.

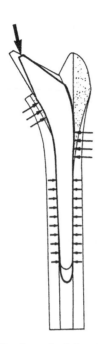

Fig.1 Press-fitted prosthesis into the femur bone

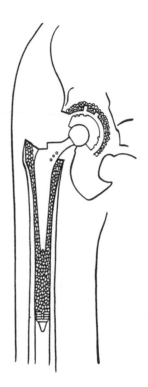

Fig.2 Prosthesis placed by the impaction grafting technique

The techniques of impaction grafting was introduced at Exeter in 1987 to overcome the problems of poor bone stock. In this technique impacted bone graft (chipped bone) is introduced into the femur and the prosthesis is inserted into the bone graft. In Fig.2 a schematic diagram showing the femur bone, bone graft and the prosthesis is presented. In a recent clinical study Gie et al (1993) have observed that this technique is very successful in all of the revision surgeries carried out. The load imposed on the prosthesis is several times more than the body weight of the patient. It is estimated that during normal level walking the load on the prosthesis is 2.5 to 3.5 times that of the body weight. Owing to the long term nature of the surgery, it is very likely that the patient will engage in light sporting events some months or years after the revision THR surgery. In light sporting events like jogging, swimming etc., it should be expected that the load on the prosthesis is even more. Further, for an average person the number of walking cycles is as high as 1×10^6 per year. The load transfer mechanism in the impaction grafted arthoplasties is very different from compared to the press-fitted prosthesis in a primary THR operation. The load from the prosthesis (see Fig.2) is first transferred into the bone graft by generating shear stresses along the prosthesis-bone graft interface and the end bearing of the prosthesis into the bone graft. The stresses in the bone graft will induce shear stresses along the femur-bone graft interface. It is imperative that there will be shear deformations within the bone graft before it mobilises its full shear strength. As a result, there will be subsidence of the prosthesis.

In this paper, we will investigate the stress generated in the different regions of the bone graft by considering it as a granular material. The types of loading considered will be both static as well as cyclic. Also the deformation of the bone graft and settlement of the prosthesis under both these types of loading will be investigated. Since the mechanical properties of this granular material, namely the bone graft, are not known. These are investigated in the next section.

2 MECHANICAL PROPERTIES OF THE BONE GRAFT

A bone graft sample was obtained by putting a freeze dried femoral head through a specially constructed bone mill. The bone graft obtained from the mill are cleaned with formalin, alcohol and water and are dried by centrifuging for 15 minutes. Also, a sample core from the femoral head is taken from which the apparent density (total weight of the bone sample/total volume) of the bone is established. Mueller et al (1966) describe methods for determining the apparent density of bone. Using a similar procedure the apparent density of the bone was estimated as 285 kg/m^3. Also, the porosity of the bone is taken as 85 % following Galante et al (1970), which gives the void ratio of the bone as 5.7. The real density of the femoral head can be obtained from the above values as 1900 kg/m^3. These values compare satisfactorily with the study of Galente et al (1970) on the physical properties of Trabecular bone.

The bone graft sample was tested in a direct shear apparatus often used to test granular materials. The tests on the bone graft were conducted at a constant strain rate of 2.5 mm/minute. In Fig.3 the shear stress-shear strain plots obtained from these tests are presented. The normal stress was changed in each experiment within a range of 13.625 kPa to 231.625 kPa. From the plots in Fig.3 we can see that the bone graft material is clearly strain hardening as the shear strain in the test progresses. In Fig.4 the shear stress at failure (or 10% strain) is plotted against the normal stress. This plot is traditionally used to identify the Mohr-Coulomb failure surface. From the data points in this figure it is clear that a linear Mohr-Coulomb failure surface can be drawn for this material. Also the friction angle of the material is obtained from the slope of the Mohr-Coulomb surface as 27.5°. It must be pointed out that there is a small intercept on the shear stress axis at zero normal stress. This intercept is interpreted as the interlocking of the material.

The mechanical properties obtained from these tests were used in the finite element analyses.

3 FINITE ELEMENT ANALYSES

The non linear finite element analyses reported in this paper were carried out using a program called SWANDYNE, Chan (1988). SWANDYNE is a generalised, fully coupled, effective stress based code for problems in geomechanics. This code was chosen for following reasons;

a) availability of a wide variety of constitutive models including the Mohr-Coulomb relation with a non-associative flow rule.

b) since cyclic nature of the forces on the prosthesis were considered, it was essential to have a code with time stepping facility. (This dynamic facility was preferred compared to Psuedo-static analysis approach of other programs).

c) it is possible to include the effect of pore fluid (bone marrow, clotted blood etc.) in the bone graft in future analyses.

3.1 Constitutive model

Two different constitutive models were used in the present analysis. The femur bone was modelled as a simple elastic material following Kang et al (1993). However, the bone graft itself was modelled as a Mohr-Coulomb type material with a non associative flow rule. It is known that the non associative flow rule is essential to capture the behaviour of granular material, Pastor et al (1985). The Mohr-Coulomb type yield surface was used primarily due to the failure surface in the experiments discussed in the previous section have fitted very closely this yield criterion. However, it is possible to use more sophisticated constitutive models in future analyses. In Table 1 the material parameters used for the femur bone and bone graft are listed.

Table 1 Material Parameters

Zone	Const. Model	Youngs Modl (kPa)	Poisson ratio	Friction Angle $^\circ$	Dilatancy Angle $^\circ$
Femur	Elastic	14.0E6	0.3	--	--
Bone graft	Mohr-Clmb*	1100	0.3	27.5	5
Prosthesis	Elastic	200E6	0.2	--	--
Femur-Bone graft Inter.	Mohr-Clmb*	110	0.3	20	5
Bone Graft-Prosth Inter.	Mohr-Clmb*	2.0E6	0.23	15	10

* Mohr-Coulomb criterion with Non Associated Flow Rule

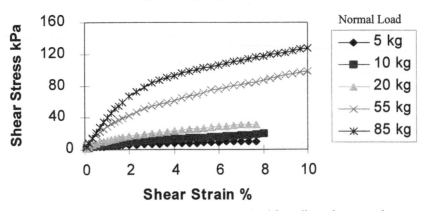

Fig.3 Stress-strain relationship of the bone graft determined from direct shear experiments

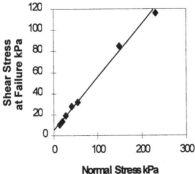

Fig.4 Mohr-Coulomb yield surface of the bone graft

introduced at the Femur-bone graft interface as well as the bone graft-prosthesis interface. In Fig.5 the finite element mesh used for all the analyses discussed in this paper is presented. In this figure all the five regions considered in the analyses are identified. The finite element mesh consisted of a total of 172 isoparametric, quadrilateral and triangular elements. The quadrilateral elements were used for the femur and the prosthesis while triangular elements were used for bone graft as shear deformations are expected in this region. Also the left hand corner node at the base of the FE mesh is fixed in both x and y directions to prevent lateral translation of the femur.

Note that in Table 1 the friction angle for the interface elements is smaller than that for the bone graft.

3.2 Finite element mesh

The total hip arthoplasty using impaction grafting involves fixing the prosthesis in the bone graft as seen in Fig.2. In this figure the femur bone, bone graft and prosthesis are clearly seen. However, in the finite element analysis interface elements were

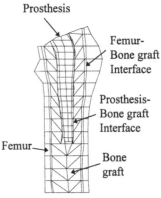

Fig.5 Finite element mesh used in the present analyses

4 FORCES ON THE FEMUR

The force transferred to the femur bone is several times that of the body weight of the patient. In the previous work of Crowninshield et al (1980) the force on the femur was taken as 200 N. Huiskies (1990) has considered a much larger force of 3000 N. In this paper four analyses were carried out in which the force applied to the femur is changed from 1000 N to 4000 N. These analyses are designated as FC-1 to FC-4. Material parameters for these analyses are taken from Table 1. It must be noted that this force is a static force under which shear deformations occur as the equilibrium is reached. Cyclic loads are considered in the next section.

The results from the four analyses are presented in Table 2. From this table it can be seen that as the force transferred to the prosthesis is increased the

settlement of the prosthesis increases. Also the lateral movements of the femur increase with load and also along the height of the femur bone. Further, the unsymmetric structure of the bone and the inclined application of the loading resulted in a difference between the movements on the left and right of the femur. The stresses just below the prosthesis increase with load but their magnitude is lower than that would be expected in a press-fitted prosthesis, Huisk es (1990). On the other hand the settlements predicted are higher than in press-fitted prosthesis. This is reasonable as the bone graft is much more compressible and will require the shear deformations to mobilise the strength to support the external load applied by the prosthesis.

The deformed mesh for the analysis carried out with a prosthesis load of 2000 N is presented in Fig.7. In this figure the undeformed mesh is overlain to show the extent of settlement of the prosthesis. From this figure it is seen that the prosthesis has suffered significant settlement. This settlement also induced shear deformations in the bone graft which induces shear stresses in the femur bone via the interface elements. The directions of major and minor principal stresses in this analysis are shown in Fig.8. From this figure it can be seen that the maximum stress rotation occurs in the right hand corner of the femur bone. Some stress rotation occurs in the left hand side of the femur. This figure suggests that the load transfer mechanism is via shear and the load is finally carried into the femur (with a bias on the right hand side owing to inclined loading and unsymmetrical geometry). This fact is reinforced by the fact that the stresses just below the prosthesis are relatively small as the load is carried via shear in the interface elements between prosthesis and the bone graft.

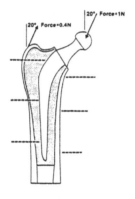

Fig.6 Angle of the load transfer into the prosthesis (after Crowninshield et al,1980)

Table 2 Results of FE analyses with a variation of prosthesis loads

Analysis	Force (N)	Prosth. Settlement (mm)	Femoral lateral Movemt. at Ht-1*(µm)		Femoral lateral Movemt. at Ht-2*(µm)		Femoral lateral Movemt. at Ht-3*(µm)		Stresses under Prosthesis (kPa)		
			Left	Right	Left	Right	Left	Right	σ_x	σ_y	τ_{xy}
FC-1	1000	1.024	-0.23	1.23	-9.64	46.75	-18.2	97.93	-4.3	-1.6	18.4
FC-2	2000	2.026	-0.46	2.47	-19.42	93.93	-36.65	196.9	-6.63	-2.5	28.1
FC-3	3000	3.303	-0.69	3.70	-29.07	140.85	-54.83	295.3	-9.88	-3.6	42.1
FC-4	4000	4.041	-0.93	4.93	-38.79	187.84	-73.16	393.8	-13.2	-4.7	56.1

*Ht-1, Ht-2 and Ht-3 refer to the base, middle and top levels of the femur bone

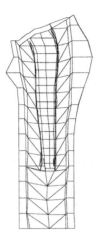

Fig.7 Deformed mesh after a
load of 2000 N was applied
to the prosthesis

Fig.8 Directions of major and
minor prinicipal stresses after
2000 N was applied

5 CYCLIC LOADING

One of the main concerns of this paper are the
additional stresses and deformations induced in the
bone graft due to light sporting activities. The
nature of the load coming onto the prosthesis will
be cyclic and its magnitude larger than the load
induced during normal level walking. Also the
load in transferred into prosthesis at an angle as
seen in Fig.6. Following Crowninshield et al
(1979) we have taken this angle as 20° to the
vertical. Also the nature of the cyclic loading is not
known (sinusoidal or otherwise). Obviously the
exact nature of the cyclic force transferred to the
femur depends on the type of sporting activity. In
this paper we have considered two types of forces
namely sinusoidal and triangular cyclic forces.

5.1 Sinusoidal loading

The sinusoidal loading is applied as a periodic
displacement of the prosthesis at the desired angle
of 20° to vertical with a frequency of 20 Hz. It is
assumed that any cyclic forces induced in light
sporting events will be well below this frequency.
The magnitude of the cyclic displacement is varied
from 0.02 mm to 4 mm in the horizontal direction
and 0.04 mm to 8 mm in the vertical direction.
Only typical results for a horizontal displacement
of 0.2mm and a vertical displacement of 0.4 mm
are shown here.

In Fig.9 the lateral movement of the femur at the
base, middle and top level are shown both to the
left hand side and right hand side. As the
prosthesis is moving up and down in a sinusoidal
manner, the femur expands and contracts laterally.
The lateral movements increase from the base to
the top of the femur. Also at all levels the right
hand side movement is larger than the left hand
side movement. There is some decrease in the
magnitude of the right hand side movement after
first few cycles owing to the compaction of the
bone graft.

In Fig.10 the stresses (sigma-x, sigma-y and Tau-
xy) in the femur at the three different levels to the
left and right of the femur are shown. The
magnitude of all the stresses increases as we move
from top to bottom. This is to be expected as the
stresses on either side of the prosthesis will be
small as very little shear is transferred in this
region. All the stress components increase from
the top level on as we move towards the base. Also
at any one level the sign of the stress components
changes from + ve to -ve as we move from left
hand side to the right hand side. The compaction
of the bone graft after first few cycles seem to
produce an upward movement of the stress
components on the right hand side of the femur.
The left hand side stress components seem to
change very little with the number of cycles. In
Fig.11 the stresses in the bone graft just below the
prosthesis and well below the prosthesis are

presented. In this figure we can see that the stresses just below the prosthesis are an order of magnitude larger than those in the region well below the prosthesis. Again this is to be expected as there will be stress distribution as we move away from the region just below the prosthesis bringing a drop in the stresses. The redistributed stresses in the bone graft cause larger stresses at the base of the femur bone.

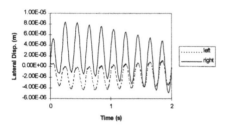

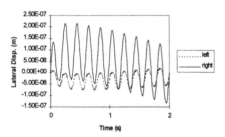

Fig.9 Lateral movements in the femur during sinusoidal displacement of 0.2mm horizontal and 0.4 mm vertical

5.2 Triangular loading

Since the nature of the cyclic loading is not known a second type of periodic displacement which varies as a triangular function was applied to the prosthesis. The magnitude and frequency of this loading was same as the sinusoidal loading so that direct comparisons of the resulting movements of the femur bone and stresses in the femur and bone graft are possible.

In Fig.12 the lateral movement of the femur at the base, middle and top level are shown both to the left hand side and right hand side. As before when the prosthesis is moving up and down in a triangular manner, the femur expands and contracts laterally following the same triangular variation. The lateral movements increase from the base to the top of the femur. Also at all levels the right hand side movement is larger than the left hand side movement. The peak lateral movements in the triangular loading are comparable to those that occurred during the sinusoidal loading. However, this may change when a more impulse type loading is applied. This aspects needs further investigation.

In Fig.13 the stresses (sigma-x, sigma-y and Tau-xy) in the femur at the three different levels to the left and right of the femur are shown when the triangular loading is applied. The magnitude of all the stresses increases as we move from top to bottom. As before, all the stress components increase from the top level on as we move towards the base. Also at any one level the sign of the stress components changes from + ve to -ve as we move from left hand side to the right hand side. The phase relationships between the three stress components changed in the low stress region near the top of the prosthesis. The shear stress Tau-xy on the left hand side is clearly 180° out of phase with the other two stress components. On the right hand side this phase change is about 90°. The reason for this phase difference is not clear and needs further investigation. In Fig.14 the stresses in the bone graft just below the prosthesis and well below the prosthesis are presented. In this figure we can see that the stresses just below the prosthesis are an order of magnitude larger than those in the region well below the prosthesis. Again this is to be expected as there will be stress distribution as we move away from the region just below the prosthesis bringing a drop in the

Stresses at top left of the femur

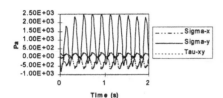

Stresses at the top right of the femur

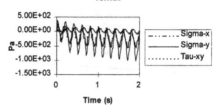

Stresses in the femur to left of mid height

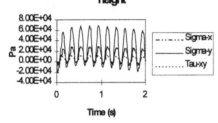

Stresses in the femur to right of mid height

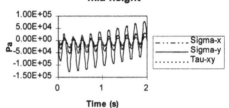

Stresses in the femur to left of base

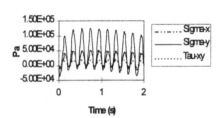

Stresses in femur to right of base

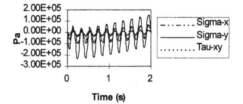

Fig.10 Stresses in the femur during sinusoidal displacement of the prosthesis

Stresses in the bone graft just below the prosthesis

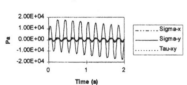

Stresses in the bone graft well below the prothesis

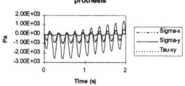

Fig.11 Stresses in the bone graft during sinusoidal displacement

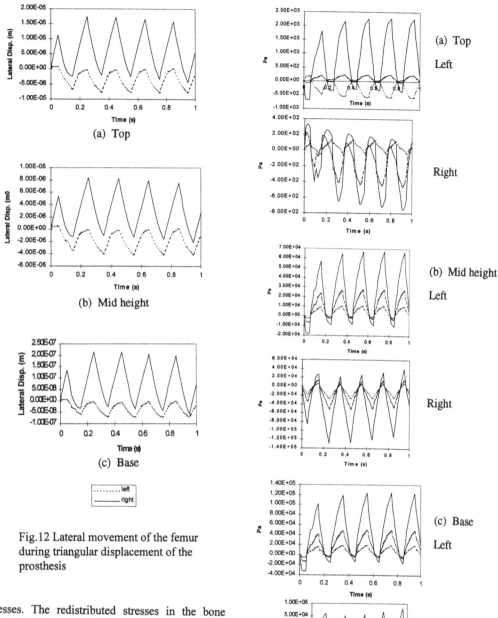

Fig.12 Lateral movement of the femur during triangular displacement of the prosthesis

Fig. 13. Stresses in the femur during triangular displacement of the prosthesis.

stresses. The redistributed stresses in the bone graft cause larger stresses at the base of the femur bone.

6 CONCLUSIONS

The technique of impaction grafting was developed at Exeter for revision total hip arthoplasty operations. This technique, while being extremely successful, depends to a large extent on the mechanical strength of the bone graft. Based

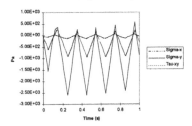

Fig.14 Stresses in the bone graft
during triangular displacement

on initial direct shear box experiments the stress-strain curve of the bone graft was determined. The data suggested that a Mohr-Coulomb type failure criterion is suitable, at least as a first approximation, to carry out the finite element analysis of the prosthesis-bone graft-femur system. Using the data from the direct shear box experiments, finite element analyses were carried out to study the stresses in the bone graft and load transfer mechanisms in the femur-impacted bone graft-prosthesis systems. The load transferred to the prosthesis was varied in the range of 1000 N to 4000 N. In this paper we carried out the finite element analyses using a non-associative form of Mohr-Coulomb yield criterion for the bone graft. Interface elements were used between the prosthesis and the bone graft as well as the femur and the bone graft. The results of the FE analyses indicate that the load is transferred from the prosthesis into the bone graft and then into the femur bone.

Sporting activities will induce in the prosthesis additional cyclic stresses. Since the nature and the magnitude of these cyclic stresses is not known, two types of cyclic loadings namely sinusoidal and triangular loadings were considered in this paper. The finite element analyses were able to predict the lateral movement of the femur due to the sinusoidal movement of the prosthesis. The predicted stress components showed an increasing trend as we move from the top of the femur to the base. Also there was a change of sign of the stress components from left hand side to right hand side of the femur. Similar observations were made for the triangular loading. However, in the triangular loading there were phase differences between the shear stress component and the other stress components. This aspect will be investigated further in future studies. Also, in the future analyses a finer FE mesh with more elements will be used.

Based on these analysis it appears that the magnitude of stresses induced in the femur in the impaction grafting technique is smaller than those induced by the press-fitted prosthesis. This is also true when there are additional stresses on the prosthesis during the light sporting events considered. However, the magnitude of settlement of the prosthesis is larger compared to the press-fitted prosthesis. This is due to the nature of load transfer mechanism which carries the load from the prosthesis via the bone graft into the femur. When the shear deformations in the bone graft occur the bone graft is able to mobilise its strength thus transferring the load into the femur bone.

ACKNOWLEDGEMENTS
We wish to express our gratititude for the helpful discussions of Prof. W.J. Gillespie, Dr. Colin Howie and Dr. Nigel Brewster of Princess Margaret Rose Orthopaedic Hospital, Edinburgh.

REFERENCES
Chan, A.H.C., (1988), A unified finite element solution to static and dynamic problems of Geomechanics, Ph.D thesis, University college of Swansea, Swansea.

Crowninshield, R.D., Brand, R.A., Johnston, R.C. and Milroy, J.C., (1980), An Analysis of Femoral Component Stem Design in Total Hip Arthoplasty, Jnl. of Bone and Joint Surgery, Vol.62-A, No.1, pp 68-78.

Galante,J., Rostoker, W. and Ray, R.D., (1970), Physical properties of Trabecular Bone, Jnl Calc.Tiss.Res., Vol.5, pp 236-246.

Gie, G.A., Linder, L., Ling, R.S.M., Simon, J.P., Slooff, T.J.J.H. and Timperley, A.J., (1993), Impacted cancellous allografts and cement for revision total hip arthoplasty, Jnl. of Bone and Joint Surgery, Vol.75-B, No.1, pp 14-21.

Huiskes, R., (1990), The various stress patterns of press-fit, ingrown and cemented femoral stems, Clinical Orthopaedics and Related Research, No.261, pp 27-38.

Kang, Y.K, Park, H.C., Youm,Y., Lee, I.K., Ahn, M.H. and Ihn, J.C., (1993), Three dimensional shape reconstruction and finite element analysis of femur before and after the cementless type of total hip replacement, Jnl. Biomed. Eng., Vol.15, pp497-504.

Kershaw, C.J., Atkins, R.M., Dodd, C.A.F. and Bulstrode, C.J.K.,(1991), Revision total hip arthoplasty for aseptic failure: a review of 276 cases, J Bone Joint Surg., Vol.73-B, pp 564-568.

Mueller,K.H., Trias, A. and Ray, R.D., (1966), Bone density and composition, Jnl. of Bone and Joint Surgery, Vol.48-A.,No.1, pp 140-148.

Pastor, M., Lueng, K.H. and Zienkiewicz, O.C., (1985), Simple model for transient soil loading in earthquake analysis, Part II: Non associative model for sands, Int. Jnl for Num. and Anal. methods in Geomechanics, Vol.9.

The Engineering of Sport, Haake (ed.)© 1996 Balkema, Rotterdam. ISBN 90 5410 822 3

The influence of body weight on ski jumping performance

Wolfram Müller & Trevor T.J. DeVaney
Institute of Medical Physics and Biophysics, Karl-Franzens University, Graz, Austria

ABSTRACT: All today's world class ski jumpers are very light. This paper gives a scientific background for the discussion concerning the underweight and anorexia problem. There is a staggering tendency to submit very young (and therefore small and light) boys to international competitions. Since ski jumping is a dangerous kind of sport the question of responsibility for the athletes arises.

A highly reliable computer model based on wind tunnel experiments and field study results can be used to explain the effects of low body weight on the ski jumping performance. The computer simulation allows one to study the impact of the athletes weight on the jump length and other characteristics of the flight. All important parameters and initial values influencing the flight path and the jump length can be varied. The angle of the ramp, the air density, aerodynamic characteristics due to positional changes of the athlete or due to improved equipment, the profile of the landing hill, the inrun velocity, the velocity of a gust blowing in the plane of the flight path from any desired direction as well as the perpendicular velocity due to the athletes jumping force. Increasing jumping force has a larger potential to compensate for higher weight at small jumping hills.

Based on these computer simulation studies suggestions can be made for changes to the ski jumping regulations in order to make it less attractive for the athletes to be extremely underweight. The unsolved question of the ski length regulation is connected with the body weight and should consider the results presented here.

1. INTRODUCTION

Reliable predictions from a computer model of ski jumping have to be based on field studies, wind tunnel measurements and a modelling concept which contains the dependancy of the lift force and the drag force to the actual flight position. We have published such a model which is based on the coupled and nonlinear equations of motion (Müller et al. I and II, 1995, 1996). The results obtained contribute to questions of safety (Müller et al. I and II, 1995, 1996), the design of jumping hills (Müller III, 1996) and to the optimization of the flight (Schmölzer 1996; Müller IV 1996).

Straumann (1927) published the first physical analyses of ski jumping in 1927. since then many scientists have studied ski jumping from different points of view. There are several papers on computer modelling of ski jumping (e.g. Denoth

1987; Hubbard et al. 1989; Remizov 1984). Recently Virmavirta and Komi (1993) published results on force measurements during take-off. Kaps et al. (1996) calculated the ground reaction forces during take off by inverse dynamics. Two analyses of the athlete's body position angles during take-off and the early flight phase have recently been published by Schwameder et al. (1995) and Vaverka et al. (1996). Wind tunnel data using a puppet and computer simulation results have recently been published by Jin et al. (1995).

Small and light athletes are advantaged today by the ski length regulation in that they have disproportionally larger skis (today jumpers are allowed to use skis 0.80 m longer than their height). Most of today's world class athletes are very light, some of them are alarmingly underweight and two cases of anorexia have been reported in the media recently (one in Austria, one in Switzerland). Here the weight and height of a representative set of world class ski jumpers is shown and compared to the weight of other athletes and untrained people. The computer simulation studies demonstrate why the ski jumpers have to be light and give the scientific arguments which the discussion on further regulation changes should be based.

2. METHODS

2.1 Reference Jump

The simulation results shown here start out from a highly realistic reference jump based on the mean position angles of 15 excellent jumps (mean jump length: 186.6 m) measured during the World Championship 1994 in Planica. The according sets of lift (L) and drag (D) areas (in m²), measured in a 5 x 5 m² wind tunnel, are: 0.2 and 0.4 (at t = 0 s, i.e. at take off), 0.65 and 0.60 (0.2 s), 0.68 and 0.58 (0.4 s), 0.77 and 0.64 (2.3 s), 0.79 and 0.73 (4.0 s), 0.78 and 0.77 (4.6s), 0.78 and 0.79 (from 5.0 s on until landing). Linear interpolation was used between. The mass of the athlete (including the equipment) was 70 kg and the air density was 1.15 kgm⁻³. This reference jump has before been described in detail (Müller I 1995, Müller II 1996).

For the simulation studies made here the profile data of three jumping hills has been used (Planica, K = 185 m; Calgary, K = 114 m; Calgary, K = 89 m) in order to demonstrate the influence of a variation of the mass m (athlete with equipment) at jumping hills of different sizes. The approach velocities v_0 used were 28.538 ms⁻¹ (Planica), 25.04 ms⁻¹ (Calgary, K = 114 m) and 22.92 ms⁻¹ (Calgary, K = 89 m), respectively. The velocity v_{p0} (perpendicular to the ramp, due to the athlete's jumping force) was 2.24 ms⁻¹ in all three cases. These protocols lead to jumps of l = 185.0 m, l = 114.0 m, and l = 89.0 m. The approach velocity v_0 and the perpendicular velocity v_{p0} are initial values for the equations of motion.

2.2 Equations of motion

$$\dot{v}_x = \left(-F_d \cdot \cos\varphi - F_l \cdot \sin\varphi\right)\frac{1}{m}$$

$$\dot{v}_y = \left(-F_d \cdot \sin\varphi + F_l \cdot \cos\varphi\right)\frac{1}{m} - g$$

$$\dot{x} = v_x$$

$$\dot{y} = v_y$$

$$F_l = \frac{\rho}{2} \cdot w^2 \cdot L$$

$$F_d = \frac{\rho}{2} \cdot w^2 \cdot D$$

$$\vec{w} = \vec{v}_g - \vec{v}$$

x, y, v_x, v_y, t, state variables
F_l ....lift force
F_d ...drag force
w....wind velocity, v_g....velocity of a gust
v....velocity of motion along the path
ρ....air density
m....mass (athlete with equipment)

The athlete's position changes during flight. Our computer model considers the dependency of the aerodynamic forces to the actual flight position.

For the simulations the program package DYNAMIX (by D. Platzer, compare with Müller et al. I 1995, and II 1996) has been used.

2.3. Parameter and initial value variations.

The influence of an altered mass m in combination with other changes of initial values or parameters has been studied. Additionally to the mass the velocity of a gust (v_g) was changed from 0 to 1 ms⁻¹. The angle of the gust ζ was 130° (i.e. blowing up the hill) in all simulations considering a gust ($v_g \neq 0$). ζ was measured from a horizontal line (x-axis) and counted positive in counter-clockwise direction (this is also true for all other angles). A change in the mass m was also combined with different values for v_{p0}, or v_0, or the ramp angle a, or the air density ρ.

3. RESULTS

3.1 Weight and height of athletes

Fig. 1a-d shows the weight and height of 59 ski jumpers, most of them participating in the World Cup (filled circles; curve A).

The data are from measurements with the Austrian skiing team (13 values), from a publication of the Int. Ski Federation (FIS, 1994, 41 values), and some from personal communications or media (5). Curve A is the according quadratic polynomial fit. This curve is below the curves B, C, D and E in

Fig. 1a, which give the mean weight for men (Soc. of Actuaries, 1959) aged from 15 to 16 years (B), 17 to 19 (C), 20 to 24 (D), and 25 to 29 (E). This curve family B-E shows that young men of a given height are lighter than older ones and indicates why many ski jumpers are very young today.

F_1 and F_2 in Fig. 1b,. indicate the zone of „desired weight", which is the weight associated with the highest life expectancy (Stat. Bull. Metrop. Life Insur. Co, 1959).

Fig. 1c compares the ski jumpers weights with the demands of a coach (SF, DRS1): Accordingly the ski jumpers body mass (in kg) should be that of his height minus 120.

In Fig. 1d the data of ski jumpers (filled circles, curve A) are compared to long distance runners (open circles, curve H) and to other athletes (gymnasts, wrestlers, boxers, judokas, shot put athletes, and 100 m runners; open squares, curve J). The values for H and J are from participants of the Olympics in Montreal in 1976 (Carter, 1982).

3.2. *Reference jumps at three jumping hills of different sizes*

For each of the trajectories shown in Fig. 2 the same lift and drag area table (see 2.1), which uses the wind tunnel data obtained with the outstanding World Cup winner and World Champion Andreas Goldberger, has been used. The approach velocities used resulted in jumps to the K-points of the three ramps, thus the jump lengths l obtained were 185.0 m (Planica, Fig. 2a), 114.0 m (Calgary, large hill, Fig. 2b), and 89.0 (Calgary, Fig. 2c), respectively. In competition, as in the computer model the jump length l is determined from a defined length at the K-point and follows the surface of the landing slope. The according lengths of the trajectories l_t were 189.4 m, 117.6 m, and 92.4 m, respectively. While the jump length l of 114 m is 128% of a 89 m jump, the step from 114 m to 185 m is much larger (162%). Presently there are no jumping hills with K ≈ 150 m. Such hills should be considered as well as competitions at very small hills (K ≈ 50 m), where the jumping force would be a predominant factor. For all jumping hills approved by the FIS the parameters can be found in the FIS Certificates of Jumping Hills.

3.3. *Simulation results*

The tables 1 to 3 give a survey of simulation results obtained when the mass m (athlete with equipment) is altered and also when additional variations are made. The tables show the results for a normal jumping hill (Tab. 1: Calgary, K = 89 m), for a large jumping hill (Tab. 2: Calgary, K = 114 m) and for a ski flying hill (Tab. 3: Planica, K = 185 m). The values of the jump length l and the landing velocity v_l and the absolute and percentage differences, when compared to the reference jumps are given. The tables demonstrate the importance of being light: low weight increases the jump length l and decreases the landing velocity v_l. Line g demonstrates that an excellent take off jump has a larger potential to compensate for higher weight at small hills (compare line G in Tab. 1, 2 and 3).

4. DISCUSSION

4.1 *The influence of the mass on the flight path*

In reality, the change of one parameter will influence the others. For instance, if a ski jumper looses weight, this might also affect his flight style. Despite this possible concatenation of effects we can learn much from variation protocols modifying a limited number of parameters, holding all others constant. The wind tunnel data used here is based on a field study from the World Championship in Skiflying, 1994. Here we use the same table of lift (L) and drag (D) values for all three ramps.

A reduction of the mass (athlete with equipment) results in reduced landing velocities. The resulting effect was greatest at the largest jumping hill (Planica, K = 185 m) and smallest at the small hill in Calgary (K = 89 m). The absolute increase in jump length was largest in Planica (6.7 m at a 5 kg reduction) and smallest in Calgary (K = 89 m; 3.5 m). However, the percentage changes in the jump length were approximately equal. An increased velocity perpendicular to the ramp at take off (v_{p0}) due to the athlete's jumping force can compensate for a higher mass. This compensation potential is more pronounced at small ramps. This must be taken into consideration when making changes to the regulations.

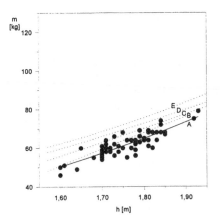

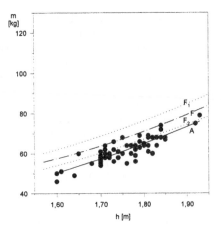

Fig. 1a
Comparison of ski jumper's weight (A) with mean weight of men (Soc. of Actuaries, 1959).
B: men aged from 15 - 16 years
C: 17 to 19 years
D: 20 to 24 years
E: 25 to 29 years

Fig. 1b
Comparison with desired weight (Stat. Bull. Metrop. Life Insur. Co, 1959)
F_1: for men with heavy body stature
F_2: for men with light body stature

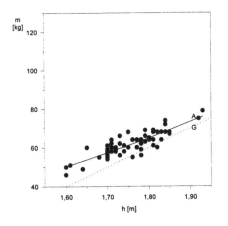

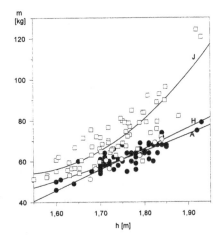

Fig. 1c
Comparison with the demand of a coach (SF, DRS1).

Fig. 1d
Comparison with other athletes (Carter, 1982)
H: long distance runners (marathon: 1; 5000 m : 4; 10 000 m: 3)
J: gymnasts (11), wrestlers (14), boxers (22), judokas (13), shot put (1), 100 m (3)

Fig. 1   Weight and height of athletes. Filled circles: ski jumpers of today

4.2 Comparison of athlete's weights

A comparison of the ski jumper's weight and height data with other athletes demonstrates, that only long distance runners are comparatively light (Fig. 1d). Other athletes, who also need muscle forces for high accelerations of the body or body segments, are also shown in 1d. The weight of ski jumpers is typically 8 kg below the „desired weight" line F in Fig. 1b and only a few ski jumpers meet the requirements formulated by the Swiss coach (SF DRS1), which is shown in Fig. 1c (curve G). Such requirements can lead to very dangerous health situations for the athletes and should be prohibited by means of an addition to the regulations. Such a regulation should consider the weight to height ratio and the severely underweight athletes should have a disadvantage. This could be in the form of penalty points. Another possibility to prevent extreme underweight could be a reduction of the permitted ski length in cases where the weight to height ratio lies below a critical line.

A reasonable regulation of the permitted ski length is absent and together with a minimum body weight to body height regulation should soon be developed. It is to be pointed out that the majority of the ski jumpers already lye either in the vicinity or above the critical area. These would not be noticeably affected by a weight to height ratio regulation. Such a weight to body height regulation would also prevent a tendency that could be observed during the last few years toward younger ski jumpers. The reason for this tendency can be explained using Fig 1a. The Curves E to B give the mean body weight in decreasing age groups. Curve B corresponds to 15 to 16 years and they are typically 10 kg lighter than those of the same height but aged between 25 and 29. Although the advantages gained in having low weight can result in a contradictory situation with extremely low weight (muscle weakness, tiredness, decreased resistance to infections, eating disorders, anorexia in extreme cases), it has to be expected that some athletes and their coaches will ignore the limits of reason. Severe health problems can be found in other sports (e.g. doping), which are difficult to solve. The question of being underweight could be solved and easily controlled. Only a small group of extremely underweight ski jumpers would be affected.

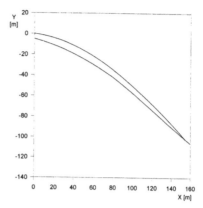

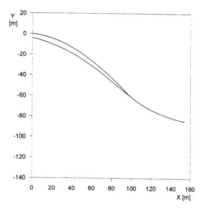

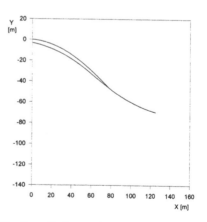

Fig. 2a-c. Reference jumps
a: Hill profile of Planica (K = 185 m)
b: Calgary (K = 114 m); c: Calgary (K = 89 m)

Table 1. Simulation results using the Planica (K = 185 m) jumping hill (profile data: a = -11.6°, b = -38.1°, b_1 = -40.1°, c = - 38.5°, H = 58.38 m, N = 107.33 m, R_1 = 125 m, R_2 = 228 m, M = 24.40 m, M_1 = 37.35 m, T = 10 m, S = 5 m).

	varied parameters and initial values											
	m [kg] / ζ [deg]	v_g [ms⁻¹]	v_{p0} [ms⁻¹]	v_0 [ms⁻¹]	a [deg]	ρ [kgm⁻³]	l [m]	Δl [m]	v_l [ms⁻¹]	$Δv_l$ [ms⁻¹]	Δl [%]	$Δv_l$ [%]
*	70						185.0	-	31.38	-	-	-
a	71						183.6	- 1.4	31.56	+ 0.18	- 0.76	+ 0.57
b	75						177.5	- 7.5	32.21	+ 0.83	- 4.05	+ 2.64
c	65						191.7	+ 6.7	30.47	- 0.91	+ 3.62	- 2.90
d	70	1 / 130					191.5	+ 6.5	30.57	- 0.81	+ 3.51	- 2.58
e	75	1 / 130					185.0	0	31.45	+ 0.07	0.00	+ 0.22
f	65	1 / 130					196.6	+ 11.6	29.61	- 1.77	+ 6.27	- 5.64
g	75		2.74				184.8	- 0.2	32.32	+ 0.94	- 0.11	+ 3.00
h	75			28.84			185.6	+ 0.6	32.33	+ 0.95	+ 0.32	+ 3.03
i	75				- 10.6		183.6	-1.4	32.30	+ 0.92	- 0.76	+ 2.93
j	75					1.2	182.3	- 2.7	31.70	+ 0.32	- 1.46	+ 1.02

* reference jump (v_0 = 28.538 ms⁻¹; v_{p0} = 2.24 ms⁻¹; ρ = 1.15 kgm⁻³; a = - 11.6°)

Table 2. Simulation results using the Calgary (K = 114 m) jumping hill (profile data: a = -11°, b = -37°, b_1 = -, c = - 35°, H = 42 m, N = 79.2 m, R_1 = 106 m, R_2 = 132 m, M = 24 m, M_1 = -, T = 7 m, S = 4 m).

	varied parameters and initial values											
	m [kg] / ζ [deg]	v_g [ms⁻¹]	v_{p0} [ms⁻¹]	v_0 [ms⁻¹]	a [deg]	ρ [kgm⁻³]	l [m]	Δl [m]	v_l [ms⁻¹]	$Δv_l$ [ms⁻¹]	Δl [%]	$Δv_l$ [%]
*	70						114.0	-	30.06	-	-	-
a	71						113.0	- 1	30.17	+ 0.11	- 0.88	+ 0.37
b	75						109.0	- 5	30.56	+ 0.50	- 4.39	+ 1.66
c	65						118.9	+ 4.9	29.41	- 0.60	+ 4.30	- 2.16
d	70	1 / 130					118.9	+ 4.9	29.46	- 0.60	+ 4.30	- 2.00
e	75	1 / 130					114.1	+ 0.1	30.07	+ 0.01	+ 0.09	+ 0.03
f	65	1 / 130					123.1	+ 9.1	28.69	- 1.37	+ 7.98	- 4.56
g	75		2.74				116.2	+ 2.2	30.79	+ 0.73	+ 1.93	+ 2.43
h	75			25.34			115.0	+ 1.0	30.82	+ 0.76	+ 0.88	+ 2.53
i	75				- 10		114.5	+ 0.5	30.73	+ 0.67	+ 0.44	+ 2.23
j	75					1.2	112.0	- 2.0	30.26	+ 0.20	- 1.75	+ 0.67

* reference jump (v_0 = 25.04 ms⁻¹; v_{p0} = 2.24 ms⁻¹; ρ = 1.15 kgm⁻³; a = - 11°)

Table 3. Simulation results using the Calgary (K = 89 m) jumping hill (profile data: a = -10°, b = -37°, b_1 = -, c = - 35°, H = 31.8 m, N = 62.4 m, R_1 = 84 m, R_2 = 138.9 m, M = 19 m, M_1 = -, T = 6.2 m, S = 3.2 m).

	varied parameters and initial values											
	m [kg]	v_g [ms^{-1}] / ζ [deg]	v_{p0} [ms^{-1}]	v_0 [ms^{-1}]	a [deg]	ρ [kgm^{-3}]	l [m]	Δl [m]	v_l [ms^{-1}]	Δv_l [ms^{-1}]	Δl [%]	Δv_l [%]
*	70						89.0	-	28.79	-	-	-
a	71						88.3	- 0.7	28.87	+ 0.08	- 0.79	+ 0.28
b	75						85.6	- 3.4	29.10	+ 0.31	- 3.82	+ 1.08
c	65						92.5	+ 3.5	28.37	- 0.42	+ 3.93	- 1.46
d	70	1 / 130					92.7	+ 3.7	28.38	- 0.41	+ 4.16	- 1.42
e	75	1 / 130					89.4	+ 0.4	28.78	- 0.01	+ 0.45	- 0.03
f	65	1 / 130					96.0	+ 7.0	27.86	- 0.93	+ 7.87	- 3.23
g	75		2.74				92.0	+ 3.0	29.47	+ 0.68	+ 3.37	+ 2.36
h	75			23.22			90.7	+ 1.7	29.47	+ 0.68	+ 1.91	+ 2.36
i	75				- 9		90.3	+ 1.3	29.36	+ 0.57	+ 1.46	+ 1.98
j	75					1.2	87.7	- 1.3	28.91	+ 0.12	- 1.46	+ 0.42

* reference jump (v_0 = 22.92 ms^{-1}; v_{p0} = 2.24 ms^{-1}; ρ = 1.15 kgm^{-3}; a = - 10°)

REFERENCES

Build and Blood Pressure Study, Vol. 1. 1959. *Society of Actuaries.* p. 16. Chicago.

Carter, J.E.L. (volume editor) 1976. *Physical Structure of Olympic Athletes (The Montreal Olympic Games, 1976).* App. C, 158-166. Basel, New York: S. Karger.

Denoth J., S.M. Luethi, H. Gasser 1987. Methodological problems in optimization of the flight phase in ski jumping. *Int J Sport Biomech,* Vol 3:404-418.

FIS World Cup Guide (1994). Publisher International Ski Federation. Oberhofen, Switzerland. 74-98.

Hubbard M., R.L. Hibbard, M.R. Yeadon, A. Komor 1989. A multisegment dynamic model of ski jumping. *Int J Sport Biomech* 5:258-274.

Jin H., S. Shimizu, T. Watanuki, H. Kubota, K. Kobayashi 1995. Desirable Gliding Styles and Techniques in Ski Jumping. *J. Applied Biomechanics,* 11: 460-474.

Kaps P., H. Schwameder, C. Engstler 1996. Computation of Ground Reaction Forces during Take-Off in Ski Jumping by Inverse Dynamics. *Book of Abstracts. 1st Int. Congress on Skiing and Science,* St. Christoph a. Arlberg, Austria, p. 44, (Proceedings in press).

Müller W. (I), D. Platzer, B. Schmölzer 1995. Scientific approach to ski safety. *NATURE* 375:p. 455.

Müller W. (II), D. Platzer, B. Schmölzer. Dynamics of human flight on skis: Improvements on safety and fairness inski jumping. *J Biomech* (in press).

Müller W. (III) 1996. Biomechanics of ski jumping- A scientific basis for the design of jumping hill profiles.

Book of Abstracts. 1st Int. Congress on Skiing and Science. p. 28, (Proceedings in press).

Müller, W. (IV) 1996. Physics of ski jumping: The lift and drag forces in the early flight phase have a pronounced impact on the performance. *Congress of the Canadian Society of Biomechanics.* Vancouver. (in press).

Remizov L.P. 1984. Biomechanics of optimal flight in ski-jumping. *J. Biomech.* 17:167-171.

Schmölzer B., W Müller 1996. The influence of lift and drag on the jump length in ski jumping. *Book of Abstracts. 1st Int. Congress on Skiing and Science.* p. 274. (in press)

Schwameder H., E. Müller 1995. Biomechanische Beschreibung und Analyse der V-Technik im Schispringen. *Spectrum der Sportwissenschaften* 1:5-36.

SF DRS1, Schweizer Fernsehen. Jan. 15th, 1996.

Straumann R. 1927. Vom Skiweitsprung und seiner Mechanik. *Jahrbuch des Schweizerischen Ski Verbandes.* 34-64.

Vaverka F., M. Janura, M. Elfmark, M. McPherson, J. Salinger 1996. Inter- and Intraindividual Variability of the Ski-Jumper's Take-Off. *Book of Abstracts. 1st Int. Congress on Skiing and Science,* p.40. (in press)

Virmavirta M., P.V. Komi, 1993. Measurement of take-off forces in ski jumping (I and II). *Scand J Med Sci Sports* 3:229-243.

Weight with the highest life expectancy 1959. *Stat. Bull. Metrop. Life Insur. Co.* p. 40. Nov-Dec.

The Engineering of Sport, Haake (ed.) © 1996 Balkema, Rotterdam. ISBN 90 5410 822 3

Comparison of the powers at the lower limb joints during walking at different velocities and their significance for a possible optimal walking velocity

W.Wang, R.H.Crompton, Y.Li & M.M.Gunther

Department of Human Anatomy and Cell Biology, The University of Liverpool, UK

ABSTRACT: This paper investigates the optimal velocity for human walking. The subjects were children and adults (age range 8-40, height range 120-185 cm and weight range 10-85 kg walking at their subjectively selected 'slow', 'normal' and 'fast' velocities. Kinematic data (joint angles and angular velocity) were obtained by photogrammetric reconstruction using two orthogonal, genlocked video cameras. Ground reaction forces were measured using a Kistler forceplate. Joint moments and mechanical joint power were then calculated by inverse dynamic analysis. Results indicated that as the forward velocity of the body centre of mass increases, the mean powers for joints vary in different ways: the mean powers for ankle and hip increase with velocity, but the mean power for the knee decreases. Thus, when the forward velocity is smaller than 2 m/s, the sum of mean powers for all joints is almost unchanged. When the forward velocity is greater than this value, the sum increases very quickly. Therefore, in order to walk as fast as possible while minimizing necessary power output, a speed should be selected as close as possible to 2m/s .

KEY WORDS: Power, velocity, optima, walking.

1. INTRODUCTION

It is a commonplace that every human has a 'comfortable' walking speed, which is neither very quick nor very slow. It is believed that humans select a optimal velocity which minimizes energy wastage while maximizing speed. How is this speed determined? Despite the large number of biomechanical studies of human gait, surprisingly few papers have addressed the issue of a possible optimal velocity for human walking. One of the few such studies is Zarrugh (1981), which suggested an optimal velocity of 1.7 m/s. on the basis of the power requirements and mechanical efficiency of treadmill walking. However, he did not take negative joint power into consideration when estimating efficiency, and it is known that negative power can play an important role in normal human gait (Winter, 1990; Schenau, 1990). Winter (1985) calculated joint power at different velocities, but did not derive an optimal velocity for human walking, and Cavagna and colleagues (1977) estimated efficiency in standing by the ratio of work/metabolic energy. Schenau (1990) and Williams (1983) both investigated efficiency using estimated joint power, although the number of assumptions needed rendered the

results rather unreliable. Nonetheless, these papers are the pioneers in this field.

In general, the sum of absolute powers for joints increases in proportion with increase of the sagittal velocity of the body centre of mass. Winter (1990); Eng and Winter (1995) and Williams(1983) noted that positive joint power is the source of propulsion for the body, while negative power may play a role in energy absorption. When the joint power output is negative, potential energy will be stored in the muscle and connective tissue surrounding the joint, and later released to provide a source for propulsion. Thus, the proportion of positive and negative power at joints is an important parameter. Thus, the sum of positive and negative power, the mean joint power, is a key to understanding how and why humans select walking velocities.

2.SUBJECTS AND METHODS

2.1.Subjects

The subjects were 4 adult males, 4 adult females and 2 children. They walked barefoot along an 10 m plywood walkway fitted with a Kistler 9281B forceplate, at self-determined slow, normal, and fast speeds . Two genlocked video cameras were set along the frontal and sagittal axes of the walkway. 28 sequences of video and kinetic data were available for analysis. Joint coordinates and ground reaction forces were assessed during one stride, at 50 Hz and 500 Hz, respectively, and filtered using an Ellip third-order filter. The statistics for subjects and speeds were as follows: 1) males: mean height 1.730 m (SD 0.02), mean weight 73.69 kg (SD 8.5), mean velocity 1.45 m/s (SD 0.38); 2) females, mean height 1.6625 m (SD 0.06), mean weight 59.65 kg (SD 7.17), mean velocity 1.525 m/s (SD 0.41); 3) children: mean height 1.29 m (SD 0.1), mean weight 19.80 kg. (10.62), mean velocity 1.405(0.26).

2.2. Methods

There are various alternative methods for computing joint power (see, e.g.: Cappozzo (1975), Janice (1995), Keith (1983), Zarrugh(1981). These methods differ primarily in (reviewed in Schenau, 1990). We shall not discuss method further here, since our goal is simply to assess the role of joint power in speed selection. Power was therefore calculated using a standard method from Winter (1990):

$$\text{Power} = \frac{1}{N} \sum \frac{Mo_i \omega_i}{\text{weight}} \qquad (1)$$

where: Mo_i is the moment; ω_i, angular velocity and N is the total number of frames. Following Newton's laws, there will be identical equations for each segment:

$$\sum F_i = ma \qquad (2)$$

and

$$\sum Mo_i + \sum F_i R_i = \beta I \qquad (3)$$

where: F_i is the joint force; Mo_i the moment; R_i the radius; β the angular acceleration and I the moment of inertia.

If these equations are applied to the sagittal plane for each segment, we can work out the three unknown variables (F_{iz} ;F_{iy} and Mo_i) from the equation groups for each segment. So, with the measured ground reaction forces, we can calculate all unknown variables centripetally from the most distal segment and joint, subject to the assumption that every segment may be treated as rigid.

The joint angles and angular velocity were obtained by stereophotogrammetric reconstruction of segment coordinates for succeeding video fields after application of the filter. Joint angles were defined as the angle between the proximodistal long axes of two adjacent segments. The origin of the local reference system was fixed at the most proximal joint, and the main reference system was defined as x: frontal, y: sagittal and z: vertical.

The moments at joints were obtained by inverse dynamic analysis (Winter, 1990). The process was as follows (1) recording ground reaction forces (2) computing from these all joint moments centripetally (3) calculation of mechanical joint power as the dot product of the joint relative velocities and moments during a stride.

The power was expressed as absolute, negative, positive and mean power. To permit comparison of different subjects, the moments and powers for joints were normalized over a full stride and expressed as a multiple of body mass.

When ω_i and M_i have the same sign, (ie, either both positive or both negative), power is positive. When they have opposite sign, power is negative. Several papers have discussed the significance of the sign of power (eg, Winter, 1990; Schenau, 1990). Here we consider positive joint power to indicate the work done by muscle groups operating at a given joint and negative power to represent the work done by other agencies, (eg, gravity). While positive, negative, absolute and mean power were all computed our discussion centres on mean power.

Power was defined as the power in the sagittal plane, averaged value over one stride. (1989). Average velocity may be obtained from a knowledge of both 3D kinematics and segment mass distribution (inertial properties were taken from Jensen, 1989), and ranged from 0.8 to 2.4 m/s. The method of calculation is as follows:

$$Y_{ct} = \sum \frac{Y_{it} M_i}{\text{weight}} \qquad (4)$$

$$V_{mc} = \frac{(Y_{cN} - Y_{c1})}{\text{time int erval}} \qquad (5)$$

where: Y_{ct} is Y_{mc} in time t; Y_{it} is distance travelled in the y direction during the time period t, and M_i is the mass of segment i.

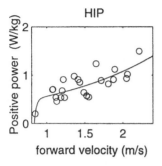

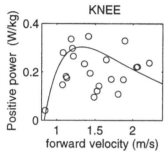

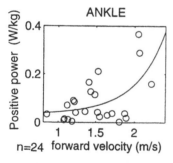

Figure 1. Positive power and forward velocity for the hip, knee and ankle.

3. RESULTS AND DISCUSSION

Since we know that both positive and negative power play an important role in walking, both must be taken into consideration. For the hip joint, it can be seen from Figure 1 that as walking velocity increases, both positive and negative power increase. But for the knee, while positive power does not change with increased speed, negative power increases with speed. For the

ankle, positive power increases while negative power decreases with speed. Therefore, total negative power increases with speed, and so does total positive power, but the latter increases more slowly (Figure 2).

Now let us consider mean power, the sum of positive and negative power. Figure 3 shows results for each joint, and indicates, as is commonly understood, that the mean power at the hip increases in proportion to the forward velocity of the body centre of mass. Thus, joint power at the hip plays a key role in propelling the body. In contrast, the mean power at the knee slowly *decreases* as forward velocity increases. This is a result of the kinematics of the knee joint. During certain phases of stance the body centre of mass has an acceleration towards the ground, and the muscle groups operating at the knee must absorb (and store) this energy. The higher the forward velocity, the greater the energy that is stored, and thus the greater will be the negative joint power. When the feet are pushing-off, the knee extends to produce positive power which contributes to the work needed for propelling the body mass. Other contributions to this work are made by muscle groups acting at the ankle and hip, and during push-off the larger contribution is from the ankle.

The mean power at the knee joint is the sum of positive and negative power at the joint.

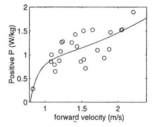

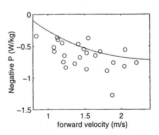

Figure 2. Positive and negative power versus forward velocity for all joints.

So, when the negative power is larger than the positive power, the mean power for the knee is reduced over the whole cycle. Because the negative power at the knee increases with velocity, while the positive power is almost unchanged, the sum of the two powers will tend to decrease slowly as velocity increases (Figure 3).

The pattern for the mean power at the ankle differs from the patterns for the hip and knee. Mean power increases slowly while the forward velocity of the body centre of mass is low

(<1.8 m/s). But it increases quickly when the velocity of mass centre is high (> 1.8 m/s). In fact, over the same period, positive power for ankle increases sharply with the velocity of the body centre of mass, while negative power decreases, but at a lower rate.

4. CONCLUSIONS

As a result of the net function of all joints, the sum of all mean powers varies as shown in Figure 4. In general, the total of mean powers is almost unchanged while the forward velocity of mass centre is less than 2 m/s, but it increases quickly at when the velocity is greater than 2 m/s. This suggests that when people are walking at a speed range below 2 m/s, they will spend almost the same energy no matter what their precise velocity. But when the range of their forward velocity is above this threshold, they will spend much more energy. This surely bears on the commonplace observation that we cannot greatly increase walking speed, but at some stage will be forced to break into a run.

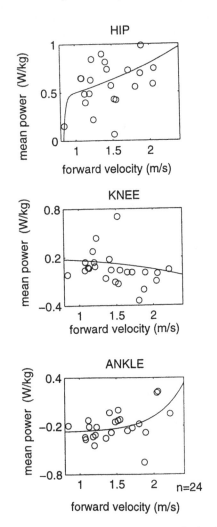

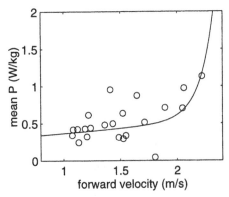

Figure 4. Mean power versus velocity for all joints.

5. ACKNOWLEDGEMENTS

This research is funded by supported by grants from the Natural Environment Research Council, Biotechnology and Biological Sciences Research Council, The Royal Society, and the Erna and Victor Hasselblad Foundation.

Figure 3. Mean power and forward velocity for the hip, knee and ankle.

6. REFERENCES

Apkarian, J. "A three-dimensional kinematic and dynamic model of the lower limb", J. Biomechanics Vol. 22,143-155 (1989)

Cappozzo,A."A general computing method for the analysis of human locomotion", J. Biomech. Vol. 8, 307-320 (1975)

Cavagna,G., Heglund, N.C., Taylor, C.R. "Mechanical work in terrestrial locomotion:two basic mechanisms for minimizing energy expenditure", Am. J. Physiol. Vol. 233(5), R243-R261 (1977)

Eng, J.J.,Winter,D.A. "Kinetic analysis of the lower limbs during walking:what information can be gained from a three-dimensional model?", J. Biomech. Vol. 28(6), 753-758 (1995)

Hreljac,A. "Determinants of the gait transition speed during human locomotion: kinematic factors", J. Biomech. Vol. 28, 669-677(1995)

Jensen,R. K. "Changes in segment inertia proportions between 4 and 20 years", J. Biomech. Vol. 22, 529-536 (1989)

Schenau, G., Van Ingen, J., Cavanagh, G. "Power equations in endurance sports", J. Biomech. Vol. 23, 865-881 (1990)

Williams, K. R. "A model for the calculation of mechanical power during distance running", J. Biomech.Vol. 16, 115-128 (1983)

Winter, D. A. "Energy generation and absorption at the ankle and during distance running." Clin. Orthop. Rel. Res. Vol 197, 147-154 (1983)

Zarrugh,M.Y. "Power requirements and mechanical efficiency of treadmill walking", J. Biomechanics. Vol. 14, 157-165 (1981)

3 Design

The Engineering of Sport, Haake (ed.)© 1996 Balkema, Rotterdam. ISBN 90 5410 822 3

The modern tennis racket

H. Brody

Physics Department, University of Pennsylvania, Philadelphia, Pa., USA

ABSTRACT: The official rules of tennis give the racket manufacturer tremendous latitude in the parameters allowed in the design of the racket. Because of this, there have been major changes in tennis frames during the last 20 years. Virtually everything that can be changed has been changed, including the racket weight, mass distribution, stiffness, length, head size and shape, string pattern, moments of inertia, and material. To some degree, this change in racket design was fuelled by the availability of composite materials which allow the designer to construct rackets with these new shapes, sizes and weight distributions. On the other hand, there was an awareness that the dimensions of the "standard" wood racket that had been in use for 50 years, were neither sacred or optimal. The modern tennis racket attempts to have the three "sweet spots" located close to the same place (close to the center of the head), have a predictable response (even for slight miss-hits by the player), provide good ball speed without too much effort, be easy to manoeuvre, and not be the cause of injury to the arm or hand.

1. INTRODUCTION

For most of this century, tennis rackets were fabricated from wood by steaming the wood to soften it, then bending, glueing, and binding it into the desired shape. Most of the racket frames weighed 14 or more ounces, and had a head that had about 75 square inches of area capable of being strung. So standardized was the 27 inch long racket with its 9 inch wide head, that the height of the net at the center (36 inches) was often checked and adjusted by measuring that height using two rackets where the 27 inch and 9 inch dimensions were added together. (The rules of tennis are spelled out in English units, such as inches, feet, ounces, etc, so the units employed in this article will also be in those units.) Because of the limitation of the strength to weight ratio of wood, to withstand the stress of stringing at high tension and the pounding that hitting a tennis ball produces, the only way to make a racket that would endure, was to increase the amount of wood in the frame (and the weight), or keep the head size small.

Over the years various players experimented with larger heads or longer rackets, but the extra weight required made then unwieldy. With the introduction of space age, composite materials, it was possible for the racket designer to change virtually any frame parameter, and still have an implement that could be swung easily, yet not fall apart. Today's rackets are lighter, stiffer, longer, thicker, wider, and have a bigger head However, there are some limitations placed on racket design by the rules of the game of tennis. For example, the rules state that the racket shall not exceed 32 inches in overall length and the strung surface shall not exceed 15½ by 11½ inches. Occasionally, a racket within these limitations comes along (such as the spaghetti strung racket) which markedly changes the way the game can be played. Then the tennis establishment modifies the rules slightly to ban such rackets. This is the fate that a racket designer faces, in the unlikely case that a new concept in rackets design can change the way the game is played.

2. THE STRINGS

The literature of tennis used to advocate high string tension to produce high ball speeds, and it is only in the last 10 to 15 years that this myth has been debunked. The tennis ball, by the rules of tennis, must lose a great deal of energy when it compresses and stores energy (very large hysteresis). The strings on the other hand, are not constrained by the rules of tennis, and can return 95% of the energy that they store. In the interaction of the ball and the strings the kinetic energy of the ball is converted into potential energy stored in both ball and string deformation. By storing a larger fraction of the incoming energy in the strings, less is dissipated, and more is returned to the ball's rebounding kinetic energy. This can be accomplished by lowering the string tension.

It is not the absolute value of string tension that determines the playing characteristics of a racket, but the string plane deformation versus externally applied force. Since the strung area of the head and the string spacing or density can be different from racket to racket, when you specify tension, it does now allow you to compare different rackets. It is the string plane deformation that determines the power, dwell time, and feel of the strings. Even though many players and manufacturers specify string tension (because stringing machines are calibrated in pounds or Newton's), there are some instruments on the market that measure string plane deformation, or the equivalent.

Most strings are made of a synthetic material, usually a type of nylon. However, as tension increases, the elasticity of these materials degrades, so animal intestines (gut) are used by players who want better performance and are willing to pay for it.

3. SWEET SPOTS OF THE RACKET

When you hit the ball at a particular location on the racket and it feels good to you, you claim the ball has hit the "sweet spot". There are actually three separate locations on the racket that might qualify for this good feeling, and they are caused by three different physical effects.

3.1 *Center of Percussion*

To the extent that a racket frame can be considered to be a rigid body, one of the sweet spots is the Center of Percussion (COP). When a ball strikes the strings at the COP, the initial shock or jar at the conjugate point on the racket (where you are holding it in your hand) is minimized. This is caused by a cancellation of the racket recoil motion (momentum conservation) by the handle's forward rotational motion (conservation of angular momentum). Clearly, the shock and jar that your hand feels will be better when the ball hits the COP than when the ball strikes a different point. The location of this point can be experimentally determined by pivoting the racket at the location where the hand is placed, using the frame as a physical pendulum, and determining the period of oscillation.

The distance between the pivot point and the COP is given by

$$COP = 9.77 * T^2$$

where T is the measured period of the pendulum in seconds and the COP distance is measured in inches.

The COP location in a modern racket is usually slightly above or not too far from the center of the head.

3.2 *The Node*

The second sweet spot is call the node, and comes about because the racket is not rigid, but flexible. When the hand held racket is struck by the ball, certain modes of oscillation can be set up. The mode that usually has the largest amplitude and feels most uncomfortable, is the lowest frequency oscillation of the free body, and it has two nodes, one near the center of the head and one near the top of the grip. For a modern racket, the frequency of the vibration of this lowest mode of vibration is between 125 and 200 Hz, depending on the racket stiffness and mass. When the ball strikes the strings close to the node that is located in the head, the amplitude of this oscillation is very small, and the hit feels "sweet". The farther the impact is from the node, the larger is the amplitude of the oscillations. The location of this node can be determined experimentally by holding the racket between two fingers at the other node and tapping the strings at various places. The node in the head will be that impact point that results in minimum vibration.

The location of the node and the location of the COP point in the head are usually not too far apart, so if a ball strikes one of them, it will be very close to striking the other.

Various mechanisms have been used to attempt to damp out the vibrations that result when the impact is not close to the node. The human hand, gripping the handle of the racket, is still one of the best vibration dampers that is available.

3.3 *The Power Spot*

The third of the sweet spots concerns the impact location on the strings that produces the maximum rebound ball speed (most power). This leads to a problem, since when a racket is swung, different parts of the racket are moving with different velocities. To find the location of the maximum power, the actual swing mechanics of the player must be known.

Then ball rebound data taken in the laboratory with the racket at rest can be transformed into the tennis court frame of reference (with the racket being swung, not translated), and the region of maximum power on the head can be established.

An electronic "speed trap" was built to determine the effective center of rotation (or pivot point location) of a swung racket, and a number of tennis players were measured for both forehand and backhand swings. From laboratory data on the ball rebound ratio (not the Coefficient of Restitution) of a free, static racket and the swing data, the power is transformed into the court frame of reference using the formula:

$$V_{hb} = e * V_{ib} + (1+e) * V_r$$

where V_{hb} is the velocity of the hit ball, V_{ib} is the velocity of the incident ball, and V_r is the velocity of the racket at the impact point and e is the ratio of ball rebound speed to ball incident speed in laboratory tests of a racket. From the data with the electronic speed trap, it is possible to replace V_r with the angular velocity of the swing multiplied by the distance of the ball impact point from an effective pivot point, or center of rotation. This will give the actual power contours of a given racket for a given swing. When this calculation is done, the power maximum often comes out closer to the throat of the racket than the center of the head, and closer to the throat than the node or COP locations.

To get the points to coincide, the racket designer then must attempt to raise the maximum power point toward the racket tip, or get the players to swing with much more "wrist" (which makes the tip of the frame move much faster than the throat). To raise the power point, the racket can be made stiffer, the balance point (center of mass) can be moved higher on the head, or the string pattern can be modified (along with the head shape). All of these have been done in the newly designed rackets.

4. RACKET WIDTH AND STABILITY

If the ball hits the strings at the location away from the principal axis (the long axis) of your racket, the resulting impulsive torque will tend to make the racket twist or turn. Since you are gripping the handle, when the racket turns or twists, so will your hand, but not your elbow, and this can lead to trouble. The amount of the racket twists and how fast it twists, depends on how hard you hit the ball, how far off axis the impact , the moment of inertia of the racket about the long axis, and the torsional stiffness of the racket. By increasing the moment of inertia about the long axis, the twisting of the racket frame on off-axis hits will be reduced. In addition, the ball rebound velocity ratio falls off very quickly as the impact point moves off-axis because the moment is relatively small. Increasing the moment about the long axis will also increase the power on off-axis hits, giving a more uniform response across the string plane.

The moment about the long axis (mr^2) can be increased either by adding weight to the side of the head or by making the racket head wider. The latter is the preferred solution, since the effect goes as the square of the width. Therefore, a 10% increase in width will lead to a 20% increase in moment. The moments of inertia of a number of different types of racket were measured, and when those values were plotted against the racket mass multiplied by the head width squared, the data lay on a smooth curve.

5. RACKET LENGTH

The standard racket length for most of this century has been 27 inches. There have been some attempts to change this, but a shorter racket does not play well, and a longer racket would work well only if it could be made light enough to allow it to be easily swung. Just extending a standard racket by a few inches resulted in a frame that was both too heavy and had too large a swing weight (moment of inertia about the butt end) for the average player to be comfortable with. Recently, rackets that are one to two inches longer (28 to 29 inches long) have appeared on the market, many of which having the same swing weight as a standard frame. These rackets give a greater reach to the player, allow the ball to be hit harder, and allow more serves to go in. Using a computer program to track ball trajectories, it is possible to calculate the improvement in first serve percentages by the use of these longer rackets. It turns out that they help all players, but their benefit is proportionally greater for the small player compared to the taller player.

Even though the official rules of tennis allow rackets up to 32 inches long, the ATP (Association of Touring Professionals) recently enacted rules limiting the length of tennis rackets to 29 inches in all ATP tournaments.

6. THE ERROR CORRECTING RACKET

The ideal racket would correct for the errors a player makes in not striking the ball at the same location on the racket every time. There is no way for a racket to correct for errors in racket angle or racket velocity at impact time, because there is no way for the racket to know what the player is attempting to do (is the shot a lob, drive, drop shot etc).

One way to correct for the errant rebound angle a ball would have if the impact is off axis, is to have a curved (parabolic) shape to the string bed. This is not a legal solution, since the rules of tennis require the strings to be in a plane. As the impact point moves away from the principal axis, the rebound speed of the ball decreases, because the racket has a small moment of inertia about that axis. To correct for that, the string spacing can be increased somewhat nearer the side of the racket head, trying to lead to a softer string plane there.

A similar problem exists along the long axis of the racket, but it is not as pronounced, since the moment of inertia involved is much greater. By moving the maximum power point up into the center of the head this problem can be overcome. This can be accomplished by the methods mentioned in section 3.3

7. CONCLUSIONS

There is no doubt that the modern tennis racket is a much better implement than the one in use 20 or more years ago. It provides more power (ball speed) with less effort on the players part. It provides more control over the direction of the ball by being somewhat forgiving of small miss-hits by the player, and it provides a predictable response. The modern racket will also last much longer in play than the old, classic wood frames that were in use through most of this century.

REFERENCES

Brody, H., "Tennis Science for Tennis Players", Univ of Pennsylvania Press, Philadelphia PA (1987)

Brody, H., "How Would a Physicist Design a Tennis Racket", Physics Today, Vol 48 No. 3, 26-31, Mar 1995.

Hatze, H., "Impact Probability Distribution, Sweet Spot, and the Concept of an Effective Power Region in Tennis Rackets", Journal of Applied Biomechanics, Vol 10, 43-50 (1994)

The Engineering of Sport, Haake (ed.) © 1996 Balkema, Rotterdam. ISBN 90 5410 822 3

The design optimisation of poles for pole vaulting

S.C. Burgess
Cambridge University Engineering Design Centre, UK

ABSTRACT: The main function of a vaulting pole that affects its design is strain energy storage. For a given pole stiffness, the main design objective is to minimise mass to enable the vaulter to maximise his running velocity at take-off. The pole mass is minimised by selection of the optimum pole material and geometry. In this paper, the optimum pole design is derived using strain energy methods. The effect of flight path requirements, production constraints and vaulter characteristics on pole design are briefly discussed.

1 INTRODUCTION

1.1 *Pole Functions*

In pole vaulting, the pole is a device which converts the kinetic energy of the vaulter into potential energy - see Figure 1. Modern poles and pole vaulting techniques are highly efficient in this energy conversion process and this means that very high vaults of more than six metres can be achieved.

The pole has three main functions:
- Energy storage function
- Pivoting column function
- Stiffness function

The reason why pole vaulting is so efficient is that the pole undergoes very large deflections to give a smooth flight path. During deformation, the pole is performing an energy storage function. This function involves a conversion of some of the

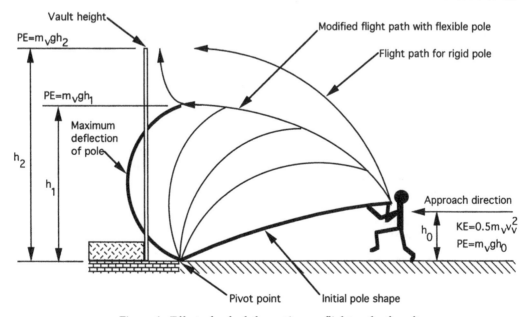

Figure 1. Effect of pole deformation on flight path of vaulter

vaulters kinetic energy into temporary strain energy during the first half of the vault. In the second half of the vault, the strain energy is released such that the pole provides an upward thrust to the vaulter.

The pivoting column function involves the pole pivoting around a fixed ground location such that the vaulter is lifted up into the air. The inclination of the pole at take-off is such that when it impacts the ground, the forward motion of the vaulter automatically provides two components of force on the end of the pole. One force is in the direction of the pole axis (x axis) and causes it to deform. The other force is perpendicular to the pole axis (y axis) and causes the vaulter to accelerate upwards.

The pole also has to provide bending stiffness in the x-z plane and torsional stiffness around its own axis (x-axis). This is because the vaulter applies significant side loads to the pole i.e. loads which are out of the x-y plane. These loads are caused mainly by the technique of vaulters to swing upside down on one side of the pole during the vault.

The energy storage function has the most influence on the design of the pole because the deformation of the pole in the x-y plane involves very large bending moments and very large stresses. The compressive stresses due to the pivoting column function and even the bending stresses due to side loads are relatively small in comparison.

1.2 Pole geometry

Figures 2 and 3 show a typical transverse and longitudinal section of a pole respectively. In the undeformed state, the pole has a hollow circular cross section, although when a pole is made with an initial radius of curvature the production process may produces a slightly elliptical shape. The internal pole diameter is usually kept constant and the external diameter varies along the length to give an optimum design. The mean diameter of the pole is typically about 40 mm which is convenient for gripping. The wall thickness varies from about 3.5 mm at the middle of the pole to about 1 mm at the end of the pole. For Olympic vaulters, the length of the pole is in the region of 5.4 m.

One of the unique features of a vaulting pole, compared to equipment used in other field sports like the javelin, is that it does not have to conform to a detailed design specification. For example, the pole material and geometry can be optimised for a particular vaulter.

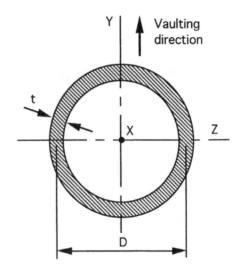

Figure 2. Transverse cross section of pole

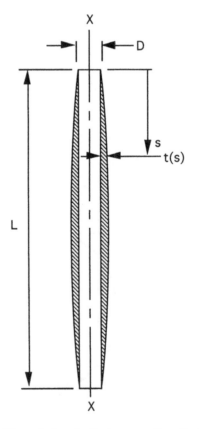

Figure 3. Longitudinal cross section of pole

2 OPTIMISATION STRATEGY

2.1 *Optimisation objective*

In order to maximise the vault height, a vaulter maximises his kinetic energy at take-off by maximising his take-off velocity. The vaulter also tries to convert this energy input as efficiently as possible into potential energy. In order to maximise the take-off velocity, the vaulter requires a pole of low mass because this causes less hindrance to running. However, in order to convert kinetic energy efficiently into potential energy the pole stiffness must be low enough to produce an adequately smooth flight path.

Sections 3-5 describe how to minimise the pole mass for a particular input of kinetic energy and a particular pole stiffness (i.e. for a particular pole deflection and flight path). The effect of changing the input kinetic energy and changing the pole stiffness will be discussed in Section 6.

The objective function which is to be minimised is given by

$$m_p \approx \pi D \rho \int_0^l t \, ds \qquad (1)$$

where m_p is the mass of the pole, D is the mean diameter [m], ρ is the density of the material [kg/m^3], t is the wall thickness at distance s [m] and l is the length of the pole [m].

The constraints are:
• Specified kinetic energy input, U_{KE} (J)
• Specified pole deformation, R_{min} (m)

The free variables are:
• Pole diameter, D (m)
• Wall thickness distribution, t = f(s) (m)

2.2 *Optimisation Principle*

To minimise the mass of an energy storage device it is necessary to maximise the average stress level when subjected to the maximum load as described by Boiten (1963). If the pole starts off straight or with a constant initial radius of curvature, then the optimum fully deformed shape must be an arc of constant radius of curvature. This is because in linear-elastic theory for a thin-walled tube, stress is proportional to the change in radius of curvature as defined by

$$\sigma_a = \frac{ED}{2} \left(\frac{1}{R_{min}} - \frac{1}{R_0} \right) \qquad (2)$$

where σ_a is the maximum allowable bending stress (N/m^2), E is the Young's modulus (N/m^2), D is the mean diameter of the pole (m), R_0 is the initial radius of curvature (m) and R_{min} is the minimum radius of curvature (m).

As well as having a constant change in curvature, the pole diameter should be maximised (wall thickness minimised) to achieve the highest value of average stress. Maximising the ratio of D/t has the effect of maximising the bending shape factor as described by Ashby (1992).

Figure 4 shows a simple model of the pole deformation. The pole is assumed to be unrestrained at its ends with an axial loading of P. At the maximum deformation, the pole has a radius of curvature of R_{min} which is an input requirement.

It should be noted that the pole is not in a state of buckling instability, otherwise there would be nothing to stop the pole fully collapsing. Therefore, P is less than the Euler buckling load. This means that it is quite difficult to deform the pole initially because the initial pole stiffness is relatively very high. To get the pole deflection started, the vaulter applies a bending moment through his two arms at the beginning of the vault. In addition, the deformation is made easier by having an initial radius of curvature in the pole.

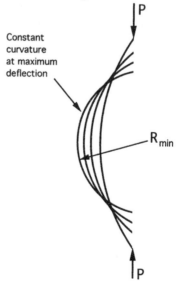

Figure 4. Model of pole loading/ deflection

3 OPTIMUM POLE DIAMETER

The pole diameter has to be maximised to maximise the average level of stress i.e. to maximise the bending shape factor. Rearranging Equation (2) gives the following relationship for D

$$D = 2\frac{\sigma_a}{E}\left(\frac{R_{min}R_0}{R_0 - R_{min}}\right) \qquad (3)$$

The maximum and optimum pole diameter is typically about 40 mm.

4 OPTIMUM WALL THICKNESS VARIATION

4.1 *Pole curvature constraint*

The wall thickness variation must be such that the maximum bending stress, at any section along the length of the pole, is equal. The pole curvature constraint determines the bending moment distribution due to the load P and, therefore, determines the optimum distribution of wall thickness as a function of P.

From linear elastic theory, the maximum bending stress along the length of a fully deformed hollow circular pole due to an applied load P is given by

$$\sigma_a \approx \frac{M_\theta D}{I_\theta} \approx 4\frac{M_\theta}{\pi D^2 t_\theta} \qquad (4)$$

where M_θ is the bending moment at an angle θ (Nm) - see Figure 5, I_θ is the second moment of area at an angle θ (m⁴) and t_θ is the wall thickness at an angle θ (m).

To achieve a constant value of maximum bending stress it is necessary to have a constant ratio of M_θ/I_θ. Since the bending moment varies and the mean pole diameter is approximately constant, it is necessary to vary the wall thickness t_θ along the length of the pole.

The optimum wall thickness t_θ at any point along the pole is found by rearranging Equation (4) and is given by

$$t_\theta = f\{M_\theta\} = \frac{4}{\pi D^2 \sigma_a} M_\theta \qquad (5)$$

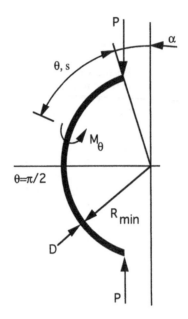

Figure 5. Pole geometry at maximum deflection

The maximum bending moment at any angle θ is given by

$$M_\theta = PR_{min}[\sin(\alpha + \theta) - \sin\alpha] \qquad (6)$$

where P is the load at maximum deflection (N).

Therefore, the optimum wall thickness distribution is given by

$$t_\theta = \frac{4}{\pi D^2 \sigma_a}\{PR_{min}[\sin(\alpha + \theta) - \sin\alpha]\} \qquad (7)$$

The wall thickness distribution for a straight pole is obtained by substituting $\theta = s/R_{min}$ into Equation (7) and is given by

$$t_s = \frac{4}{\pi D^2 \sigma_a}\left\{PR_{min}[\sin(\alpha + \frac{s}{R_{min}}) - \sin\alpha]\right\} \qquad (8)$$

The maximum wall thickness at the centre of the pole is given by

$$t_s = \frac{4}{\pi D^2 \sigma_a}\{PR_{min}[1 - \sin\alpha]\} \qquad (9)$$

Therefore, the optimum wall thickness distribution is a sine function with an amplitude that is proportional to the load P - see Figure 6. The load P is itself a function of the energy input. In practice, it is not practical for the thickness to decrease to zero at the ends of the pole, otherwise local buckling will take place. The wall thickness has a minimum value of approximately 1 mm over a section of about 400 mm in length. In addition, the end of the pole which impacts the ground is usually toughened and built up locally with kevlar fibre.

4.2 Energy storage constraint

The energy storage constraint determines the load P which itself determines the amplitude of the wall thickness distribution. The larger the energy input, the larger the load P will be and the larger will be the required wall thickness.

The stored strain energy in the fully deformed pole is equal to the difference between the kinetic energy of the vaulter at take-off and the potential energy of the vaulter when the pole is fully deformed

$$U_{SE} = \frac{m_v v_v^2}{2} - m_v g(h_1 - h_0) \qquad (10)$$

where U_{SE} is the strain energy in the fully deformed pole (J), U_{KE} is the kinetic energy input by the vaulter at take-off (J), m_v is the mass of the vaulter, g is the gravitational constant (m/s^2), h_1 is the height of the fully deformed pole (m) and h_0 is the initial height of the vaulter (m).

The maximum load P is determined by equating the strain energy storage requirement (Equation (10)) with the total strain energy stored in the fully deformed pole. The strain energy in the pole is determined by integrating the product of the bending moment and angular rotations throughout the pole i.e.

$$U_{SE} = \frac{R_{min}}{2EI} \int_{\alpha}^{\pi-\alpha} M_\theta^2 d\theta$$

$$= \frac{(R_0 - R_{min})}{R_0} \int_0^{\frac{\pi}{2}-\alpha} M_\theta d\theta \qquad (11)$$

Therefore, from Equations (6) and (11) the maximum load P is given by

$$P = \frac{R_0}{(R_0 - R_{min})} \frac{U_{SE}}{R_{min} C_1} \qquad (12)$$

where

$$C_1 = \int_0^{\frac{\pi}{2}-\alpha} [\sin(\alpha + \theta) - \sin \alpha] d\theta$$

$$C_1 = [\cos \alpha + \left(\alpha - \frac{\pi}{2}\right) \sin \alpha] \qquad (13)$$

From Equations (7) and (12) the optimum wall thickness distribution is given by

$$t_\theta = \frac{4}{\pi D^2 \sigma_a} \left\{ \frac{R_0 U_{SE}[\sin(\alpha + \theta) - \sin \alpha]}{(R_0 - R_{min}) C_1} \right\} \qquad (14)$$

Therefore, the wall thickness distribution is a sine function whose amplitude is proportional to the strain energy input.

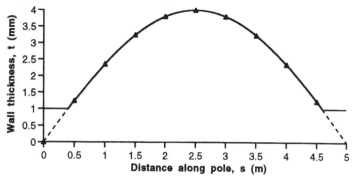

Figure 6. Optimum wall thickness variation

5 OPTIMUM POLE MASS

By inserting the stiffness and energy constraints into the objective function (Equation (1)) the free variables (D and t) are eliminated and the objective function defines the optimum pole mass in terms of material properties and input constraints.

The longitudinal cross sectional area is given by

$$\int_0^l t\,ds = 2R_{min} \int_0^{\frac{\pi}{2}-\alpha} t_\theta\,d\theta$$

$$= \frac{8R_{min}}{\pi D^2 \sigma_a} \int_0^{\frac{\pi}{2}-\alpha} M_\theta\,d\theta$$

$$= \frac{8R_{min}}{\pi D^2 \sigma_a} \{PR_{min}C_1\} \qquad (15)$$

Inserting Equations (3) and (15) into the objective function (Equation (1)) gives

$$m_p = (2U_{SE})(2)\left(\frac{\rho E}{\sigma_a^2}\right) \qquad (16)$$

Therefore, for the specified curvature constraint, the pole mass is a function of the material properties and the strain energy input.

The objective function can also be written in terms of the maximum load P by inserting Equation (12) into (16) i.e.

$$m_p = (2P)\left(2R_{min}\left[1-\frac{R_{min}}{R_0}\right]C_1\right)\left(\frac{\rho E}{\sigma_a^2}\right) \qquad (17)$$

Therefore, the pole mass is a function of the material properties, the change in curvature and the maximum load. Equation (16) can also be derived by considering the volume of material necessary to store an amount of strain energy. For example the energy storage capacity of an ideal spring bar element in tension is given by

$$U_{SE} = \frac{Fx}{2} = Al\frac{\sigma_a \varepsilon}{2} = \frac{m}{2}\left(\frac{\sigma_a^2}{\rho E}\right) \qquad (18)$$

where F is the force applied (N), x is the deflection (m), A is the cross sectional area (m^2), l is the length, (m), σ_a is the material allowable stress (N/m^2), ε is the material strain, m is the mass of the spring element, ρ is the material density (kg/m^3) and E is the Young's modulus of the material (N/m^2).

The mass of a non-ideal spring is given by

$$m = (2U_{SE})\left(\frac{1}{\lambda}\right)\left(\frac{\rho E}{\sigma_a^2}\right) \qquad (19)$$

where λ is the fractional energy storage efficiency as defined by Boiten (1963).

Therefore, from Equations (16) and (19) the efficiency of the pole design considered in this paper is 0.5 (i.e. 50%). The inefficiency is due to the low stresses near the neutral axis of the hollow circular section. This type of section has a maximum spring efficiency of 0.5 for bending as shown by Ashby (1992).

6 ROUTES TO MASS MINIMISATION

6.1 Optimum material

In Section 5 it was shown that the best material is the one with the highest value of the index

$$M = \left(\frac{\sigma_a^2}{\rho E}\right) \qquad (20)$$

where σ_a is the maximum allowable stress, E is the Young's modulus and ρ is the material density. Materials with the highest value of M are ceramics and glass but these are too brittle in tension and torsion respectively. Rubber also has a high value but is not practical because of a high loss coefficient and also because the low elasticity would not meet the pole stiffness requirements. The best practical material is carbon fibre reinforced plastic (CFRP) which is slightly ahead of GRFP.

6.2 Optimum transverse shape

In Section 5 it was shown that the efficiency of hollow circular poles is approximately 0.5. To improve the energy storage efficiency, the shape can be changed to an ellipse which has an energy storage efficiency of between 0.5 - 0.75. The ellipse has to be aligned such that the minor axis is in the plane in which the bending moments are acting. The efficiency depends on the ratio of height to width with the highest efficiency achieved with the highest ratio. Therefore, the use of an ellipse could result in a mass saving of up to 25%.

One problem with an elliptical section is that it would be difficult to fabricate. Another problem is that a change of section would be required where the vaulter grips the pole. An additional problem for analysis is that an ellipse would be more prone to being deformed (i.e. increased elliptical shape) when subjected to bending loads. Therefore, the initial fabricated ellipse would have to compensate for the distortion that would take place.

6.3 *Optimum wall thickness variation*

As shown in Section 4, it is possible to vary the wall thickness such that the maximum bending stress is constant along the outer edges of the pole. This enables a lower volume of material to be used to store a given amount of strain energy compared to a uniform wall thickness where the average maximum stress is relatively low. Modern vaulting poles do already exploit this route to mass minimisation.

One problem in having a varying wall thickness is that it is more difficult to produce than a constant wall thickness. Figure 7 shows a typical production method for fabricating a carbon fibre pole. A sheet of pre-preg carbon fibre in the shape of a "D", approximately half a sine function, is rolled into a pole. The pole has a small step in the thickness which forms a spiral from each end of the pole to the centre. However, this step is very small and can be neglected in the bending analysis.

6.4 *Optimum initial radius of curvature*

Having an initial radius of curvature makes the initial pole stiffness lower and, therefore, has the advantage of making the initial part of the flight path smoother. Having initial curvature also makes the net change in curvature less which allows a larger pole diameter. Having a larger pole diameter has the advantage of making a higher level of average stress and lower mass.

Some modern poles do have a small amount of initial curvature although it may be that greater initial curvature could be advantageous.

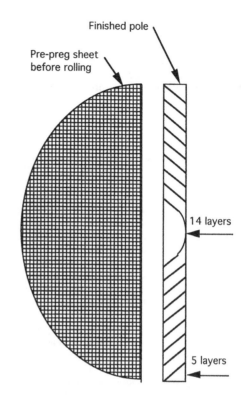

Figure 7. Typical production method

6.5 *Characteristics of vaulter*

The characteristics of the vaulter affects the input kinetic energy and this affects the pole mass as shown by Equations (10) and (16).

For example, a heavy runner has more kinetic energy and, therefore, for a given running speed will require a heavier pole. Vaulters who have a very high take-off speed, like Olympic vaulters, also have a high value of kinetic energy and so also need a relatively heavy pole.

6.6 *Stiffness requirement*

The amount of deformation of the pole has a significant effect on the pole mass. For example, if the pole does not deform very much then a large amount of the vaulters kinetic energy is converted directly into potential energy with not very much put into deforming the pole - see Figure 1. Therefore, a light pole can be achieved by having high stiffness. However, if the pole does not deform very much then the flight path of the vaulter is not smooth but involves a sharp upward acceleration at the beginning of the vault. This sudden acceleration would cause a significant energy loss and could be more significant than the gains in energy due to the lighter pole. Therefore, there is a fundamental trade-off between pole mass and pole stiffness.

7 CONCLUSIONS

Modern vaulting poles are very efficient devices. The materials currently used - CFRP and GFRP - are the best materials currently available for the device. Modern poles also have an optimum variation in wall thickness along the length of the pole which makes each transverse cross-section work equally hard. The circular hollow section currently used is also an efficient design feature although there is potential for mass reduction by changing to a more oval shape. Having an initial radius of curvature also gives potential for mass reduction and a smoother flight path.

The pole stiffness has a very significant impact on the pole mass. As the pole stiffness decreases, the pole mass increases because more strain energy is required to be stored in the pole. However, a low stiffness pole gives a smoother and more efficient flight path. Therefore there is a fundamental trade-off between pole mass and stiffness. It would be useful to combine a study of the pole design and flight path efficiency in order to establish an optimum combination of stiffness and mass.

ACKNOWLEDGEMENTS

The author acknowledges helpful discussions with Mr K. Winter (Olympic pole vaulting coach) and Dr P. Weaver of the Cambridge EDC. The author also acknowledges the financial support of the EPSRC for the Cambridge Engineering Design Centre.

REFERENCES

Boiten, R.G. 1963. Mechanics of instrumentation. Proc. I. Mech. E. 177, 269.
Ashby, M.F. 1992. Materials selection in mechanical design. Pergamon Press.

The Engineering of Sport, Haake (ed.) © 1996 Balkema, Rotterdam. ISBN 90 5410 822 3

Shuttlecock design and development

Alison J.Cooke
Department of Engineering, The University of Cambridge, UK

ABSTRACT: This paper presents the main findings from research on the aerodynamics and mechanics of shuttlecocks. It also describes how the knowledge has been used to develop a technical design process and a new synthetic shuttlecock for Dunlop Slazenger International Ltd..

1 INTRODUCTION

Badminton is now played in over 123 countries and by an estimated 4 million people in England alone. Feather shuttlecocks have always been the "elite" projectiles since the origin of the game in the mid-19th century but the invention of the cheaper more durable synthetic shuttlecock in the 1950s gave the game a wider appeal. Today, synthetic versions account for 60 per cent of shuttlecock sales and are used mostly by amateurs. Although they are comparable in price, they last about three or four times longer than traditional feather shuttlecocks.

Manufacturers have yet to produce a synthetic shuttlecock that exactly mirrors the flight of the feather shuttlecock during the game and they are continually striving to reach that goal. The differences between them are subtle. Players talk of a lack of control when playing with synthetic models - quite a disadvantage when the shuttlecocks can travel at velocities in excess of 150 mph. Players also point out that they "travel faster through the air and so favour the attacking player who likes to smash" (Cooke 1994).

This paper firstly summarises the main elements of a four year programme of fundamental research on shuttlecock aerodynamics and mechanics which explains some of the players' comments. The research took the form of a comparative study between feather and synthetic shuttlecocks (Cooke 1992). It then explains how the research was put to commercial use by describing the commercial background, the development of a technical design process and the new product which was launched

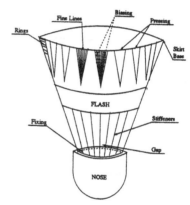

FIGURE 1: Geometry of Synthetic Shuttlecock

into the badminton market in the summer of 1995 (Dixon 1995).

2 SHUTTLECOCKS

Feather shuttlecocks are made of a hemispherical cork or plastic nose with 16 feathers attached to it, and weigh around 5 grams. The spines of the feathers are glued into holes in the cork or plastic and fan out behind the nose to form a cone. The manufacturing process is necessarily very labour-intensive and hence most feather versions are now made in the Far East.

Synthetic shuttlecocks have the same shape but replace feathers with an injection-moulded plastic skirt which uses a special blend of nylons (see Figure 1). The International Badminton Federation issues

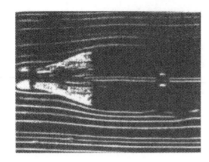

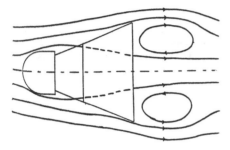

FIGURE 2:   Flow over Feather Shuttlecock

guidelines on shuttlecock size, mass and shape but the range of dimensions for each parameter is large enough to mean that shuttlecocks manufactured within the guidelines can still behave very differently, both aerodynamically and mechanically.

3 MAIN AREAS OF RESEARCH

The main areas of research are discussed below and the practical implications for players are noted in boxes.

3.1 *Aerodynamics*

3.1.1 *Flow Visualisation*

The flow over the shuttlecocks was investigated using three techniques:

- smoke flow visualisation,
- surface static pressure measurements
- wake velocity measurements.

The shuttlecock is a bluff body and the predominant drag mechanism is base drag. The flow separates at the downstream end of the nose and some passes through the gap behind the nose

resulting in an axial jet of air downstream from the gap. There is an annular stagnation region in the wake behind the shuttlecock skirt (see Figure 2).

The introduction of the gap in the shuttlecock skirt was found to increase the drag coefficient. This is explained by the entrainment of stagnant wake air by the strong axial jet emerging from the gap in the shuttlecock base. This generally reduces base pressure and hence increases the drag.

3.1.2 *Aerodynamic Coefficients*

The drag, lift and pitching moment coefficients, based on shuttlecock skirt diameter, were measured for four production shuttlecocks (1 feather, 3 synthetic) using a custom-made, sting-mounted, three-component strain gauge load cell which fitted inside the shuttlecock nose and allowed free rotation in the wind tunnel. Results were obtained for Reynolds numbers of up to 200,000 (equivalent velocities of up to 45 m/s) using this technique.

At lower Reynolds numbers, where the forces were below the resolution of the load cell, a terminal velocity experiment was conducted in a water column, producing aerodynamic coefficients down to a Reynolds number of 13,000 (equivalent to a velocity of 3 m/s).

It was found that, above velocities of approximately 23 m/s, the drag coefficients for some synthetic shuttlecocks were lower than those for the feather shuttlecock. Image processing of the shuttlecocks from films of the experiments in the wind tunnel showed that this drag reduction was due to skirt deformation. It was particularly noticeable with synthetic shuttlecock skirts with a deeper "pressing" or fluting as the skirt then tended to close up causing lower frontal area. The feather shuttlecock is generally stiffer and so it does not suffer from this problem.

Note: The drag coefficient reduction for the synthetic shuttlecocks explained why players feel that the synthetic shuttlecocks travel faster and fly further.

The lift coefficients were not significantly different between the production models. In all cases the coefficients increase with increasing incidence.

The pitching moment coefficients were not significantly different between the production models. In all cases the position of the aerodynamic centre is behind the centre of gravity (which gives the shuttlecock its stability during flight).

3.2 *Mechanics*

Various design parameters were compared, including: mass, moments of inertia, centres of gravity, angular damping factors. The most interesting phenomenon was found to be caused by the shuttlecock's rotation about its own axis. A gyroscopic phenomenon was shown to be present.

The shuttlecock flies at incidence. The restoring moment which occurs as a result of the incidence, together with the shuttlecock rotation, can induce a gyroscopic precession and a subsequent sideways "drift" (see Figure 3). This was demonstrated experimentally and modelled using gyroscopic theory. It is more noticeable with a feather shuttlecock because it was found to have a higher rotation about its own axis (up to twice that for synthetic models).

Note: Drifting is noticed by players particularly with the feather shuttlecock during high clears (when the incidence is likely to be greatest).

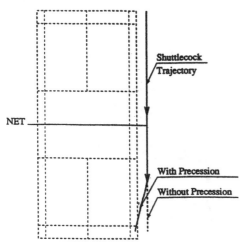

FIGURE 3:   Plan of Precessional Drift

This drifting leads to a secondary precess which was also identified experimentally. This causes the shuttlecocks to "nose-dive" after the trajectory vertex (i.e. turnover).

Note: "Nose-diving" is also noted by players using the feather shuttlecock.

3.3 *Ballistics*

A computer program was written to predict the linear and angular motion of the shuttlecock in two dimensions (i.e. in the plane of its trajectory) using the measured aerodynamic and mechanical coefficients. An experiment was also constructed to measure shuttlecock trajectories for imposed initial conditions.

The measured trajectories were found to be in agreement with trajectories predicted by the computer program (within experimental error). The computer program was then used to simulate various badminton shots.

Note: It was shown that the synthetic shuttlecocks were flying further than their feather counterparts.

4 COMMERCIAL BACKGROUND

The Carlton Sports synthetic shuttlecock factory is part of the Dunlop Slazenger International Group. Bill Carlton, the entrepreneur, formed the company in the 1950's, and masterminded the development of the world's first injection moulded shuttlecock. This was the product development project on which the profitability of the company was to rest for four decades.

The more recent story of product development is less dramatic. Throughout most of the 1980's, Carlton Sports' product range had been largely unchanged. Individual new products had been launched periodically but the success rate of new product launches was poor. The state of the synthetic shuttlecock business at the beginning of 1994 could be summarised as follows:-

- a tired product range, suffering from quality problems
- a declining market share
- a management which had lost confidence in the company's ability to develop new products
- increasing manufacturing costs
- little in-house design knowledge

It was therefore decided to use the elements of shuttlecock research described above and in Cooke 1992 to facilitate the development of a structured design process for shuttlecocks. Such a design process was intended to ensure that the company would not lose any of its technical product expertise in the future and it was hoped that it would provide a methodology to transfer this expertise into profitable products.

A project plan was devised which not only developed a new design process, but also developed a new shuttlecock product using a radical new production process. The cost of the total project approached £2 million.

The next section provides an overview of how the technical design process was developed. All the knowledge about product performance was combined. This included the theory of shuttlecock performance described above (Cooke 1992), the company's empirical knowledge and comments from players.

5 TECHNICAL DESIGN PROCESS

A shuttlecock's characteristics can be described at four different levels (see Figure 4):

- design parameters: centre of gravity, total mass, drag coefficient, etc.
- design features: geometry and material.
- test performance: describing the performance through objective experiments.
- play performance: the subjective judgements of badminton players.

provided information on aspects of play performance which had previously been purely subjective. Some translation between design parameters and design features was also possible: for example, centre of gravity can be calculated from geometry and material.

The production process was also fundamental in determining the design features. The company's process knowledge was considered a "black art": empirical rules had been established historically by correcting product faults using the process controls. The rules provided the missing link between certain fine details of geometry and performance. The theory of the design parameters structured the "black art" and enabled the transfer of key messages from production into the design sphere, i.e. design for manufacture.

The final stage in developing the technical design process was to simplify the knowledge into one coherent unit. The shuttlecock performance depends almost entirely on the properties of the shuttlecock skirt (i.e. design features) and so the key objective was therefore to provide a structured approach to its definition. The final methodology is described pictorially by Figure 5.

The above design process was used to produce the new product which was launched into the synthetic shuttlecock market in the summer of 1995. The next section describes the key design features which differentiates it from previous models.

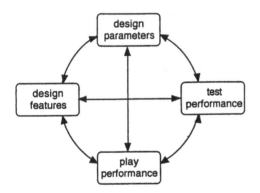

FIGURE 4: Characterising Shuttlecocks

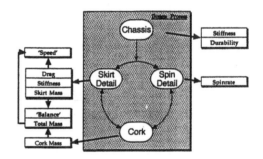

FIGURE 5: Shuttlecock Design Process

This model provides a clear way of assessing the design knowledge and the ways of applying it profitably. The ultimate aim is to be able to specify the design features required to achieve desired play performance, which is considered to be the key customer requirement.

Cooke 1992 defines the design parameters and, from them, the predicted test performance. This

6 THE NEW PRODUCT

6.1 *Compensation for Drag Coefficient Reduction*

The new production process eliminates the requirement for a secondary "pressing" operation. The shuttlecock skirt is moulded in a single process and the plastic is stress-relieved in its final shape during the cooling phase of the process. This ensures that the skirt shape and hence the aerodynamic performance is more predictable.

The fluting has been designed to emulate the effect of the individual feathers. It is shallow and therefore resists skirt collapse at high speed and therefore drag coefficient reduction. This reduces the tendency for the new product to fly further than a feather shuttlecock.

6.2 *Rotational Improvement*

The fluting is asymmetric and the pattern of lines on the skirt (or "biasing") is such that the rotation or spin of the new product is much greater and therefore more similar to the feather shuttlecock (which rotates rapidly). This ensures that the gyroscopic behaviour, i.e. drifting and nose-diving, emulates the feather shuttlecock more accurately.

7 CONCLUSION

Fundamental research on the theory of shuttlecock design has been integrated with the "black art" of the production process to provide the company with a simple to use and durable technical design process. A new product has been developed which more closely emulates the flight of the feather shuttlecock. It has been launched into the market place using a radical new injection moulding production facility.

REFERENCES

Cooke, A.J. 1992. *The Aerodynamics and Mechanics of Shuttlecocks.* PhD Thesis, Cambridge University

Cooke, A.J. 1994. *The Flight of the Shuttlecock.* New Scientist March 1994.

Dixon, J. & A. J. Cooke 1995. *Managing Product Design: A Case Study from the Consumer Sports Industry.* Product Design Seminar 1995, Teaching Company Directorate.

The Engineering of Sport, Haake (ed.) © 1996 Balkema, Rotterdam. ISBN 90 5410 822 3

Design and prototype manufacture of a composite bicycle frame

D. Katsanis & S. M. Grove
Department of Mechanical Engineering, The University of Plymouth, UK

ABSTRACT: This paper reports the design and construction of a prototype cycle frame intended for sprint racing. Preliminary materials selection and detailed specification of composite ply lay up sequence were based on the results of computer-aided laminate analysis and on static tests on representative coupons. The frame design was based on the overall requirements of aerodynamics (but without detailed calculation), and the need for a low sideways area concentrated close to the rear wheel. An important design consideration was that the cycle should accept standard fittings (wheels, forks, etc.).

Initially, the bicycle was conceived as a "one off". Moreover, the work was undertaken in the context of an undergraduate student project, so that very little funding was available for tooling. It was decided, therefore, to manufacture the frame by hand lay up on a foam core. This is necessarily a laborious process requiring hand finishing, but is a low risk route, allowing continual monitoring of quality. Wet lay up (dry fabric and liquid resin) was used, and vacuum bag processing gave good laminate consolidation. Aluminium bottom bracket shell and rear drop-outs, and carbon fibre tubes for the head tube and seat tube, were bonded to the foam core and integrally laminated in the frame.

The finished frame was subjected to a series of static tests to verify its stiffness. These included vertical compliance, seat tube bending and overall torsional stiffness. The results were "benchmarked" against two commercial frames and found to be of satisfactory performance. A detailed finite element model was under development at the time of writing this paper.

A complete bicycle was assembled and raced by a member of the National Team of Greece in the 1994 World Championships. The rider succeeded in setting a new national record in the sprint event. Since then a number of bikes made for riders of national teams of Greece, Italy, Russia, Estonia and Latvia.

The paper also considers the manufacturing options for series production of cycle frames using more advanced processing technology such as resin transfer moulding (RTM). Further design studies are in progress on alternative raw materials with the aim of producing a 1 kg road racing frame.

1. INTRODUCTION

The advantages of fibre-reinforced polymer composites in sporting and leisure applications are well documented [1-3]. These include high specific strength and stiffness and good fatigue resistance. The ability of polymer composites to be moulded, rather than fabricated from stock material such as tube or sheet, allows lightweight "monocoque" structures to replace conventional frameworks. This in turn gives the designer considerable freedom not only to save weight but also to explore possible aerodynamic advantages.

Surprisingly, perhaps, advanced composites have only made a significant impact on bicycle design and construction in the last few years. The fundamental diamond-shape frame, fabricated from steel tube, has changed little since the late 19th century [3]. Within the last 20 years, some attempt has been made to improve strength-to-weight ratios by using aluminium and titanium tubes. A more radical departure from

tradition was the production of a cast magnesium alloy frame in the 1980s [4]. The first use of advanced composites appears to have been in the mid 1970s, when Exxon constructed bicycle frames from carbon fibre-reinforced tubes and aluminium lugs [5]. Classical triangular frames have been constructed as one-piece composite components [4, 6].

One reason for this conservatism was that, prior to 1990, the governing body of international racing (the Union Cycliste International) specified that "...the bicycle frame must consist of a main triangle with 3 tubes measuring a maximum 40 mm in diameter...". At this time, the only route to improved aerodynamic efficiency was to flatten the tube cross-section. When the rule was relaxed in 1990, the new design freedoms were exploited most notably by Mike Burrows, the creator of Lotus Engineering's "Sport" pursuit cycle, on which Chris Boardman won the Olympic gold medal in 1994 [7, 8].

Despite the success of the monocoque design in time trial events, bicycles for "mass start" racing are generally still based on traditional tubular design, albeit with non-circular aerodynamic tube cross-sections or relatively minor design changes such as the omission of the seat tube. In track sprint racing, designs appear to be particularly conservative, and many bicycles are still based on steel tubes.

The overall objective of the work described in this paper was to produce a monocoque composite bicycle frame for sprint racing - this was the first such frame produced for international competition. The authors emphasise that design, manufacture and testing were all carried out as an undergraduate student project, with limited time and resources. Aerodynamic studies and detailed structural analyses (see [7, 9] for example) were both beyond the scope of the project, although work is continuing in this area at the University of Plymouth.

2. DESIGN

2.1 *Overall Considerations*

The original objective was to construct a composite frame with a maximum weight of 2 kg which would be as stiff as a conventional steel sprint frame. Weighing 79 kg, track sprint riders are on average 9 kg heavier than endurance athletes. . 1 (from [10]) shows the large power output possible during short term sprint racing - these are 3 or 4 times greater than in longer races. As a consequence, maximum speeds are typically 20 m/s (see Fig. 1), and cornering on a tight track results in accelerations of 2-3g. Although sprint events are of very short duration (typically 10-15 s), the short term demands placed on the bicycle frame are thus higher than in other forms of racing. Conventional steel sprint frames weight about 2.1 kg - this is some 15-20% heavier than a high quality road bicycle frame.

The importance of aerodynamics results from the high speeds attained in sprint racing. Generally the power requirement is proportional to the product of drag force and velocity, so an increase of speed from say 30 to 40 km/h would require the rider to double their power output [10]. Aerodynamic analysis was beyond the scope of this project, but it was expected that a monocoque frame based on a cruciform shape would give some advantage over a tubular design.
The conceptual design considerations are summarised as follows:

Aerodynamic, cruciform monocoque construction.

Low sideways area, concentrated close to the rear wheel. As discussed in [7], it is important for stability that the centre of pressure resulting from cross winds is as far as possible to the rear of the cycle's centre of gravity.

Minimum distance between head tube, bottom bracket and rear wheel.

Use of conventional fittings (wheels, forks, etc.).

Conformity to UCI regulations.

It should be noted that the decision to incorporate standard fittings rules out the adoption of the

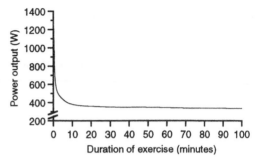

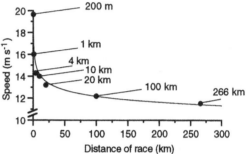

Figure 1. Graphs from [10] showing the large power output possible during short term sprint racing.

"monoblade" concept as used in the Lotus Sport bicycle, and requires two rear chain stays.

2.2 *Materials Selection*

Intermediate modulus carbon fibre (Toray M40J)/epoxy resin on a PVC foam core was selected as the most suitable material for the main structural beams of the frame. This was used as a 200 g/m2 unidirectional prepreg and as a 200 g/m2 twill woven fabric in wet lay up with SP Systems Ampreg 20 epoxy resin in all parts of the structure.

Laminate construction was investigated using GENLAM analysis software [11].

A pragmatic approach to structural analysis was taken, relying on a combination of simplified models and component testing. The main beam and chain stays were analysed by representing them as tapered beams of elliptical cross-section. In an attempt to maximise the longitudinal bending stiffness of the beam while simultaneously minimising the weight, 9 plies of unidirectional fabric was used at $0°$ (i.e. parallel to the beam axis) at the top and bottom of the beam - the sides of the beam carried only 3 plies of the same reinforcement. This resulted in a non-uniform laminate thickness, varying between from 1.2 mm to 0.4 mm.

A typical equivalent static load was calculated by assuming a 80 kg rider can negotiate a 20 m radius bend at 75 km/h. With a factor of safety of 1.5, a load of 3000 N is obtained. It was assumed that the load was shared equally between the main beam and the two chain stays.

A simplified analysis of the beams indicated that, if regarded as cantilevers under a concentrated end load, the maximum beam deflections were just less than 0.9 mm. The chain stays in particular are required to undergo minimum deflection due to potential problems with misalignment of the chain.

3. MANUFACTURE

For production of a "one off" prototype, no investment was available for tooling. It was therefore decided to construct the frame from wet lay up with vacuum bagging over a foam core.

The core was constructed from PVC foam of density 40 kg/m3. The main frame and chain stays were cut and shaped separately by hand then bonded in place with epoxy resin. The rear drop-outs were made from 8 mm flat aluminium alloy plate (7020 T651).

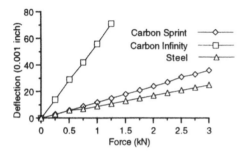

Figure 2. Vertical compliance test.

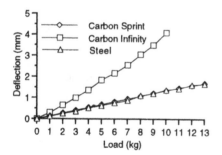

Figure 3. Torsion test.

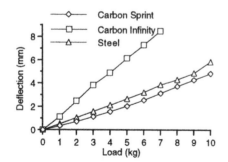

Figure 4. Seat post bending test

The bottom bracket shell was made of 2017 T3 alloy. All alloy parts were wrapped in a layer of woven glass mat/epoxy resin to avoid galvanic corrosion. The seat tube and head tube was made by wrapping a single layer of woven glass/epoxy followed by layers of woven carbon/epoxy to a total thickness of 1 mm around a aluminium mandrel of appropriate diameter. All inserts were fitted to the jig and secured in place. Then the prepared foam core was bonded to these parts using epoxy resin adhesive. The foam core proved sufficiently rigid to assemble the crucial parts (rear wheel, crank arms and chain). Slight misalignments were corrected at this stage. At the same time, the front fork was built, using steel tubes, and aligned on a special jig.

The fibre layers were placed in two separate operations. Each side was laminated completely on the foam core while it was on the alignment jig and vacuum bagged for consolidation. The whole structure was covered with a portable curing chamber and the temperature was raised to 60°C for 10 hours.

After the cure cycle was complete, the bag was removed. Surface irregularities due to folds and creases in the vacuum bag were removed by the application of a lightweight filler and careful sanding.

4. TESTING

In view of the simplified design approach used in this project, the stiffness of the composite frame was compared directly with two other frames. These were a conventional steel tube mass production frame and a commercial carbon fibre monocoque frame designed for road racing - these are labelled as "sprint", "steel" and "road" in the following figures. The carbon fibre road frame was of similar geometry to the prototype sprint frame developed here, except that it had a single "monoblade" chain stay.

Three static tests were carried out as detailed below.

4.1. *Vertical Compliance*
In this test the frame was simply supported at the head tube and rear drop outs. A load cell was positioned vertically above the seat post. Deflection was measured by a dial gauge positioned beneath the bottom bracket. Test results are shown in Fig. 2.

4.2. *Seat Tube Bending*
Here, the frame was laid on its side, clamped rigidly only at the bottom bracket. The frame was loaded by hanging weights from a steel rod inserted in the seat tube. These weights acted at a distance of 75 cm from the bottom bracket, with deflections being measured by a dial gauge at 63 cm from the bracket. Test results are shown in Figure 3.

4.3. *Torsional Stiffness*

To compare torsional stiffness, the frame was again supported at the bottom bracket as above with an additional support placed under the head tube. Weights were hung from a steel rod passing through the head tube at a distance of 23 cm from the support, and deflection was measured by a dial gauge 18 cm from the support. Results are shown in Fig. 4.

Generally, the prototype sprint frame displayed similar stiffnesses to the steel frame. The greatest difference was in the vertical compliance test, where the triangular steel frame was some 40% stiffer than the sprint frame. The carbon fibre road frame was considerably more flexible than the other two in all tests, and in view of its larger deflections it was not taken to the same loads.

The agreement between measurement and calculation for the vertical compliance test is remarkably good - as seen from Fig. 3, the actual deflection at a load of 3000 N was 0.91 mm, compared with a theoretical value of 0.87 mm.

5. PERFORMANCE

After completion in the summer of 1994, the sprint frame was ridden in competition by Greek National Team member Lambros Vassilopoulos. In the World Cup (Toulon, France) he set a new national sprint record. He also competed in the World Championships in Sicily, in which he finished 10th and again broke the national record.
A similar frame was made for the Italian National Team and ridden by the former tandem sprint world champion Gianluca Capitano. He achieved 7th place in the kilometre event in the 1994 World Championships. Both riders reported the frames to be very rigid, stable and fast.

The Greek national team have now a total of 7 frames for its riders. The Estonian national team has two, Latvia one, Russia two and Italy one. Enquiries have also been received from Argentina, South Africa USA and some well known British riders.

A number of international level athletes have now raced this frame in sprint events.

Some of the results are:

1994
Lambros Vassilopoulos, Greek National Team
Twice new national sprint record, 10th in sprint event in World Championship in Sicily, Italy

Gianluca Capitano (twice World champion), Italian National Team
7th in kilometre event in World Championship in Sicily, Italy

1995
Lambros Vassilopoulos, Greek National Team
New national sprint record, 4th in sprint event in World Cup in Cottbus, Germany

Greek National Team (Lambros Vassilopoulos, G. Himonetos)
3rd place in Olympic sprint event in World Cup in Cottbus, Germany

Gianluca Capitano (twice World champion), Italian National Team
2nd in kilometre event in World Cup in Athens, Greece
1st in Olympic sprint event in World Cup in Manchester

Erika Salumae (twice Olympic champion, twice World champion)
3rd in sprint event in World Cup, Manchester, 3rd in the sprint event in the World Championship in Bogota, Colombia.

6. FURTHER DEVELOPMENTS

6.1 *Manufacture*

In view of the commercial interest being shown in this product, there is clearly a need to adopt a faster, less labour-intensive manufacturing route. An obvious candidate is RTM (resin transfer moulding). This process involves loading a closed mould with dry reinforcement and injecting a pre-catalysed thermosetting resin under internal vacuum and/or external pressure. Resin injection pressures are usually in the range 1 to 5 bar - this means that low cost composite tooling is appropriate. The process has several environmental advantages over contact moulding, and is able to produce composite components of high tolerance and good surface finish. For larger volume production, the process is particularly suitable for automation [12].

Several techniques are available for the production of hollow or cored structures such as the bicycle frame. A foam core may be shaped by hand or machine, or moulded in a dedicated tool. Dry fibres are wrapped around the core - this may be a manual process, or involve more sophisticated techniques such as filament winding or braiding. The complete "preform" is then loaded into the tool for resin injection and curing. Hollow structures, including a bicycle frame [6], have been produced by wrapping the reinforcement around an inflatable rubber bag instead of a foam core - after moulding this is deflated and removed. A proprietary process has recently been demonstrated for the manufacture of large hollow aerogenerator blades and a furling yacht boom [13].

6.2 *Design*

Test results suggest that there is scope for improved torsional stiffness of the frame. This is now being achieved in development by the addition of more layers at ±45o to the beam axes. This is being accompanied by a more sophisticated structural analysis, using finite element software. There is also a need to reduce the overall frame weight. It is believed that this can be achieved by relatively simple design changes, including the use of a lower density foam and more careful sizing of the aluminium inserts. It is estimated that 300-400 g can be saved in this way. Optimal design and manufacture are expected to realise further structural efficiencies, making the overall design target of a 1 kg composite frame with the same stiffness as a 1.7 kg steel frame easily achievable.

ACKNOWLEDGEMENTS

The authors would like to thank Andy Lewis, Mike Stringer and Tim Searle for their practical input to this project.

REFERENCES

1. *"Composites Takes to the Slopes"*, Advanced Composites Engineering (Design Council, London), Spring 1988, pp 8-11; "Sporting Composites", Ibid., January 1990, pp 7-9.
2. Haines, R.C. *"Volume Production with Carbon Fibre Reinforced Thermoplastics"*, Plastics and Rubber Processing and Applications, 5 (1), 1985, pp 79-84.
3. Easterling, K.E. *"Advanced Materials for Sports Equipment"* , Chapman and Hall, 1993.
4. Glaskin, M. *"Bicycles"*, Engineering, 229, February 1989, pp 78-79; "Magnesium Makes a Better Bike", Ibid, April 1987, pp 235-6.
5. Redcay, J. *"Composite Contemplations"*, Bicycling, June 1987, p 192.
6. Jun, E.J., B.S. Kim and C.R. Joe. *"Composite Monocoque Frame for Bicycles, Design and Fabrication"*, Proc. 9th International Conference on Composite Materials (1994), VI, pp 208-215.
7. Hill, R.D. *"The Design and Development of the LotusSport Pursuit Bicycle"*, Proc. Institution of Mechanical Engineers - D - J. Automobile Eng., 207 (D4), 1993, pp. 285-294.
8. Parsons, N.G. *"Development of the Composite Structured LotusSport Bicycle"*, Proc. 14th International SAMPE European Chapter Conference, October 1993, pp 133-148.
9. Lessard, L.B., J.A. Nemes and P.L. Lizotte. *"Utilization of FEA in the Design of Composite Bicycle Frames"*, Composites, 26 (1), 1995, pp 72-74.
10. Burke E.R., *"The Science of Cycling"*, Human Kinetics Books, Illinois, 1986.
11. Tsai, S.W. *"Theory of Composites Design"*, Think Composites, Ohio, 1992.
12. Bland, R.J., A.R. Harper and S.M. Grove. *"Cost Effective Automation for Low to Medium Volume RTM Production"*, Proc. 50th Annual Technical Conference, SPI Composites Institute, Cincinnati, January 1995.
13. *"RTM Hollow Mouldings"*, Reinforced Plastics, April 1994, pp 34-37; "Hollow RTM Becomes a Booming Business", Ibid., January 1995, pp 20-23.

The Engineering of Sport, Haake (ed.) © 1996 Balkema, Rotterdam. ISBN 90 5410 822 3

The design and development of a shuttlecock hitting machine for training badminton players at all levels of the game

J.E. Morgan

Department of Mechanical Engineering, University of Bristol, UK

ABSTRACT: This paper describes some of the criteria that have to be considered when designing a prototype on-court machine for hitting or throwing feather shuttlecocks at badminton players. A large number of design concepts were initially identified, but many of these were rejected on safety grounds bearing in mind that the training machine will operate in a public sports hall environment, and hopefully will be used by juniors just starting to learn the game, as well as top class players.

1 BACKGROUND

1.1 In recent years, sport sophistication, the advancement of materials technology, dedicated microprocessor circuits and design ingenuity, have combined to produce machines that enhance sport training to the extent that in many sports both coaches and players regard such equipment as a fundamental aid for game improvement. Unfortunately, as of yet, no such machines have been developed to improve the racquet skills of badminton players. Despite both real and imaginary improvements made to the game in terms of using, for example, composite frame materials, space-age string materials, and incredibly complicated finite element design packages in, say, the production of badminton racquets, badminton coaches still spend a considerable amount of their time in "feeding" shuttlecocks to players in order to develop consistent shot reproduction.

2 RAISON D'ÊTRE

2.1 One of the major problems facing badminton coaches is the ability to repeatedly feed shuttlecocks to a player at a constant rate, constant speed, and into a precise place, where an error of only a few inches may well nullify the coaches desire to force the player into respondingin a particular way. Although with

practise a coach may achieve a high degree of consistent feeding, the fact that he/she is often required to deliver precisely positioned shuttlecocks to a player means that much of his/her time and concentration is pulled away from observing exactly what the player is doing and instead centres on his/her own feeding accuracy. In addition the feeding operation usually means that the coach is always on the other side of the net from the player, a situation which may often diminish the ability of the coach to effectively work with the player. In reality, due to human nature, absolute consistency of hand feeding will never be achieved and as a result a player being coached may be forced to make unwanted corrections in stroke play, thus hampering the repeated practice of specific shots. The availability of an automated shuttlecock feeding machine should therefore be an invaluable aid to all badminton coaches since they would then be able to devote all of their time to the player, secure in the knowledge that shuttlecocks can be repeatedly delivered into exactly the same place in the court every shot.

3 THE ENIGMA OF THE SHUTTLECOCK

3.1 The mechanical hitting (or throwing) of balls for sports, such as table-tennis balls, squash balls, tennis balls, cricket balls and even footballs only concern hitting symmetrically round, relatively hard, and often elastic balls, which by the nature of their associated sports have been designed and manufactured to be violently abused during their use. For example, footballs are violently kicked, cricket balls and golf balls are violently hit by fast moving, solid surfaces, and even tennis balls are more or less almost indestructible to the extent, for example, that they can be soaked in sea water and covered in abrasive sand, and even in this condition still fulfil their function in the playing, of say, beach cricket - a popular English pastime during the cold summer months The symmetry of these balls combined with the fact that when hit they have no orientation requirement (i.e. no up or down) means that the design of a device to mechanically hit or throw them is at least, in the conceptual stage, relatively simple to achieve. In addition, after impact, all of these balls follow more or less a parabolic flight path, as described by classical collision and dynamic theory. On the other hand, a non-badminton playing design engineer could be forgiven for thinking that the feather shuttlecock, with its extraordinary shape and component parts, was in fact designed to prevent anybody making a successful mechanical device for hitting it.

3.2 A feather shuttlecock, with its ridiculous asymmetric shape, comprises a small cork base (c. 2.5 grams) and a relatively large feather skirt (also c. 2.5 grams), the feather skirt being made up of about 16 right hand wing goose feathers held together with string and glue. Because of the disporportionate front end (cork base) loading, in flight the centre of pressure of the shuttlecock is always well behind the shuttlecock's centre of gravity. Thus in a game of badminton, the shuttlecock will normally be travelling cork base first toward one of the players. This player will try to return the shuttlecock over the net and towards the other end of the court by hitting the cork base of the shuttlecock with his/her racquet. The shuttlecock starts to travel in its new direction, but initially with the feather skirt leading. However, as a result of its non-aerodynamic shape, the shuttlecock almost instantaneously rotates through 180 degrees to assume its normal stable flight configuration i.e. cork base leading and the feather skirt trailing. Experiment has shown that at low velocities, after relatively gentle racquet impact, turnover of a shuttlecock occurs after the shuttlecock has travelled about 200 mm. At high velocities, typically resulting from more powerful shots, turnover may not be completed until the shuttlecock has travelled about 800 mm.

3.3 Due to the overlap of the feathers making up the skirt of the shuttlecock, and the fact that only right hand wing feathers of a goose are used in its construction, when travelling through the air the shuttlecock spins around its axis of symmetry and also fractionally continually curves to the right during flight. In addition its total allowable weight is strictly controlled. To remain within the laws of the game the shuttlecock must only weigh between 4.74 and 5.50 grams (73 and 85 grains). From momentum considerations it can be shown that for a racquet (or hitter) of average mass, the shuttlecock will leave the racquet at about twice the racquet's linear velocity. This approximate doubling of speed has been confirmed by experiment. For example, by using stroboscopic photography, it has been observed that a racquet impact speed of 60 km/hr produces an initial shuttlecock speed of just over 100 km/hr. However, it has also been identified that about half the shuttlecock's initial velocity is lost during the shuttlecock's turnover manoeuvre. Thus, since turnover typically occurs in the first few hundredths of a second after hitting, once turned over, and travelling in its normal stable flight orientation, the shuttlecock can be considered to have an effective initial velocity roughly equal to the linear velocity of the hitter at impact. In this normal flight orientation vortex drag from the feather skirt slows the shuttle down, but at an exponential rate. It is this drag effect created by the feather skirt, that largely makes the game what it is, and part of the result of this drag is that the trajectory of a shuttlecock is not the normal parabolic path, observed say when a spherical ball is hit, but instead follows a "parachute" trajectory (Fig. 1). Needless to say if the shuttlecock is hit anywhere else other than its cork base, instantaneous terminal damage to the shuttlecock will probably occur.

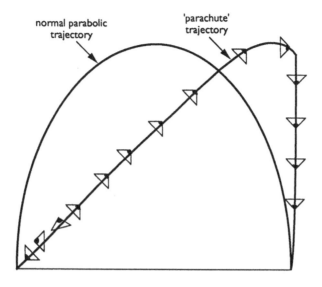

normal parabolic trajectory

'parachute' trajectory

FIG. I

3.4 The exponential deceleration of a shuttlecock in flight means that for any particular court condition it almost impossible to hit a shuttlecock much further than its design distance. The two main factors which govern how far a shuttlecock will travel after being hit are the velocity of the hitter and the weight of the shuttlecock. Of these two variables, it is without doubt the weight of the shuttlecock that has the greatest effect in determining the maximum distance achieved.

3.5 Shuttlecocks are "speed rated" by identifying their weights measured in grains. The rules of badminton say that a shuttlecock has correct flight/distance characteristics when, if hit with a firm underarm stroke, it travels between 12.64 m and 13.10 m. (The length of a badminton court is 13.4 m.) Thus in hot climates, say at high altitude, where the air is relatively thin, a firmly hit 73 grain shuttlecock might have sufficient weight to travel the required distance of about 13m. On the other hand in a cold climate, with humid dense air, it would be difficult to hit the same 75 grain shuttlecock much further than just over half the court length (c. 7 m), and a heavier 85 grain shuttlecock might be required in such extreme "slow" conditions if the regulation flight distance of the shuttlecock was ever to be achieved. Generally speaking every extra grain of weight

increases the distance a shuttlecock will fly by about 200 mm.

3.6 The only reason why badminton players are able to hit regulation shuttlecocks the full length of the court at all, is that by developing their hitting technique they are able to produce incredibly high racquet velocities at the moment of impact. Generally this is accomplished by not only swinging the racquet arm through the hitting arc, but in addition, adding a velocity vector to the arm's rotational component by bringing the wrist through in a high speed flicking movement at impact. Top class players manage this combined action, and as a result can produce very high initial shuttlecock velocities, reported to be about 300 km/h (~80 m/sec). The rest of the badminton playing public try to emulate this sort of action and speed, only to find that the maximum shuttlecock speeds they can achieve are often much slower, one result of which is that the opposition on the other side of the net keep returning the shuttlecock back into play. A number of experiments have been carried out, using racquets and mechanical hitters to measure racquet/hitter speed and shuttlecock flight distance. The results from a number of different experiments are shown in Fig. 2.

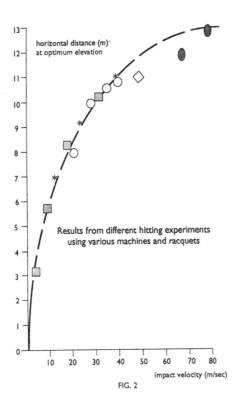

horizontal distance (m)·
at optimum elevation

Results from different hitting experiments
using various machines and racquets

impact velocity (m/sec)

FIG. 2

4   PROBLEMS AND ATTEMPTED SOLUTIONS

4.1   Figure 2 identifies one of the most fundamental problems that needs to be overcome if a mechanical device is ever going to successfully hit shuttlecocks the full length of a badminton court. The problem is simply that as the velocity of the hitter increases, the extra distance that the shuttlecock travels decreases exponentially. It can be seen from Fig. 2, for example, that a hitter speed of 40 m/sec will propel a shuttlecock roughly 11 m, whereas a speed twice this (80 m/sec) only produces an extra 2 m distance, and a speed twice this (160 m/sec) might only produce a further 0.5 m distance. In considering possible designs for mechanical hitters, the required hitting speeds need to be put in perspective with what might be thought reasonable within a safe sporting environment. For example 40 m/sec is ~140 km/hr - in car terms too fast even for motorway driving, 80 m/sec is ~300 km/hr - and again in car terms, the sort of speed only seen in flat out formula 1 racing, . . . and so on.

4.2   A small number of mechanical shuttlecockthrowers/ hitters have been made by

various interested parties. Some years ago Carlton Sports Co. Ltd. produced a small machine that impacted the shuttlecock with a released leaf spring. The machine was reliable but only propelled the shuttlecock very short distances. It is understood that the machine ended its days in storage. Reinforced Shuttlecocks Ltd. (RSL) at one time possessed a "gun" that fired shuttlecocks using compressed air. The shuttlecock travelled about 3/4 the length of a badminton court, about 10 m. However, a fundamental problem with using compressed air to blow shuttlecocks is that the shuttlecock never assumes a realistic flight pattern and tends to wobble through the air as the entraining air jet travels with it. It might be that such an air gun could be useful for shuttlecock manufacturers to test shuttlecock consistency but it would appear to be unsatisfactory as a propulsion method for a training machine since the flight characteristics of the shuttlecock are so different from those that actually occur when playing.

4.3   Patent searches revealed that although there were a number of patents existing for cricket ball, tennis ball hitters etc. no U.K. patents were discovered relating to shuttlecock throwers or hitters. A Russian patent has been subsequently identified which apparently patents a conceptual design where a racquet is fixed to a rotating arm and the shuttlecock is fed into the path of the racquet by free falling from a position above the racquet There was no indication in the patent that any attempt had been made to realise this design in practice.

5   SOME POSSIBLE DESIGN STRATEGIES

5.1 A number of design concepts were examined as part of a feasibility study. Among many other ideas, the principal ones considered included: (i) using a rotating racquet (continuous rotary movement or powered only through the hitting arc by a spring etc.); (ii) using a rotating solid hitter, (again using a continuos rotary movement or powered only through the hitting arc by a spring etc.); (iii) a catapult device to launch a hitting surface against the shuttlecock, or to launch the shuttlecock itself; (iv) a throwing machine incorporating some sort of lever arm; and (v) a blowing machines using compressed air to directly propel the shuttlecock or used in a pneumatic system to accelerate a hitter against the shuttlecock.

5.2 Prototype models were made for all of the conceptual systems described. However, after trials a number of these ideas were discarded, principally for one of two reasons. The main problems encountered were that the devices either produced unrealistic flight characteristics, or that the stored energies and the mass accelerations necessary to produce the required hitting velocities were so large that they were considered to be potentially unsafe for incorporating into a machine that was to be used repeatedly in a typical sports training environment.

5.3 Thus, it has been shown that very high hitting velocities need to be generated if the shuttlecock is to be propelled a distance of about 13m, and, in addition, any available increase in hitting velocity may in fact produce only a disproportionately small increase in distance obtained. Experiments have shown that very high hitting velocities can be achieved using stored energy systems, where a sudden release of elastic energy is harnessed to obtain these velocities. However, after evaluating a number of stored energy machines it was considered that their high elastic to kinetic energy

requirements created potentially unsafe operating conditions, especially if attention is given to what might happen if a component part of the machine were to fail. Therefore, mainly with this consideration in mind, a decision was made to concentrate on developing a machine using a continuos rotary hitter, where high accelerations and decelerations of component parts could hopefully be kept to within reasonable limits. After further evaluation it was decided to consider a principal design where a hitter was attached to the periphery of a rotating wheel.

6   SIZE OF MACHINE

6.1 The size of the hitting wheel plus hitter will to a large extent determine the general size of the machine. Not surprisingly, one design specification for the machine is that it should be transportable, and in particular that it should be able to fit into the boot of an average sized car. Clearly, therefore, all things being equal, the smaller the better. Unfortunately, however, reducing the size of the hitting wheel plus hitter brings with it severe penalties. Figure 2. has shown that ideally hitter speeds of 80m/sec need to be achieved if the shuttlecock is to be hit the full length of a badminton court. However, the smaller the hitting wheel diameter, the faster the wheel has to run to achieve this peripheral hitting speed. More importantly, with any increase in wheel speed, less and less time becomes available between complete revolutions of the hitter in which to allow placing of the shuttlecock from a position where none of it is physically in the path of the hitter to a position where it is exactly in the path of the hitter and orientated for correct hitting. For the condition of a constant hitting velocity of 80 m/sec, Figure 3 shows a graph of hitting wheel (plus hitter) diameter against r.p.m. of the wheel, and also therefore the time for the wheel to complete one revolution, i.e. the maximum loading time envelope available to introduce the shuttlecock into the hitter's path.

6.2 During the development of the badminton machine a number of alternative hitting wheel diameters were tested. The smallest wheel used had a diameter of 0.2 m. From Fig. 3 it can be seen that to achieve a hitting velocity of 80 m/sec this wheel needed to be rotating at c. 7000 rpm and had a shuttlecock loading time envelope of ~1/100th of a second The largest wheel used had a diameter of 0.75 m. This wheel needed to be

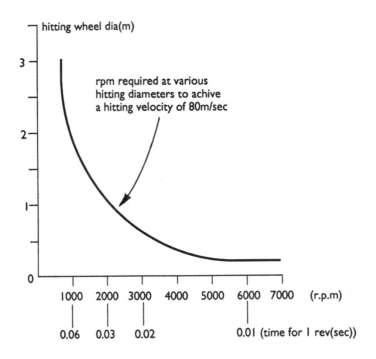

hitting wheel dia(m)

rpm required at various
hitting diameters to achive
a hitting velocity of 80m/sec

(r.p.m)

0.06   0.03   0.02   0.01 (time for 1 rev(sec))

FIG. 3

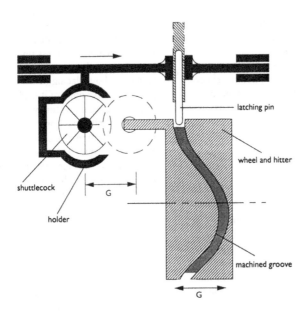

latching pin

wheel and hitter

shuttlecock

holder

G

machined groove

G

FIG. 4

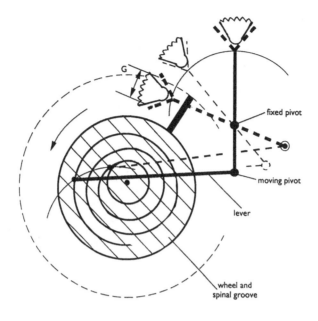

G

fixed pivot

moving pivot

lever

wheel and
spinal groove

FIG. 5

rotating at about 2000 rpm and had a shuttlecock loading time envelope of ~ 3/100ths of a second. Although larger wheel diameters appear to gain by operating at relatively low rotational speeds, increase the effective loading time envelope, and in addition imply an inherently safer machine than one incorporating high speed rotating components, the size of a very low speed machine becomes impracticable if the machine has also to be transportable. For this reason an arbitrary upper limit of 0.7 m for the diameter of the wheel plus hitter was identified as offering a good compromise between final machine size and wheel rotational speed.

7 LOADING THE SHUTTLECOCK

7.1 Experiments have shown that allowing the shuttlecock to free fall into the path of a rotating hitter is not practical if consistent shuttlecock trajectories are required. Therefore some method of placing and holding the shuttlecock in the correct position and orientation with respect to the the hitter needs to be identified. After considering a number of theoretical concepts, two designs for feeding the shuttlecock into the path of the hitter were developed.

7.2 The first of these (Fig. 4) involved mounting the hitter on to the side of a rotating wheel and pulled a shuttlecock into the path of the hitter by latching a pin into an eccentric groove formed in the circumferential surface of the wheel.

7.3 The second system (Fig. 5) involved mounting the hitter radially on the circumference of the wheel containing a spiral groove machined into one of its side surfaces. With the wheel rotating, when a pin at the wheel end of a pivoted lever was latched into the spiral groove the lever system pulled the shuttlecock into the path of the hitter.

7.4 For both systems, within one revolution of the wheel, the shuttlecock loading mechanism has to move the shuttlecock in its holder from a position where it is missed by the rotating hitter, to a position where it is correctly orientated for hitting. However, once the shuttlecock has been hit out of the machine, the hitter on coming around again passes through the empty holder without touching it. Thus, there is no urgency to remove the holder and the return of the empty holder back to its starting (and loading) position can be carried out at a relatively slow rate.

7.5 The most recently built machine has utilised the second method described for loading and hitting the shuttlecock. Either method, however, should be capable of performing satisfactorily. In both systems, however, perhaps the most important design criteria is to use the very light mass of the shuttlecock to its full advantage It would be unfortunate to design a holder of relatively high mass, and high inertia, in order to facilitate moving a shuttlecock weighing only a few grams, since high accelerations and decelarations of the holder may need to be accomplished within a very small loading time envelope. However, nothing is as simple as it may at first seem. The original machine incorporated a one-piece, relatively heavy, difficult to machine, polymer shuttlecock holder. The holder worked well, but as a simplification the polymer holder was replaced with a very simple aluminium wire frame" holder. This holder, which was very cheap and easy to make, seemed perfectly satisfactory, until it was discovered that at high hitting velocities, typically above 40 m/sec, the wind generated by the rotating hitter was blowing the shuttlecock out of the holder before it could be hit. A problem that had not occurred when the heavier "closed" polymer holder was used.

8   THE SITUATION AT PRESENT

8.1 The machine at present seems to operate extremely well. However, not surprisingly hitting a shuttlecock ~13 m can be difficult to achieve depending on shuttle condition, weight etc. For the 0.7 m. diameter wheel chosen, and with safety considerations in mind, the machine will easily hit a shuttlecock~11 m., a perfectly satisfactory distance for all training requirements. High lobs to the back of the court can be produced with the machine , say, 2 m or 3 m from the opposite side of the net. Fast flat shots from the base line, travelling just over the top of the net are no problem. Perhaps, however, the greatest benefit of the machine is to simulate the short serve, dropping an inch over the net and landing just in the front of the service court. This is one of the most difficult shots to train a player to respond to and equally difficult to set up with a coach feeding to a player. Even the best coaches find it impossible to repeatedly short serve and get it right every time. The machine, however, will produce exactly such a shot over and over again,

with of course the benefit that the coach can concentrate on watching the player rather than worrying about getting his/her feeding technique absolutely right.

9   FUTURE WORK

9.1 The design of the machine has a little way to go before it can be considered as good as it can be In particular, one aspect that has been looked into over the last few years is the development of a multiple shuttlecock feeding facility which once loaded will deliver , say, 40 or 50 shuttlecocks completely automatically into the hitting machine. At the last count, over a hundred different design concepts had been identified for devices to achieve this end. By the end of this year it is anticipated that a successful prototype of one optimised system will be in full operation with the present machine set up.

ACKNOWLEDGEMENTS

The author would like to thank F.S.T. Ltd and R.S.L. Ltd for their advice and enthusiasm for the project.

The Engineering of Sport, Haake (ed.) © 1996 Balkema, Rotterdam. ISBN 90 5410 822 3

How to win the America's Cup: Optimum control of a yacht having an optimally designed sail

T. Sugimoto

Saitama Institute of Technology, Japan

ABSTRACT: This study presents how to maximize the speed made good to windward: the problem mixed the optimum sail design with the optimum sailing strategy. The sail is designed to produce the maximum aerodynamic thrust subject to constraints on the heeling-righting moment equilibrium and the maximum sail lift. Then two control parameters, "reefing" and "flat", are introduced. The analytic expression is derived for the aerodynamic contributions due to these parameters. For wide varieties of the angle between the true wind and the sailing course, the best choices of control parameters are obtained. The reefing parameter is found to be inactive; the flat parameter becomes active if the sailing course is directed less than 60 degrees to the true wind.

1. INTRODUCTION

If you want to win the America's Cup, or the New Zealand's Cup, you have to build a fast yacht. In this paper we propose the best way to get and use a sail for a fast yacht.

A yacht is a very sophisticated sailing-craft that can run up to windward without any sort of engines. The windward sailing is possible owing to the main sail that yields the aerodynamic thrust toward the sailing course. Therefore the optimum sail design is one of the indispensable factors to get a fast yacht.

At the same time we should note that a racing yacht must have a good *path* performance from the start to the goal. Optimum design is, however, not almighty for this purpose, because the design is optimum only for required conditions of a *point* performance. Therefore another factor of a fast yacht is to consider the way of enhancing a path performance. A possible remedy is optimum control. We can expect a good sailing performance even off the design conditions by controlling and tuning the sail optimally.

This is my answer to the question how to get a fast yacht: 1) to design a sail optimally; 2) to use that sail always optimally. In this paper, the section 2 describes the theory from the opti-

mum design to the optimum control. The section 3 presents the numerical results and discussion. The section 4 summarizes our findings.

2. THEORY

2.1 Steady-state sailing to windward

Figure 1 shows schematically the steady-state sailing to windward. The essential part of sailing mechanics is to understand the velocity triangle and the equilibria of aero-hydrodynamic forces and moments. Let us introduce the velocity ratio λ, or the non-dimensional yacht speed, defined by

$$\lambda = V/W, \qquad (1)$$

where V and W denote the craft speed and the true wind speed, respectively. Then we can write the relations of the velocity triangle in the following:

$$U/W = \sqrt{\lambda^2 + 2\lambda \cos \theta + 1}, \qquad (2)$$

$$\sin \phi = \sin \theta / \sqrt{\lambda^2 + 2\lambda \cos \theta + 1}, \qquad (3)$$

$$\cos \phi = (\cos \theta + \lambda) / \sqrt{\lambda^2 + 2\lambda \cos \theta + 1}, \qquad (4)$$

where U, θ and ϕ denote the relative wind speed, the wind angle and the relative wind angle, respectively.

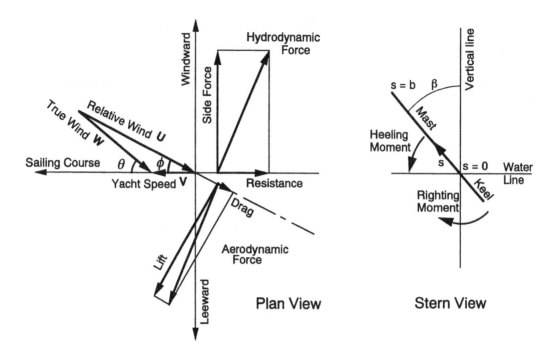

Figure 1. The velocity triangle and the equilibria of forces and moments

Although a yacht has six degrees of freedom, the most important are three of them: 1) the equilibrium between the aerodynamic thrust and the hydrodynamic resistance; 2) the equilibrium between the aerodynamic and hydrodynamic side forces; 3) the equilibrium between the heeling and righting moments. The aerodynamic thrust and side force are given by (lift)$\sin\phi-$ (drag)$\cos\phi$ and (lift)$\cos\phi+$ (drag) $\sin\phi$, respectively.

Later we will consider these equilibria to formulate the optimum sailing strategy.

2.2 Optimum sail design

This subsection reviews the known result brought by [1], and also shows the derivation of the aerodynamic characteristics, which is a new result. First we introduce the optimum sail design that maximizes the aerodynamic thrust subject to constraints on the heeling-righting moment equilibrium and the maximum sail lift. The synthesis is based on the lifting-line theory. The non-dimensional thrust C_T, maximum lift C_{Lmax} and heeling moment C_H are written in terms of the unknown circulation γ:

$$C_T = \mathcal{R}\{1 - \frac{1}{2}(1 + \sin^2\phi)\beta^2\} \int_0^1 \gamma(s)\mathrm{d}s\sin\phi$$
$$-\{C_{D0} + \mathcal{R}\int_0^1 \gamma(s)\alpha_i(s)\mathrm{d}s\}\cos\phi, \quad (5)$$

$$C_{Lmax} = \mathcal{R}\int_0^1 \gamma(s)\mathrm{d}s, \quad (6)$$

$$C_H = \mathcal{R}\cos\phi\int_0^1 \gamma(s)s\mathrm{d}s, \quad (7)$$

where $\mathcal{R}$, β, C_{D0} and α_i denote the aspect ratio of a sail, the heel angle, the friction drag coefficient of the sail and the induced angle of attack, respectively. The coordinate s runs up the mast, and is made non-dimensional by use of the mast height b^*. Application of the variational principle to this problem leads to the solution. The optimum circulation is given by

$$\gamma(s) = \frac{12C_{Lmax}}{\pi\mathcal{R}(3 - 2\mu^*)}$$
$$\times(\sqrt{1 - s^2} + \mu^* s^2 \log\left|\frac{1 - \sqrt{1 - s^2}}{1 + \sqrt{1 - s^2}}\right|), \quad (8)$$

$$\alpha_i(s) = \frac{3C_{Lmax}}{\mathcal{R}(3 - 2\mu^*)}(\frac{1 + 2\mu^*}{\pi} - 2\mu^* s), \quad (9)$$

112

where μ^* is a loading factor, which takes the value in $[0, 1/2]$. If μ^* is zero, the optimum circulation coincides with the elliptic loading. If μ^* is $1/2$, the optimum circulation becomes Jones' solution [3]. Other integrated values are in the following:

$$C_T = \{1 - \frac{1}{2}(1 + \sin^2 \phi)\beta^2\}C_{Lmax} \sin \phi$$
$$-\{C_{D0} + \frac{C_{Lmax}^2}{e^* \pi \mathcal{R}}\} \cos \phi, \quad (10)$$

$$C_H = \frac{4(1 - \mu^*)}{\pi(3 - 2\mu^*)}C_{Lmax} \cos \phi, \quad (11)$$

$$e^* = \frac{(3 - 2\mu^*)^2}{3(4\mu^{*2} - 4\mu^* + 3)}. \quad (12)$$

The aerodynamic characteristics of the optimum circulation are summarized by its lift curve slope. We can derive its analytic expression from the Kutta-Zhukovski theorem. If we assume the lift curve slope of the wing section as 2π, the theorem is written in the form:

$$\gamma(s) = \pi c(s)(\alpha_{emax}(s) - \alpha_i(s)), \quad (13)$$

where c and α_{emax} denote the chord and the maximum effective angle of attack, respectively. The effective angle of attack designates the geometrical angle of attack taking account of the zero lift angle. Substitution of (9) into (13) and integration of the equation leads to the result:

$$\frac{C_{Lmax}}{\mathcal{R}} = \frac{2\pi}{\mathcal{R}}\{\overline{\alpha_{emax}}$$
$$-\frac{3C_{Lmax}}{\mathcal{R}(3 - 2\mu^*)}(\frac{1 + 2\mu^*}{\pi} - 2\mu^*\overline{s})\}, \quad (14)$$

where

$$\overline{\alpha_{emax}} = \int_0^1 c(s)\alpha_{emax}(s)\mathrm{d}s / \int_0^1 c(s)\mathrm{d}s, \quad (15)$$

$$\overline{s} = \int_0^1 c(s)s\mathrm{d}s / \int_0^1 c(s)\mathrm{d}s. \quad (16)$$

Solving (14) with respect to C_{Lmax}, we have

$$C_{Lmax} = \frac{2\pi\overline{\alpha_{emax}}}{1 + \frac{6}{\mathcal{R}} \frac{1 + 2(1 - \pi\overline{s})\mu^*}{3 - 2\mu^*}}. \quad (17)$$

Equation (17) provides the relation between the total lift and the geometric angle of attack. We should note that the three-dimensional effect is inclusive in this equation.

2.3 Introduction of control parameters

In this subsection we shall introduce two control parameters, "reefing" and "flat", after [2] and [3].

The reefing parameter, denoted by r, represents the effect due to shortening the mast height. The one and only, neccessary assumption is the similarity of geometry: if the mast length b^* is shortened to b, then we can expect $c^*(s/b^*) = c(s/b)$. The definition of r is given by

$$r = b/b^* \leq 1. \quad (18)$$

This parameter affects the sail performance aerodynamically as well as geometrically. The following are trivial:

$$S/S^* = r, \quad (19)$$
$$\mathcal{R}/\mathcal{R}^* = r, \quad (20)$$

where S denotes the sail area. Quantities with asterisk denote those on the design conditions. From (17) and (20) one can derive the formula for the lift curve slope $C_{L\alpha}$:

$$C_{L\alpha} = \frac{rC_{L\alpha}^*}{1 - \frac{1 - r}{2\pi}C_{L\alpha}^*} \approx rC_{L\alpha}^*. \quad (21)$$

The flat parameter, denoted by f, represents the effect due to reducing the angle of attack, twist and camber of the sail. My definition is given by

$$f = \overline{\alpha_e}/\overline{\alpha_{emax}}^*. \quad (22)$$

If $f = 1$, then a sail produces its maximum lift; if $f = 0$, then the lift is zero. The lift coefficient is given by use of f and r:

$$C_L = C_{L\alpha}\overline{\alpha_e} = rC_{L\alpha}^*\overline{\alpha_{emax}}^* f = frC_{Lmax}^*. \quad (23)$$

Other aerodynamic coefficients shall be given in terms of C_L.

2.4 Optimum control problem

In this subsection we will formulate the basic equation of the optimum control problem: how to maximize the yacht speed by controlling the optimally designed sail.

Using the expression of aerodynamic characteristics described in the previous subsection, we can derive the equilibria of the steady-state sailing to windward. The aerodynamic thrust is given by

$$\frac{1}{2}\rho_a U^2 S[\{1 - \frac{1}{2}(1 + \sin^2\phi)\beta^2\}C_L \sin \phi$$
$$-(C_{D0} + \frac{C_L^2}{e^* \pi \mathcal{R}}) \cos \phi], \quad (24)$$

113

where ρ_a denotes the air density.

On the other hand, the hydrodynamic resistance is given by

$$\frac{1}{2}\rho_w V^2 S_k(C_F + C_W + \frac{C_S^2}{e_k \pi \mathcal{R}_k}), \qquad (25)$$

where ρ_w, S_k, C_F, C_W, e_k, $\mathcal{R}_k$, and C_S denote the water density, the keel area, the skin friction coefficient, the wave drag coefficient, the wing efficiency of the keel, the aspect ratio of the keel, and the hydrodynamic side force coefficient, respectively.

The equilibrium of side forces is given by

$$\frac{1}{2}\rho_a U^2 S[\{1 - \frac{1}{2}(1 + \sin^2\phi)\beta^2\}C_L \cos\phi$$
$$+ (C_{D0} + \frac{C_L^2}{e^*\pi \mathcal{R}})\sin\phi] = \frac{1}{2}\rho_w V^2 S_k C_S. \qquad (26)$$

In the equation above we neglected the effect of heel to C_S. To show the reason we shall estimate orders of related terms by introducing ϵ as the magnitude of smallness. The leeway, or the angle of attack at the keel, δ is assumed to be of order ϵ, while the heel angle β is of order $\epsilon^{1/2}$ in our analysis [1]. Although the hydrodynamic side force is affected by the heel to the factor $(1 - \sin^2\delta \sin^2\beta)^{1/2}$, the trigonometric term in this factor is of negligible order ϵ^3.

The equilibrium of heeling-righting moments is given by

$$\frac{1}{2}\rho_a U^2 Sb \frac{4(1-\mu^*)}{\pi(3-2\mu^*)}C_L \cos\phi = \frac{1}{2}\rho_w V^2 S_k d C_{M,\beta}(0)\beta, \qquad (27)$$

where d denotes the keel draft. The righting moment coefficient is denoted by C_M, and $C_{M,\beta}(0)$ denotes its first derivative at $\beta = 0$.

Solving (26) with respect to C_S, we can eliminate C_S from (25) by substitution. In a similar manner, we can eliminate β from (24) by use of (27). Dividing the equation by $\rho_a W^2 S^*/2$, we arrive at the non-dimensional form of thrust:

$$\sqrt{\lambda^2 + 2\lambda\cos\theta + 1}$$
$$\times \left[\{1 - \frac{1}{2}(\lambda^2 + 2\lambda\cos\theta + 1 + \sin^2\theta)\right.$$
$$\times \frac{4(1-\mu^*)}{\pi(3-2\mu^*)}(\frac{b^*}{d})\frac{\cos\theta + \lambda}{\lambda^2\sqrt{\lambda^2 + 2\lambda\cos\theta + 1}}$$
$$\times \frac{fr^3\sigma^* C_{Lmax}^*}{C_{M,\beta}(0)}\}fr C_{Lmax}^* \sin\theta$$
$$\left. - (C_{D0}^* + \frac{f^2 r C_{Lmax}^{*2}}{e^*\pi \mathcal{R}^*})(\cos\theta + \lambda)\right], \qquad (28)$$

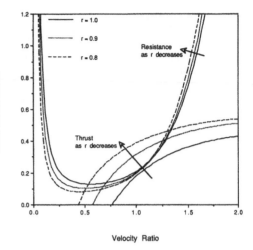

Figure 2. Effects due to "reefing" upon trim velocities

where

$$\sigma^* = \frac{\rho_a S^*}{\rho_w S_k}. \qquad (29)$$

The non-dimensional resistance is written in the form:

$$\frac{1}{r\sigma^*}[(C_F + C_W)\lambda^2 + \frac{r^2\sigma^{*2}}{e_k\pi \mathcal{R}_k}\frac{\lambda^2 + 2\lambda\cos\theta + 1}{\lambda^2}$$
$$\times \{fr C_{Lmax}^*(\cos\theta + \lambda)$$
$$+ (C_{D0}^* + \frac{f^2 r C_{Lmax}^{*2}}{e^*\pi \mathcal{R}^*})\sin\theta\}^2], \qquad (30)$$

The trim velocity of a yacht is given as a solution of the equation (28) = (30). In this equation the unknown is λ, while r and f are control parameters to be adjusted. All the rest should be given quantities. Parameters on the design conditions are shown with asterisk.

3. NUMERICAL RESULTS AND DISCUSSION

Numerical calculation of (28) and (30) has been done for the following conditions:
$\rho_a/\rho_w = 0.00125$, $S^* = 280\ [m^2]$, $\mu^* = 1/2$, $\mathcal{R}^* = 8.3$, $S_k = 2.8\ [m^2]$, $e_k = 0.8$, $\mathcal{R}_k = 11$, $b^*/d = 8.5$, $C_{Lmax}^* = 1.2$ at $\alpha_{emax}^* = \pi/12\ [rad] = 15\ [deg]$, $C_{D0}^* = 0.01$, $C_F = 0.01$, $C_W = 0.005\lambda^4$, $C_{M,\beta}(0) = 1.4$.

The loading parameter μ^* was chosen so that the thrust becomes the largest. According to (28)

114

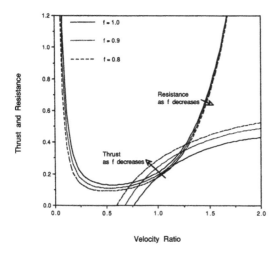

Figure 3. Effects due to "flat" upon trim
velocities

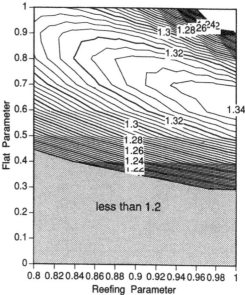

Figure 4. Trim velocity contour in the parameter
plane at $\theta = \pi/6$

it is obvious that larger μ^* produces larger thrust. Therefore the largest, possible value for μ^* is $1/2$.

Figure 2 and 3 show how the control parameters affect the trim velocities at the wind angle $\theta = \pi/6$. In both figures the non-dimensional resistance and thrust are plotted against the velocity ratio λ for r and f specified in the legend.

If both r and f are equal to unity, there is no equilibrium. As one reduces either r or f, there appear two equilibria. The trim at the smaller velocity is, however, unstable, because there exists no restoring force off this trim [1].

Two parameters affect the stable trim velocities in different manners. As f is reduced, the resistance is reduced too. The reduction in f results in the reduction in lift and hence side forces. This reduces the resistance in the end. As is shown in (30), the dominant term in the resistance is not related to the side force but the wave drag in proportion to λ^4. Therefore the effect due to f is not so large.

The reduction in r, however, comes to the contrary conclusion. The r effect by way of the side forces is similar to that of f, but the reduction in r increases the resistance. This is because the reduction in r results in the reduction in sail area. This effect is described by the term $(r\sigma^*)^{-1}$ appearing just out of the bracket in (30). Since the wave drag dominates the resistance, the resistance is magnified in inverse proportion to r.

Although the equation (28) = (30) is strongly nonlinear with respect to λ, it is analytic. There-

fore, to solve the equation, we have used the Newton-Raphson method with relaxation. The scheme, starting from the large, initial value of λ with given r and f, converges fairly fast to a trim velocity, if it exists. To assure if the solution converged to the higher trim velocity, we have checked the sign of the first derivative of the equation.

Figure 4 shows the contour of the higher trim velocity in the control parameter plane at $\theta = \pi/6$. One can see that there is the globally maximum trim velocity near $(r, f) = (1, 0.65)$. To get to the global optimum, we have used the following method: 1) to start calculating λ on rather coarse meshes in rf plane; 2) to find the temporal optimum; 3) to calculate λ on finer meshes in rf plane near the temporal optimum; 4) to repeat refining a mesh and calculating λ until significant digits of the optimum λ becomes four.

Figure 5 shows the polar plot of the optimum trim velocities obtained by the method of solution described above. In this numerical example the speed made good to windward is 1.17 times faster than the true wind velocity at 34.5 degree wind angle.

Figure 6 shows the control strategy for the performance in fig. 5. The result is rather simple: the reefing parameter r is equal to unity, while the flat parameter f becomes less than unity for the wind

115

angle smaller than about 60 degrees.

The change of the strategy is well illustrated by fig. 7. This shows the relation between the heel angle and the wind angle. The heel angle increases almost monotonically, as the wind angle decreases from 90 to around 60 degrees. To the contrary the heel angle decreases, as the wind angle decreases less than 60 degrees. This change implies the following strategy: 1) to use the maximum lift at the wind angle larger than 60 degrees; 2) to reduce the heel angle at the wind angle less than 60 degrees. For this purpose, "reefing" is not so effective as "flat" to reduce the heel angle.

4. CONCLUSIONS

In this paper we proposed the analytic method to derive the optimum sailing strategy by use of an optimally designed sail. The nature of two control parameters, "reefing" and "flat", was investigated based on the lifting-line theory. Numerical example showed the detail of the strategy: 1) the reefing parameter is inactive; 2) the flat parameter becomes active at the wind angle less than 60 degrees.

ACKNOWLEDGEMENT

My interest in this study arose as a result of suggestive talk with Prof. P. S. Jackson, University of Auckland, on the occasion of my visit there in 1994.

REFERENCES

[1] SNEYD, A. D. & SUGIMOTO, T. 1994. *Influence of a Yacht's Heeling Stability to Optimum Sail Design* . Res. Rep. Ser. 2 No. 33, Dept. Maths & Stats, University of Waikato, New Zealand, pp. 19

[2] KERWIN, J. E. 1978. *A Velocity Prediction Program for Ocean Racing Yachts*. MIT Rep. 78-11, MIT, USA, pp. 17

[3] JACKSON, P. S. 1996. Modelling the Aerodynamics of Upwind Sails. *J. Wind Engng & Ind. Aerodyn* . to appear

[4] SUGIMOTO, T. 1993. Analysis of Windward Performance of Sailing Crafts. *Theo. & Appl. Mech.* **42**, pp. 271-275

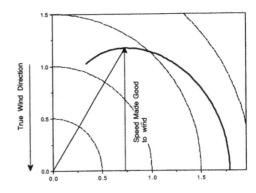

Figure 5. Polar plot of optimum trim velocities

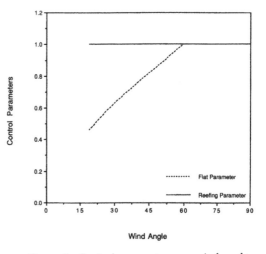

Figure 6. Control parameters vs. wind angle

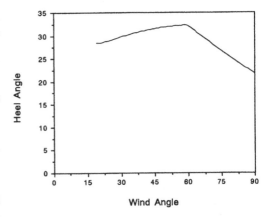

Figure 7. Heel angle vs. wind angle

4 Instrumentation

The Engineering of Sport, Haake (ed.)© 1996 Balkema, Rotterdam. ISBN 90 5410 822 3

Accuracy of kinematic data collection, filtering and numerical differentiation

Peter Dabnichki, Serdar Aritan, Mike Lauder & Dimitrios Tsirakos
The Manchester Metropolitan University, Crewe & Alsager Faculty, Biomechanics Research Group, UK

ABSTRACT: The study represents an experimental methodology for enhancing the calibration procedure in order to increase the accuracy and reliability of the kinematic data collection in biomechanics studies. The frequency response function of the on-line system was studied from tests with controlled input frequency. A parallel synchronised feedback signal with higher sampling frequency was designed to measure the true values of the kinematic derivatives.

1. INTRODUCTION

The kinematic data collection is a very popular method for analysis of sport movement both during field competition and in laboratory conditions. The finite precision of technical measuring systems is a source of a largely stochastic type of errors affecting the kinematic derivatives. Normally in sport analysis the kinematic data is the only possible source of information and through mathematical procedures certain important dynamic characteristics such as acceleration, force and speed are derived. In sporting competitions when the subject should not and cannot be disturbed by measuring devices, film or video analyses are the only methods that can be implemented by the researcher interested in quantitatively measuring performance. One needs prior knowledge of the error to implement proper filtering procedure. An obvious target is the noise generated by the system itself - they can be obtained via controlled experiments implementing parallel and more accurate data collection. The purpose of the presented study was to propose simple ways for deriving system's noise characteristics and possible application to "system sensitive filtering".

2. METHODS AND EXPERIMENTAL DESIGN

The series of experiments was designed such that a controlled and repeated rotational movement was created. Strict control of the frequency of movement was facilitated allowing one to obtain the system's response function.

An electrical servomotor (RS 715-106) was used for the motion generation. Its speed was controlled and measured independently using frequency to voltage converter with a sampling frequency of 1000 Hz. These data were recorded through a parallel data channel of the on-line (ELITE, BTS, Milan, Italy) system via A/D converter (Figure 1). This control independently recorded the period of revolution in order to check the accuracy of the collected kinematic data from the on-line system.

Five independent groups of experiments were conducted each examining a different parameter of interest. The parameters were selected based on the results from a series of carefully conducted pilot experiments. Each experimental group had variable number of identical trials depending upon the step increment and the limits of the parameter of interest. For each trial within the experimental groups, five successful data sets were recorded. The kinematic and time sampling frequencies were 100 Hz and 1000 Hz respectively for all experiments.

2.1. Variable distance between the camera and the object

The distance between the camera and the rotating motor was varied throughout the first series of experiments with an increment of 0.250 ± 10^{-3} m. The starting point was the minimum possible distance (2.510 m) when the system was still able to encompass the whole calibration area and the maximum distance was 5.510 m.

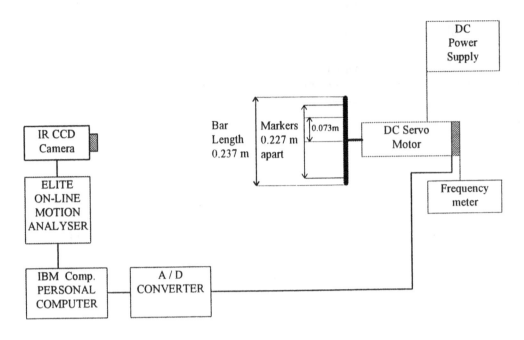

Figure 1. Experimental set-up

2.2. Variable distance object - calibration plane

A principal recommendation for 2-D data collection is that the moving object stays in the calibration plane throughout the recorded motion. This sequence of experiments was designed and performed in order to check the influence of the distance between the plane of motion and the calibration plane. The calibration plane was kept stationary but the motor was moved three times forward and three times backward by an increment of 0.050 ± 10^{-3} m from the calibration plane.

2.3. Variable ratio calibration area / area of the movement

An obvious factor which is expected to affect the accuracy is the size of the calibration area. Clear indications how this size affects the accuracy of the collected data were not found in the literature. The experiment was conducted with the motor in a fixed position (in the centre of the total calibration area) and the calibration area expanded gradually from 0.300 x 0.300 m (4 x 4) to 0.900 x 0.900 m (10 x 10 markers). This was achieved by expanding the calibration area by one column of markers horizontally followed by one row of markers vertically anti-clockwise.

2.4. Variable motor speed

The influence of the frequency of movement on the accuracy of data collection was assessed by varying the frequency of the motor revolution. Thirteen different frequencies were produced from 2.710 Hz to 28.571 Hz.

2.5. Position of the moving object in the calibration area

To check the effect of the object's position in the calibration area the following experiment was performed. The area was divided into nine equally sized sub-areas (chess board) and each time the motor was stipulated in the centre of each of these sub-areas. Each of these sub-areas encompassed 16 (4x4) equidistant markers giving square calibration area with sides of 0.300 m. The sub-areas were defined according to their position on the calibration frame i.e. Left - Upper, Middle and Lower, etc.

2.6. Data processing and reduction

As a result of the experiments described above a data set consisting of a total of 280 kinematic data files consisting of 500 frames and 280 sampling files of 5000 points each detecting the period duration for every revolution that were produced. The synchronised data set provided independent data with higher accuracy and means of checking the stationariness of the process. A Wilcoxon matched pairs test (CSS : Statistica, 1991) was performed on the raw data (computed length between the two outer markers) from each of the sub-groups of trials from experimental groups to determine which trials were homogenous. Homogenous trials were then merged to provide a full set of raw data for each manipulated parameter within each of the groups of experiments. From these sets of data the average length was then computed for each sub-group and from this the frequency distribution of the error deviations from the estimated mean value was determined. The Kolmogorov-Smirnov one sample test (Hoel, 1971) was then performed on each set of frequency data to determine if the distribution of the error differed significantly from the theoretically normal distribution. The significance level was set at p⟺ 0.05 and the procedure helped to establish that the process was ergodic. In practical terms, the system was stable and the components were not affected by the duration of the work. All statistical analyses were carried out using the CSS Statistica software package (StatSoft Inc., 1991).
'Biased' error (Bendat and Piersol, 1971) was calculated by the equation:

$$b\left(\hat{x}\right)[\%\,] = \left\{ \frac{E\left(\hat{x}\right)}{x} - 1 \right\} \times 100 \qquad (1.1)$$

where $\hat{x}$ is the estimator (calculated value) of the real length x and $E(\hat{x})$ is the mathematical expectation.

3. RESULTS

The global results of average difference in length, variance in length and biased error in length for within each group of experiments are presented in Table 1.The data produced cannot be submitted in full as its size is very big. The results presented are very concise and further information if required could be obtained from the authors.

The distribution of the error in the kinematic data was a matter of special interest as some filtering procedures assume that the error is normally distributed. So these type of procedures could be

sometimes very sensitive to the error distribution (Craven and Wahba, 1979). As stated by (Hoel, 1971) the Kolmogorov-Smirnov test has "striking advantages" over the Chi squared test that its choice is very natural.

Table 1. Experimental results for the different experimental groups (2.1-2.6).

Sub-group	Mean length margins (mm)	Variance in length (mm) within the group	Biased error in length (% of real length)
MC	222.80 - 225.30	0.66 - 5.46	-1.9 to -0.8
MM	210.68 - 241.68	1.66 - 4.61	-7.2 to 6.5
CE	222.71 - 223.45	0.55 - 1.13	-1.9 to -1.5
VS	226.26 - 226.37	1.15 - 20.9	-0.9
MP	224.57 - 226.90	0.88 - 3.29	-0.04 to 1.4

The Kolmogorov-Smirnov one sample test showed three of the fifty one sub experiments to have error distributions that differed significantly from the theoretical normal distribution. These methodologies were CE16, CE56 and MC276 (p< 0.05). It could be hypothesised that these differences were attributable to some unforeseen factor that influenced the data collection. However these findings were deemed not to be of significance to the conclusions of the present study.

The experiments showed also that the data collection set-up affects more the noise characteristics as these differences are further expressed with the increase of the number of the of data points (see Figure 2).

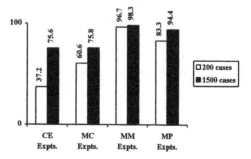

Figure 2. The percentage of significant differences between trials within each experimental group when using 200 and 1500 data points.

Another surprising finding with definite practical implications was the threshold frequency of the system. This frequency was well below the Nyquist value and the results are presented in Table2.

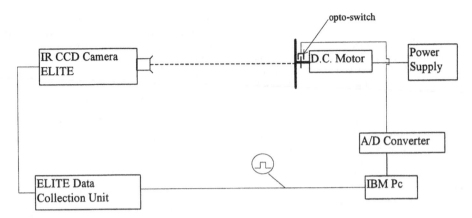

Figure 3. Proposed experimental set-up

Table 2. Frequency limits of on-line system.

Sampling Frequency	System Limit	Nyquist Frequency
25	3.1	12.5
50	4.6	25
100	11.6	50

4 DISCUSSION

The present study considered the kinematic system as one prime unit and does not hypothesise the source of error and its distribution within the system in the different stages of data derivation and processing. This approach is sufficiently flexible and is therefore applicable to different optical systems.

The results of this study are deemed to become an important part of a "system sensitive" filtering procedure. They will be used to reduce the influence of the system generated noise on the kinematic derivatives. However the presented approach has been successful in establishing the error in the displacements but has two major disadvantages in the estimation of the error in the kinematic derivatives:

- the produced data is too smooth.
- the instantaneous velocity is practically unknown.

Further improvements of the design are possible via measuring the angular positions via mounting an encoder with an optical switch. The design of the device has been accomplished (see Figure 3) and some preliminary results were obtained. The advantage of this approach is that more "rough" running of the motor is possible and the instantaneous velocity could be estimated with a much higher degree of accuracy.

5 CONCLUSION

The chosen parameters in the present study represent a set of independent statistical variables. The error distribution in each methodological experimental set represents the relevant marginal distribution. The joint density of the error distribution could be expressed as a product of these marginal distributions:

$$p\left(d_c, d_{cp}, a_r, p_{cp}\right) = p\left(d_c\right) p\left(d_{cp}\right) p\left(p_{cp}\right) p\left(a_r\right) \quad (1.2)$$

where:

p   the density function of the error distribution;

d_c camera distance;

d_{cp} distance from the object to the calibration plane;

p_{cp} position of the object in the calibration plane;

a_r ratio between the area of the movement and the calibrated area.

The marginal distributions could be assumed with a limited accuracy to be normal distribution. This limited accuracy is good enough for the majority of the practical applications to sport.

The major results of the present work could be summarised as follows:

- The frequency upper limit of the system is well below the Nyquist limit and should be obtained as part of the calibration procedure.
- When two-dimensional data collection is conducted the accuracy for a set of possible planes of motion parallel to the calibration plane should be studied in advance. A simple example is walking - the data would differ largely for the left and right part of the body.
- The error distribution is sensitive to relatively

small changes in the camera distance and the position of the object within the calibrated area.

- The on-line system showed very good accuracy in these close to ideal test conditions and was not affected by the input frequency. The results from the study suggested that the system generates wide band white noise.
- The frequency response function of the system is independent from the input frequency and this fact can be utilised in a robust filtering procedure.
- The density of the error distribution was found to be a multiparametrical distribution and can be represented as a product of the marginal distributions.

The presented methodology is not system dependant and was successfully applied with some minor modifications to underwater video data collection. Minor modifications will be required for 3-D studies.

6. ACKNOWLEDGEMENT

The authors would like to express their gratitude to Dr Diane Taktak for her valuable comments and to Mr Gerald Wright and Mr Barry Shuttleworth for the technical support.

7. REFERENCES

Craven P, Wahba G., 1979, Smoothing noisy data with spline functions. *Numer Mathem* 31: 377-403.

CSS (Complete Statistical System): Statistica, 1991, Vol. 1-3, Stat Soft, USA.

Bendat J.S, Piersol A.G., 1971, *Random data: Analysis and measurement procedures*. John Wiley & Sons, New York.

Hoel P.G., 1971, *Introduction to mathematical statistics*. Fourth edition; John Wiley & Sons, New York.

The Engineering of Sport, Haake (ed.)© 1996 Balkema, Rotterdam. ISBN 90 5410 822 3

Measuring the longitudinal force during javelin release

Simon Iwnicki
The Manchester Metropolitan University, Faculty of Science and Engineering, Department of Mechanical Engineering, UK

Peter Dabnichki & Serdar Aritan
The Manchester Metropolitan University, Crewe & Alsager Faculty, Biomechanics Research Group, UK

ABSTRACT: This paper presents the design of a simple transducer for measuring the longitudinal force acting on a javelin during a throw.

1 INTRODUCTION

The biomechanical analysis of the javelin throw has been a subject of interest for sport scientists, coaches and sport biomechanicists. Many studies have been devoted to the analysis of athletes' performance. Normally all these works [1,2] are based solely on kinematic data as these data were obtained during competition when any interference with the athletes is strictly prohibited. The kinematic data has normally been obtained using cine cameras with a sampling frequency of up to 200 Hz. The cine data is insufficient as a source of information and, to estimate the forces involved one needs to adopt different models based on certain assumptions for the work of the human joints. This approach faces three major problems:

- the data contains errors or noise with unknown frequency and amplitude which is a function of a variety of factors. Proper filtering is very difficult.
- the forces are estimated based on assumption that the system passes a sequence of equilibrium states (quasistatic approach).
- the sampling frequency limit is not sufficiently high and this frequency is not constant due to mechanical problems in filming.

The difficulties associated with the first problem are discussed in details elsewhere [3]. The second problem results from the data collection method and cannot be overcome. The way forward seems to be combined data collection of both force and kinematic data in laboratory conditions using appropriate devices. Such data will be very valuable for modeling purposes as it allows synthesis of an optimal pattern of movement and verification of model results.

Software for this purpose is available but the lack of reliable force data makes generation of realistic computer models almost impossible. It is possible to increase the sampling frequency but the cost of a suitable camera and other equipment is rather high and rises exponentially with the increase of the sampling frequency.

The work presented here outlines a simple force transducer for measurement the longitudinal force acting on the javelin during release. The importance of these data for a proper analysis and understanding the process of the throw is briefly discussed.

2. METHODS

2.1. *Design and calibration of the prototype force transducer*

The research work being carried out at Manchester Metropolitan University required the accurate measurement of the force exerted by the thrower's hand on the javelin (and thus by the javelin on the thrower's hand) in the axial direction only. In order to measure this axial force a collar has been designed so that it can slide on the javelin shaft at the point where the normal grip is. The thrower grasps the collar instead of the shaft and the force of the throw is transmitted through the transducer to the javelin.

The collar is a turned aluminum cylinder with a shield to support the transducer (figure 1). The transducer consists of a 15 mm x 4 mm rectangular section steel beam which passes through the shaft of the javelin and is supported at either side by the sliding collar.

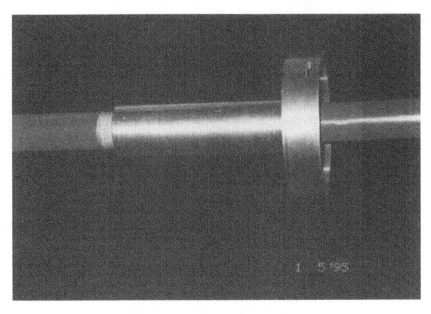

Figure 1. The sliding collar on the javelin.

The beam is therefore in 4 point bending with the javelin making up the two inner point forces and the collar providing the outer two point forces. This design means that the javelin being used must be modified with the incorporation of a slot on either side wall at the point of the grip.

Electrical resistance strain gauges are fitted to either side of the steel beam and connected as two legs in a wheatstone bridge. In normal operation, one gauge measures compressive strain on one side of the beam and the other measure tensile strain. As change in strain is directly proportional to change in the force between the javelin and the collar it is possible to calibrate the transducer to measure this force [1].

In order to design the transducer a maximum level of the force to be measured was required. The maximum acceleration of a javelin during a typical throw was calculated from data collected from synchronized infra-red cameras and found to be 363 ms^{-2}. Using the mass of the javelin this equated to an accelerating force of 268 N. The maximum design force for the transducer was set to 500 N to avoid overload damage.

A preliminary static calibration was first carried out up to the 300 N and the resulting calibration curve is shown in figure 2. The results indicate that the device shows the required linearity across the force range required.

A dynamic test was then set up to ensure that the response remained linear at frequencies up to 15 Hz. This is the maximum frequency at which data is required as an input to models of the human arm being developed. The shaft of the javelin was replaced by an electromagnetic shaker and the collar clamped in vee blocks. A force transducer was also fitted between the shaker and the device so that the input force could be accurately measured. The dynamic calibration is shown in figure 3. (a list of equipment used in the calibration tests is given in appendix 1) The system was excited across the frequency range of interest and the output from the strain bridge and the force transducer were compared.

It can be seen that the device is linear between 2. and 15. Hz with a maximum error of 5%. Between 2. and 20. Hz the device has a maximum error of 11%.

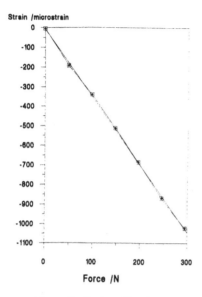

Figure 2. Static calibration curve.

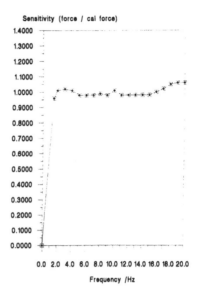

Figure 3. Dynamic calibration curve.

2.2. *Testing and modification of the force transducer*

The force transducer, once calibrated, was assembled into the modified javelin and the output was connected to an A to D sampling card in a PC (details of hardware and software in appendix 1). The connecting wires were allowed to trail behind the javelin with enough slack to allow standing throws to be made.

The device seems to function well with the results conforming to the pattern and level of forces that would be expected from previous calculations based on photography [2]. Several modifications to the design of the collar and electrical system were however suggested during the trials.

The collar proved not to be long enough to allow the thrower to achieve a normal grip and the transducer shield also interfered with a comfortable hand position. The collar has since been extended and part of the shield cut away. The collar is still not perfect and its mass of 380 g may still be too high to allow the throw to be unaffected by the measuring device. Effectively this is an increase of 46% although the design of the device is such that it should be considered as an additional mass to the thrower's arm. The other drawback is that the device changes

the center of mass and would effect the balance and the length of the throw. However both the throw distance and the magnitude of the transverse force are not of interest in this stage.

As the magnitude of the longitudinal force was only required during the period before (and immediately after) release, the wires connecting the transducer to the computer have been shortened and quick release connectors fitted. This reduces the chance of the thrower being distracted by the trailing cable.

2.3. *Data sampling*

Signals from the transducer were collected from a standing throw. A standing throw was chosen to provide a sample input as it probably represents around 70% of the best throw for a good athlete. The device has been designed to allow a real throw without too much hindrance using light wires with suitable support points

The standing throw also has several advantages for modeling purposes:

- the movement is closely controlled and synchronized data from the force platform can be collected.
- If synchronized kinematic data is required the calibrated volume will be relatively small and the accuracy higher.
- No vibrations are introduced by the running pattern of the athlete.
- The thrower can have a good control over the javelin.

With some minor modification the method could be applied to three step throws where experienced throwers could reach 85% of their best results. The force data collection could be synchronized with kinematic data to provide a better view of the release technique of the athletes and the required strength.

3 DISCUSSION OF THE RESULTS

Several sample throws were performed and a sample plot of the output is given in figure 4.

The first positive peak for each throw represents the release force and the negative peaks represent the impacts with the safety barrier. The sampling frequency of the analogue-digital converter was 1kHz. this is an adequate sampling speed but greater detail may be available if a higher rate was used.

The duration of the throw as can be seen from the same figure is in the range of 8ms to 14ms and the point of greatest interest around the actual release is a fraction of that. Taking into consideration that a real throw is significantly faster one can judge that a sampling frequency for the force transducer of up to 10 kHz is highly desirable. The calibration would also need to be carried out fior these higher frequencies to confirm that the device remains linear.

This fact emphasizes even further the necessity of direct force measurement. This result also shows clearly that the standard kinematic collection frequency is insufficient for obtaining a detailed picture of the release.

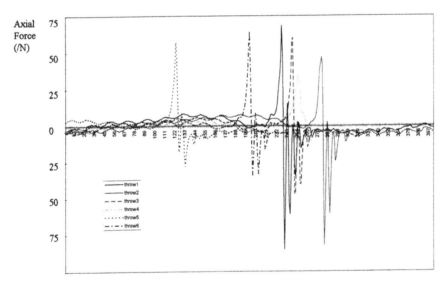

Figure 4. Test results from the javelin force transducer.

The device as already mentioned adds additional weight to the javelin (up 46% for the prototype) and to that extent the accelerations involved are expected to be lower than in an unmodified javelin. It ought to be pointed that the extra weight due to the collar should be considered as an increase of the weight of the throwing arm as opposed to increase to the javelin weight. This mass does not affect the force measured between the collar and the javelin.

Additionally, the increase of the air resistance due to the shape and size of the structure has been neglected. Further modifications are planned and best results could be achieved in conjunction with javelin manufacturers. However for modelling purposes and for estimation of the required strength the device is sufficiently good.

4 CONCLUSIONS

The device presented here offers a reliable tool for measurement in laboratory conditions. It could be used for special strength tests of javelin throwers and allows synchronized video and force data collection. Such a comprehensive data set would be beneficial for a better understanding of the dynamics of the javelin throw and could help to improve the reliability of the computer modelling currently being developed.

REFERENCES

[1]   Fiander M. Bennets J. and Ngo T.
        'Determining the throwing force of a javelin'
        Internal project report Manchester
        Metropolitan University   May 1995

[2]   Best R.J. Bartlet R.M. and Morris C.J.
        'A three-dimensional analysis of javelin
        throwing technique' 1993,11,315-328
        Journal of Sports Sciences

[3]   Dabnichki P., Lauder M., Aritan S. and
        Tsirakos D., (1996) Accuracy of the
        kinematic data collection, filtering and
        numerical differentiation. In: Haake, S. (ed)
        Proceedings of the International Conference
        of The Engineering of Sport. Rotterdam:
        A.A.Balkema.

APPENDIX 1

EQUIPMENT USED DURING CALIBRATION AND TESTING

Strain gauges:	100Ω copper-Nickel gauge factor 2.0
Force transducer: (B&Kcal 20-9-93) (used for calibration)	B&K 8200 120-3753 4.03 pC/N
Charge amplifier:	B&K 2635
PC data capture:	IBM Compatible 486DX 33 MHz RTI-815 ADC 12 bit ± 1LSB

The Engineering of Sport, Haake (ed.)© 1996 Balkema, Rotterdam. ISBN 90 5410 822 3

Experimental study of the perception of body position in space

Ingo Kobenz, Trevor T.J. DeVaney & Wolfram Müller
Institute of Medical Physics and Biophysics, Karl-Franzens University, Graz, Austria

Walter Habermann
ENT University Clinic, Graz, Austria

Michael Samastur
Institute of Applied Statistics and System Analysis, Joanneum Research, Graz, Austria

ABSTRACT: The perception of the body position in space has been investigated here. A special apparatus for the investigations has been developed in order to study persons being rotated through both the transversal and the sagital axes. A computer controlled motor enabled smooth movements with variable angular acceleration and with speeds between 0.04°/s and 400°/s, on the one hand, enabling determination of the position perception and on the other hand, simulation of the angular accelerations found in sport e.g. gymnastics, trampolining, etc.

Each test person was clinically tested for spontaneous nystagmus and head shaking nystagmus, speed of reaction, Unterbergs step test, standing test (by Romberg) before the experimental investigation on the apparatus. Test persons showing pathological indications were excluded from further tests.

Our experiments up to now have concentrated on both 0° and 30° anteflexion of the head relative to the body and all experiments were conducted with the test person having their eyes closed in a dark room. Those who took part were between 20 and 25 years old. (n = 20, 15m, 5w)

Test persons were investigated in horizontal and vertical positions using 5 differing angular velocities about the horizontal axis. All tests started in both cases at 0° relative to the position under investigation. The test person was then informed that this was the starting position and they would be pendeled through the starting position and should push a button on one of the grips when they found themselves to be in the vertical respectively horizontal position.

Test persons perceived the starting position with a high degree of accuracy. Singularly large errors in estimating the position of 0° were to be found. The presented data represent reference data for the investigation of athletes and the blind.

It is well known to coaches and athletes that the individual perception of the body configuration in space often differs markedly from the real position. Associated questions are subject of further investigations.

1 INTRODUCTION

The environment is perceived through interactions with the subject which stimulate sensory organs through sensory nerves. These stimulations are then integrated in the sensory central nervous system and result in sensation. Sensory stimulations are directly received without influence or evaluation by the subject. Normally the meaning of the sensory signals is evaluated by the subject using sensory information, learned or innate.

This evaluation of sensation results in perception [Schmidt 1976]. Those parts of the sensory system that receive information and transduce it to suitable signals are called receptors. A sense receptor is a single cell that transduces mechanical energy, chemical energy or light into sensory signals [Gibson 1973]. Four systems of orientation perception are important for the orientation sense in space, these are centrally combined and evaluated using: the optical, the acoustic, the proprioceptive and the peripheral vestibular systems [Scherer 1984a].

For the process of steering and regulating the act of movement all four systems are equally important, work closely together and are complimentary to each other. Krestownikow [1953] describes this process as a „single complex movement receptor" [Rieder 1977a].

For the maintenance of balance, apart from the peripheral vestibular receptor cells there are receptors in the muscles, the tendons, the diaphragm, the viscera as well as in the skin [Jonkees, Scherer 1984b]. This proprioceptive system is similar to the suspension system of the otoliths in that it delivers information constantly that is centrally evaluated. Cyberkinetically the vestibular system is to be seen as a complicated system with many differing functions [Groen 1961]. A weighting of the individual factors involved in the control and calculation of the sense of balance has at the moment not been reliably achieved [Rieder 1977b]. Investigations involving the importance of the optical sense in orientation in space are at the moment limited.

In the literature, investigations of the difference between the blind and the normally sighted in relation to their orientation ability in space has been described [Worchel 1950, Bittermann & Worchel 1953, Landau 1981, Stones & Kozma 1976]. These investigations are limited to either the orientation ability between fixed points in a room that are learned and then tested for the ability to find the points again or for the identification of forms through touch [Strauss & Marmor 1976].

In other investigations objects had to be brought into particular positions. Bittermann & Worchel used in their experiments the „rod and frame test" , in which the blindfolded proband had to bring a rod into the vertical or horizontal position and the variation from the desired position was measured. The values obtained from both of the groups investigated were compared. At first no difference could be seen but then as the probands were tilted sidewards the blind probands showed a significantly better result to the normally sighted.

Investigations concerning pilots and astronauts, involving the influence of gravity on human orientation ability are known [Clark & Stewart 1969, Schöne 1964, Guedry 1992]. Furthermore investigations concerning the orientation ability of pilots during rotation as well as the measurement of the limits of perception during rotation [Clark 1972a]. A few of the above mentioned papers dealt with the influence of sight on the orientation ability and in none of the investigations the blind were compared with the normally sighted. In one experiment the velocity threshold was measured with a rotatetable flight trainer cockpit. However, the velocity threshold for the situation with a dim peripheral field on the screen of the trainer side window was significantly lower than the threshold in the dark [Huang & Young 1981].

In this paper we have investigated the effect of the partial or total absence of visual cues for the estimation of ones orientation in space. A comparison was made between blind subjects and normally sighted subjects that have been blindfolded in their perception of their body position in space. The aim of this paper is to obtain new information in the field of spatial orientation, where in particular the orientation ability of the blind and the almost blind is compared with that of the normally sighted.

2 METHODS

2.1 Probands

Probands for this investigation were unpaid volunteers. The main reason of the probands for participating in this study was pure interest. The blind were scholars of a school for the blind or members of the Austrian Association of the Blind. In this paper we refer to this blind/partially blind group as „blind group". 33 persons of which 22 (7 female, 15 male) were normally sighted and 11 that were blind (7 female and 4 male) of which 3 were blind from birth. 50% of the blind probands could be classified as being totally blind. The average age of the probands was 30. The youngest proband being 16 and the oldest being 56. Of the 33 persons tested, nobody had to be excluded from analysis due to the preceding clinical tests.

2.2 Standard clinical test

Data [body-weight, height, training frequency, medication, vertigo, shoe size, sporting activities (including diving, flying, parachuting), date of birth, sex, smoking habits, alcohol use, tiredness, tinnitis, deafness (in one or both ears, partial or total), including examination of the ear drum] for each proband was collected with the help of a data formula. Afterward each were clinically examined using the following tests:

- Step test (Unterberg), as a test of the dynamic sense of balance [Scherer 1984c]
- Standing test (Romberg) as a test of the static sense of balance [Scherer 1984d]
- Identification of the presence of spontaneous nystagmus or head shaking nystagmus [Scherer 1984e]
- Reaction test (Fetz), was carried out for normal sighted probands only [Fetz 1978]

Probands that according to these tests differ from the norm were excluded from further investigation. All probands were blindfolded during the Unterberg test, the Romberg test and during the test on the rotation device. Additionaly the investigation room was darkened and noises held to a minimum. Both standard clinical tests and the investigation of spatial perception on the rotation device took altogether 60 minutes.

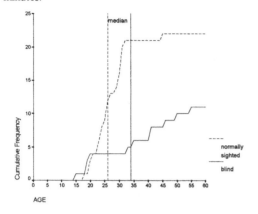

Figure 1. Cumulative frequency function for age of all probands showing also the difference in the median between the blind and the normally sighted.

2.3 Investigation of spatial perception on the rotation device

Experiments were carried out firstly in the horizontal position using velocity 1 (v1) followed by velocity 2 (v2). Both velocities were repeated with the proband's head tilted forwards 30°. The complete procedure was repeated in the vertical position.

The data relating to the head tilt will be published in a future paper.

The rocking motion of the rotation device is controlled by computer and before the start of our test the degree of swing out for each rock of the device was randomly selected. This then held constant for all of the probands. The light pointer (see below) is coupled with the signal button and a push of this button emits an acoustic signal as feedback for the proband.

The task was described to the proband with the following words: (Here translated into English) „You will be rocked back and forth. Between the changes of direction you pass through the starting position ten times. Your task is to press the signal button when you believe that you have reached the vertical

(or horizontal) position. At the beginning of each experiment you will be brought into a position that you have to recognise when passing through it during the rocking of the device. When this position is reached you press the button located on the handle. These instructions are valid for all experiments (both speeds in both orientations)."

Basic position: the person under investigation lies on their back on the rotation device, head and trunk are in the same plane [Schöne 1980]. The legs are in a stretched position. The hands hold handles mounted on the shoulder supports. The probands upper-body is fastened using a four point belt and his legs are secured with two further safety belts. The proband is blindfolded using opaque goggles.

The experiment used two angular velocities v1 (0.95°/s) and v2 (4°/s). Each angular velocity was used five times in each direction. The degree of swing of the table varied with each swing and lies between 9° and 17° for v1 and between 16° and 23° for v_2. The first experiment started in the horizontal with an angular velocity v1 and the proband was rocked with the first swing in the direction in which he faced. The same procedure was followed in the vertical position. The proband signalled reaching the reference position (horizontal/vertical) through a signal button on the shoulder support handle. Measurements were made of the difference to the reference position and that of the position signalled.

2.4 The rotation device

The rotation device consists of a supporting framework and a motorised axle upon which a table

Figure 2. Rotation device illustrating motor and axle orientation.

is mounted. Probands lying on their back on the axle mounted table can be turned about their transversal axis. The influence of gravity is normal to the axle's axis of rotation and normal to the transversal axis of the proband. The motor driving the axle is controlled using a PC enabling reproducible angular velocities, angular accelerations and distances of movement [Compax Programm Editor for Compax; Version 1.5; © 1990,91,95 Hauser Elektronik GmbH.].

The angular velocities reachable lye between 0.04°/s and 400°/s. A constant angular acceleration can be chosen. Probands are held in a nearly identical position through the use of variable mountings. Balance weights at the foot and head ends as well as mounted on the reverse of the table enable the table to be balanced. A light pointer has been mounted to the rear side of the table to enable the determination of the position of the table at all times. Using the light pointer an optical signal is emitted for determination of the actual position of the table.

Figure 3. Detail of the rotation device table.

2.5 *Statistical methods*

We used the absolute deviation from the reference position in degrees as a measure of error. The distribution of the absolute deviation was significantly different from the normal distribution. So we applied the non-parametric Wilcoxon-rank-sum test (two-sided). We tested the Null-hypotheses of equality of the central tendency.

We call p-values smaller than 0.05 significant and p-values smaller than 0.01 highly significant. A common form to show results graphically is the box plot-form. It is used to show the median* and the interquartile range (IQR=upper quartile** - lower quartile***), further the outliers and the extreme values. 50% of all values are within the box. Extreme values are bigger than the upper quartile plus 3 times the IQR, and the outliers are the upper quartile plus 1.5-3 times the IQR. Statistical analysis was performed with the Statistical Package of the Social Sciences (SPSS 6.0 for Windows).

* 50% of all data are lower and 50% are bigger than the median.
** 75% of all data are lower and 25% are bigger than the upper quartile.
*** 25% of all data are lower and 75% are bigger than the lower quartile.

3. RESULTS

We investigated the estimation accuracy concerning both the horizontal and vertical reference positions. More than 50% of all estimates deviated less than 5° from the reference positions, the median was 2.45° A total of 2633 estimates were made and evaluated, 876 from the blind and 1747 from the sighted.

Analysing the horizontal and the vertical position separately, using the data for both, the blind and the sighted, led to the results shown in Fig.4. The estimation of the vertical position did not differ significantly from the estimation of the horizontal (the median for the horizontal was 2.54°, for the vertical 2.45°)

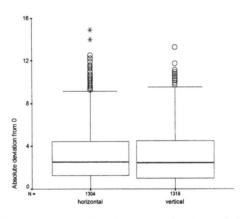

Figure 4. Comparison between estimations of the horizontal and vertical reference positions.

134

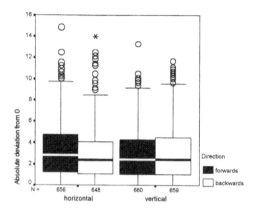

Figure 5. Comparison of forward and backward rotations for both angular velocities from the horizontal and vertical reference positions for all probands.

Figure 5 shows the difference between the two directions of rotation relating to the estimation of the vertical and horizontal reference position by all of the probands. The estimation of the horizontal reference position was highly significantly better during backward rotations than during forward rotations (p<0.01), while the estimation of the vertical reference position did not show any difference between the forward and backward rotations.

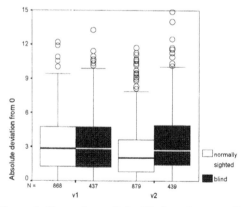

Figure 6. Comparison of the blind to the normally sighted probands in all experiments at the two velocities used.

A separat analysis of the data for differences in the two groups (blind and normally sighted) results in Fig.6. The estimation precision of both groups at v1=0.95°/s did not differ significantly, whereby at v2=4.0°/s a highly significant difference could be found (p<0.001)

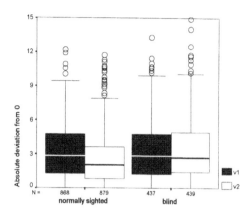

Figure 7. Comparison of the blind to the normally sighted at two angular velocities.

In Fig.7 the data from Fig.6 has been rearranged to make comparison of the two velocities easier. The box plot shows a highly significant better estimation of the reference positions by the normally sighted at v2, which is the higher velocity. No significant difference could be found with the blind.

4 DISCUSSION

The obtained results should be interpreted considering that it was not possible to obtain a truly random sample for the blind as well as the sighted groups. All volunteers available have been investigated. There was an inhomogeneity in the two experimental groups concerning the age structure. The blind were on average older than the sighted under investigation (Fig.1). Despite this no significant correlation between age and reference position estimation could be found in the groups investigated.

Fig.4 shows that the estimation precision is increased significantly at the higher velocity (v2) for the normally sighted. Thus we can conclude that the perception of the reference positions is a function of the angular velocity. Further investigations should be made to study this dependancy over a larger range of velocities and also for other reference positions (e.g. horizontal, faced downwoard or vertically in the head down position). To extend these investigation the probands also should be positioned on their side on the rotation device, to study the effects when being rotated sideways (axis of rotation parallel to the sagital axis).

A significant increase in the accuracy of the estimation of the reference position could not be found with the blind. For reasons unknown to us the blind group investigated seems to be unable to make use of sensory information available to the normally sighted at the increased velocity.

REFERENCES

Bittermann M.E. & Worchel P. 1953. The phenomenal vertical and horizontal in blind and sighted people. *American journal of psychologie* 66. pp.598-602

Clark A.H. & Stewart J.D. 1969. Effects of angular acceleration on man. *Aerospace medicine*

Clark A.H. 1972a. Comparison of the sensitivity to rotation of pilots and nonpilots. Aerospace medicine

Clark B.1967b.Thresholds for the perception of angular acceleration in man. *Aerospace medicine* 5, 38. pp.443-450

Fetz F.1978. *Sportmotorische Tests*, Inn Verlag, Innsbruck,pp.51-55

Gibson J.J. 1973. *Die Sinne und der Prozeß der Wahrnehmung*, H.Huber Verlag, p.65

Groen J.J. 1961 Problems of the semicircular canal from a mechanico-physiological point of view. *Acta oto-laryng.*, Suppl.163, pp.59-67

Guedry F.E. 1992. Perception of Motion and Position Relative to the earth. *Annals New York Academy of Science*

Huang J. & Young L.R.1981. Sensation of rotation about a vertical axis with a fixed visual field in different illuminations and the dark. *Exp Brain Res* 41, pp.172-183

Jonkees L.B.W. Physiologie und Untersuchungsmethoden des Vestibularsystems, in: *HNO-Heilkunde in Praxis und Klinik*, Bd.5, Berends J.

Krestownikow A.N. 1953. *Physiologie der Körperübungen*. Berlin, p.106

Landau B. 1981. Spatial knowledgement and geometric representation in a child blind from birth, *Science* 213

Rieder H. 1977a. *Bewegungslehre es Sports, Sammlung grundlegender Beiträge II*, Hoffmann Verlag, Schorndorf

Rieder H. 1977b. Bewegungslehre des Sports. p.28 *Sammlung grundlegender Beiträge* II, Hofmann Verlag, Schorndorf:

Scherer F.,1983. *Sport mit blinden und sehbehinderten Kindern und Jugendlichen.* Schorndorf: Hoffmann

Scherer H., 1984a. *Das Gleichgewicht* II, Springer Verlag 114

Scherer H. 1984b. *Das Gleichgewicht* II: 114

Scherer H. 1984c. *Das Gleichgewicht* I: 23-24

Scherer H. 1984d. *Das Gleichgewicht* I: 20

Scherer H. 1984e. *Das Gleichgewicht* I: Cap.5

Schmidt R.F. 1976. *Grundriß der Sinnesphysiologie*, p.4, Berlin: Springer

Schöne H. 1964. On the role of gravity in human spatial orientation. *Aerospace medicine*

Schöne H. 1980, *Orientierung im Raum*. p.128 Wissenschaftliche Verlagsgesellschaft: Stuttgart 1980

Stones M.J. & Kozma A. 1987. Balance and age in the sighted and blind. *Arch Phys Med Rehab*, Vol. 68: pp.85-89

Strauss & Marmor 1976. Human perception and performance. *Journal of experimental psychologie*

Worchel P. 1950. Space perception and orientation in the blind. *Psychological Monographs*: 65, 15

The Engineering of Sport, Haake (ed.)© 1996 Balkema, Rotterdam. ISBN 90 5410 822 3

Instrumentation of the Concept II ergometer for optimization of the gesture of the rower

P. Pudlo, F. Barbier & J.C. Angue

Laboratoire d'Automatique et de Mécanique Industrielles et Humaines, Valenciennes, France

ABSTRACT: The aim of this paper is to describe the external efforts at the contact between the rower and the environment i.e. the hands, feet and bottom. The simulation tool is a Type C Concept II ergometer. For each of the contact points the design of the instrumentation is divided into four stages. These are (1) method of measurement, (2) integration of the method chosen with the ergometer while protecting the rower ergonomics, (3) the device to locate the force plates in the experimental global system of reference and (4) experiments performed on a French regional level rower. The rower is 22 years old, male, has a height of 1.86m and a weight of 82 kg. Finally, the processing of the force plate signals is described.

1. INTRODUCTION

The aim of this research is to optimise the use of the Type C Concept II ergometer. This optimization is intended to be complete despite inherent difficulties (Hatze, 1993). Indeed according to Baumann (1993) general optimization procedures of the human gesture that could be efficiently used do not exist. Nevertheless Hatze (1976) proposed a first procedure for complete optimization of the human movement.

The first work, to perform this optimization, consisted of measuring kinematic variables, external efforts and in computing unobservable variables, such as articular efforts. The analysis of these different variables is done in collaboration with trainers. Articular efforts are computed from an iterative calculation code (Barbier 1994). The inputs of this program are on the one hand the kinematic positions of anatomical points and on the other hand, the external efforts and their application points. The acquisition of the positions of anatomical points is carried out thanks to a SAGA3 opto-electronic system. External efforts, are measured with 3 LOGABEX force plates and 2 ENTRAN mono-dimensional force transducers.

This article presents the instrumentation for a Type C Concept II Ergometer. This ergometer, despite many controversies, (McBride 1991: Duchesnes, 1991), is commonly the rower's gesture simulation device.

2. THE INSTRUMENTATION ADOPTED

The instrumentation must allow us to measure the external efforts at 5 points: the 2 hands, the 2 feet and the bottom. Quintic splines are used to smooth the different curves found and each curve is expressed as a percentage of cycle.

2.1 At the level of hands

The instrumentation is made for each hand and consequently, the dissociation between right hand and left hand has to be achieved. Two miniature mono-dimensional force transducers, functioning in compression and traction, are used. Their measurement range is 1000 N with a measurement error of ± 5N has been found to be sufficient (Hartman, 1993). Indeed, Hartman (1993) has noticed that the maximal force peak for the two hands, during the first strokes, does not exceed 1350 N. Two new handles have been built (figure 1); each handle is conceived so that the hand cannot develop couples. The handle is made from steel and the hands are protected by a grip.

The force transducers measure only the force intensity. Consequently, two markers are used and fixed on the stiff part of each handle which permit the definition of the direction vector of the force (figure 2). The force vector is computed from this normalised directrix vector and the measured intensity.

The position of the application point of the force generated on the handle by the hand, is deduced from the positions of markers C and D (figure 3). This computation is achieved because the wrist of the high level rower remains rigid during the stroke (FFSA, 1991).

The rower's gesture amplitude is preserved by the lengthening of the ergometer which is equal to the size of the new handles (figure 4).

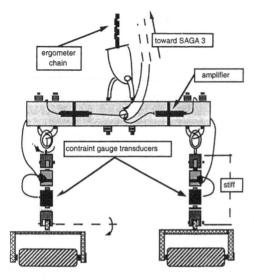

Figure 1. New handles for the ergometer

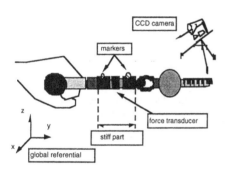

Figure 2. Determination of the force vector

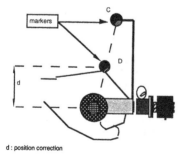

d : position correction

Figure 3. Determination of application point

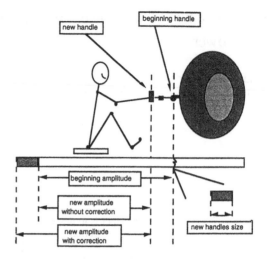

Figure 4. Lengthening of the ergometer

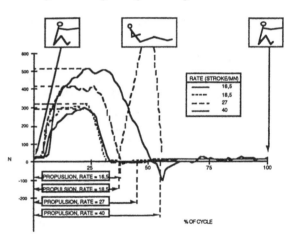

Figure 5. The force developed by the right hand

The analysis of the curves in figure 5 shows that the force peak increases with rate. The period of propulsion also increases with rate. Therefore, the decrease of stroke time is mainly due to the reduction of recovery time. During the recovery phase at 40 strokes per minute, the force reaches -100 N. Thus, the movement of the rower is faster than the return system of the ergometer.

2.2 At the level of feet

Instrumentation has to be developed out for each foot and, as with the hands, instrumentation has to dissociate the right foot from the left. Two LOGABEX force plates have been used. Their measurement range is 465 daN for Fx and Fy, 865 daN for Fz, 80 daN.m for Mx and My and 180 daN.m for Mz. The theoretical error is 1% of the measurement range for each of these components. They are fixed on a steel base forming a rigid frame with the ergometer (figure 6). The mass of the frame guarantees the stability of the ergometer during experimentation.

Two new stretchers have been created which are fixed to the 2 force plates (figure 6).

Finally, a device is conceived and fixed on each stretcher before or after the experimentation (figure 6). It allows identification of the position of the force plate in an experimental global reference system. The processing required to compute the external efforts in the global reference system are described in section 3.

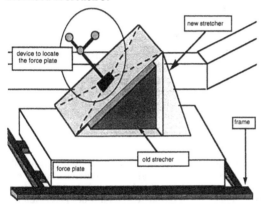

Figure 6. The new stretcher

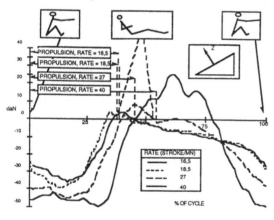

Figure 7. The force developed by the right foot in relation to the z-axis of the stretcher

Figure 7 shows the force developed by the right foot for the z-axis of the stretcher. The negative peak force is maximal for the larger rate. The percentage of the cycle between the first cancelling out of this force and the end of the propulsion section increases with the stroke rate. Therefore, the force on the handle increases and is the result of hands action and trunk action.

2.3 At the level of the bottom

The bottom is the last rower's contact point with the environment. It is not necessary to measure the external efforts for this point in order to compute the articular efforts at the hip level, but it has been carried out to verify the computed articular efforts at the hip and to test different seat forms. A LOGABEX mini force plate with dimensions 114 mm diameter, 45 mm total height and with a mass 0.335 kg, is put under the rower's seat. Its measurement range is 200 daN for Fz, 50 daN for Fx and Fy, 6 daN.m for Mz and 5 daN.m for Mx and My. The theoretical precision is 1% of the measurement range for the different components.

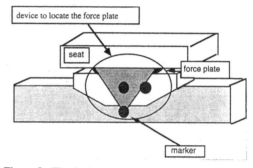

Figure 8. The device to locate the force place under the seat

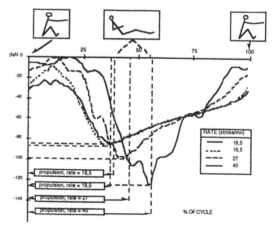

Figure 9. The force developed by the bottom in relation to the z-axis of the seat

To locate this force plate in the global reference system, a device is conceived and fixed under the seat (figure 8). The process to compute external efforts in the global reference system is presented in section 3.

Figure 9 shows the force developed by the bottom for the z-axis of the seat. The negative peak force decreases with the stroke rate. When the rowing rate is equal to 40 per minute, the force peak reaches 120 daN. As the rower's weight is 82 daN, the maximum push is approximately 40 daN. In all cases, the peak is reached before the end of the propulsion. The percentage of the cycle where this force is smaller than the rower's weight increases with the stroke rate.

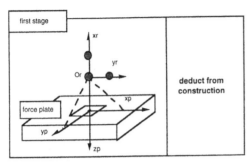

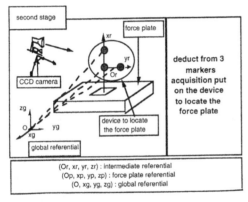

(Or, xr, yr, zr) : intermediate referential
(Op, xp, yp, zp) : force plate referential
(O, xg, yg, zg) : global referential

Figure 10. The two stages required to locate the force plate.

3. PROCESSING ASSOCIATED WITH THE FORCE PLATES

The ergometer is static during experimentation. Consequently, two types of force plates are used. The first static force plates are placed under the new stretchers. These force plates have a fixed position in the space during the experimentation, whereas the force plate under the rower's bottom is mobile and moves together with the rower seat.

For these two types of force plate the processing performed proceeds to a double change of reference system (figure 10).

The first change of reference system (first stage, figure 10) changes the coordinates for the force plate reference system to those for the device to locate the force plate. This is called the intermediate reference system. The resultant matrix is constant for the two types of force plates considered and is valid for all experimentation. Indeed, it is deduced from the construction adopted. The second reference system change (second stage, figure 10) transforms the

intermediate reference system to the global reference system. For the static type of force plate, the result matrix is constant and is valid so that the position of the ergometer is not modified. For the dynamic type, however the resultant matrix is calculated for each acquisition. The quality of the final result matrix is conditioned largely by the precision of the kinematic data produced by the opto-electronics system. In the present case, the observed error is 2.4% and 4.5% of the real distance in dynamic and static modes respectively (Barbier, 1994).

The principle of the calculation of passage matrix is presented by (Nikravesch, 1988). Finally, the forces measured by the mobile force plate must undergo a proportional readjustment to the acceleration of the seat.

4. CONCLUSION

The current research describes the development of instrumentation to measure the external efforts of a rower in relation to his environment in order to optimise the gesture of the rower. Once this initial phase is complete the apparatus will be used to analyse the articular efforts of the rower.

REFERENCES

Barbier F. 1994, Modélisation biomécanique du corps humain et analyse de al marche normale et pathologique. Application à la rééducation, Thèse de doctorat, Université de Valenciennes, Numéro d'ordre 94-16.

Baumann W. 1993, Round Table-1 : Optimisation of sports performance, International Society Biomechanics XIV Congress, p 28-29.

Duchesnes C. J., Riethmuller M. L., Nicol A. C., Paul J. P. 1991, Kinematic comparison of on-water and specific ergometer sculling, Abstracts, International Society of Biomechanics XIII Congress, p 61-62.

FFSA, Formation de cadres FFS Aviron-Document Direction Technique Nationale-Edition.

Hartmann U., Mader A., Wasser K., Klauer I. 1993, Peak force, velocity, and power during five and ten maximal rowing ergometer strokes by world class female and male rowers, Int J Sports Med, Vol 14, Supl. 1, pS42-S45.

Hatze H. 1976, The complete optimization of human motion, Math. Biosci. 28, p 99-135.

Hatze H. 1993, Reasons why progress in the optimization of sports performance is painfully slow, International Society of Biomechanics XIV Congress.

McBride M. E. 1991, Does the concept II rowing ergometer accurately simulate the biomechanics of rowing, Abstract, International Society of Biomechanics XIII Congress, p711-712.

Nikravesh P. E. 1988, Computer aided analysis of mechanical systems, ISBN 0-13-164220-0, Prentice-Hall inc., p 164-165.

The Engineering of Sport, Haake (ed.) © 1996 Balkema, Rotterdam. ISBN 90 5410 822 3

Three-dimensional kinematic analysis of upper extremity in the soft-tennis forehand drive

Lin-Hwa Wang, Hong-Wen Wu, Yi-Wen Chang & Fong-Chin Su
Physical Education Office & Institute of Biomedical Engineering, National Cheng Kung University, Tianin, Taiwan

Kuo-Cheng Lo
Physical Education Section, Kung Shan Institute of Technology and Commerce, Tianin, Taiwan

ABSTRACT: There was currently no good agreement for the pattern of forehand drive since soft-tennis contestants had a big variety of objects to imitate or learn. The goals of this study were (1) to establish a three-dimensional kinematic model in upper extremity to compute the angular movement of the trunk, shoulder, elbow and wrist joints, and (2) to compare the difference of athletes between the national representatives and college team so as to evaluate the training effect of soft-tennis forehand drive. The results showed that the national athletes and college athletes had similar angular movement patterns. The average ranges of motion of the national athletes shoulder joint were 56 degrees for flexion/extension, 53 degrees for adduction/abduction, and 45 degrees for axial rotation. For college tennis players, shoulder joints performed 70 degrees flexion/extension, 72 degrees adduction/abduction, and 65 degrees axial rotation. The range of motion for shoulder was smaller for national representatives, compared to college representatives (p=0.003). For the elbow, two groups had the almost same range of motion. To the contrast, national tennis players performed larger wrist flexion/extension movement (56 vs. 50 degrees).

1. INTRODUCTION

Soft-Tennis is one of ball games the earliest and well-developed in Asia. There are pre-swing, back-swing, forward-swing, impact, follow-through and finish phases in each cycle of forehand drive. However, currently there is no good agreement for the pattern of forehand drive since soft-tennis contestants have a big variety of objects to imitate or to learn. Also, the different methods of forehand drive regionally exist due to different environments. It is ambiguous both for novices and professionals to practice soft-tennis. Therefore, it is necessary for the contestants and coaches to scientifically obtain an elegant pattern and reasonable range of soft-tennis forehand drive in upper extremity as a consult through three-dimensional motion analysis system. A complete knowledge of soft-tennis with the consideration of mechanics and physiology would provide the basic guidelines and rationales for training the beginners to gain an optimal pattern of soft-tennis forehand drive.

There have been many investigators to study tennis forehand drive since 1949[1,2,3,8]. Recently,

the research of kinematics and kinetics were performed with the means of picture and video. However, they were limited to two-dimensional or incomplete three-dimensional studies[4,5,6,7,9]. The goal of this study was (1) to establish comprehensive kinematic model in upper extremity to compute the angular movement of the trunk, shoulder, elbow and wrist joints, and (2) to compare the difference of athletes between the national representatives and college team so as to evaluate the training effect of soft-tennis forehand drive.

2. METHODS

Marker Set

A set of thirteen reflective markers was placed on selected anatomic landmarks unilaterally on the subject as shown in Figure 1. The selected anatomic landmarks include: processus xiphoideus (A), incisura jugularis (B), the 7th cervical vertebra (C), acromion process (D), medial and lateral epicondyles (E, F), radial and ulnar styloid processes (G, H), the 2nd and 4th metacarpal heads (I, J), anterior superior iliac spine (K, L), and posterior superior iliac spine (M).

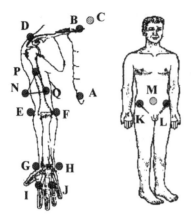

Fig 1. Markers set

Model

A three-dimensional biomechanical model of the upper extremity has been developed using above described marker set. In this model, the trunk, pelvis and arm were treated a five segment linkage system. The four segments consist of the truck, pelvis, upper arm, forearm, and hand. The model was further simplified by assuming: each segment was treated as a rigid body.

Co-ordinate Systems

All the image data were collected in a laboratory coordinate system which was calibrated at the beginning of the experiments. However, more coordinate systems would be included for the need to solve the kinematic problem. In total, six coordinate systems were established in the procedure of analysis so as to calculate the spatial trajectories and kinematic parameters of the movements.

1. Laboratory coordinate system: Defined by the initial calibration.
2. Trunk coordinate system: As shown in Figure 1, points A, B and C could be used to build an orthogonal coordinate system. AB was normalized as Z unit vector, $\hat{z}$. The crossproduct of BC and AB was normalized into y unit vector, $\hat{y}$. X unit vector, $\hat{x}$, could be obtained by a crossproduct of $\hat{y}$ and $\hat{z}$.
3. Arm coordinate system: Points D, E, and F were used to build this orthogonal coordinate system whose origin was point D. Point O was the midpoint of points E and F. OD was normalized as Z unit vector, $\hat{z}$. The crossproduct of DE and OD was normalized into

X unit vector, $\hat{x}$. Y unit vector, $\hat{y}$, could be obtained by a crossproduct of $\hat{z}$ and $\hat{x}$.
4. Forearm coordinate system: Points E, F, G and H were used to build this orthogonal coordinate system whose origin was the midpoint O of points E and F. EF was normalized as Y unit vector, $\hat{y}$. The crossproduct of HO and $\hat{y}$ was normalized into X unit vector, $\hat{x}$. Z unit vector, $\hat{z}$, was the crossproduct of $\hat{x}$ and $\hat{y}$.
5. Hand coordinate system: Following the similar procedure as in forearm coordinate system, points G, H, I and J were used to build this orthogonal coordinate system.
6. Pelvis coordinate system: Points K, L, and M were used to build this orthogonal coordinate system. KL was normalized as Y unit vector, $\hat{y}$. Point P was the midpoint between points K and L. The crossproduct of KL and $\hat{y}$ was normalized into Z unit vector, $\hat{z}$. X unit vector, $\hat{x}$, was the crossproduct of $\hat{y}$ and $\hat{z}$.

Kinematics

The angular movements of hand with respective to forearm, forearm with respective to arm, arm with respective to trunk and trunk with respective to pelvis were defined as the motion of wrist, elbow, shoulder joints and trunk, respectively. Euler angles were used to describe the orientation of a distal segment reference frame relative to a proximal segment reference frame. The first rotation about the y axis represents the flexion/extension angle (α). The second rotation about x' axis represents adduction/abduction angle (β). The third rotation about the z'' axis represents segmental axial rotation (γ). The transformation matrix of distal segment reference frame relative to proximal segment reference frame is:

$$\mathbf{R}_{yx'z''}(\alpha,\beta,\gamma) = \left[\Gamma_{ij}\right]_{3\times3}$$
$$= \begin{bmatrix} S\alpha S\beta S\gamma + C\alpha C\gamma & S\alpha S\beta C\gamma - C\alpha S\gamma & S\alpha C\beta \\ C\beta S\gamma & C\beta C\gamma & -S\beta \\ C\alpha S\beta S\gamma - S\alpha S\gamma & C\alpha S\beta C\gamma + S\alpha S\gamma & C\alpha C\beta \end{bmatrix}$$

The three Euler angles of distal segment reference frame relative to proximal segment reference frame were calculated using the following equations:

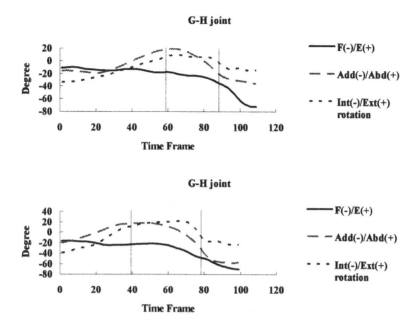

Fig 2. The angular movement of G-H joint for Group A (upper) and Group B (lower)

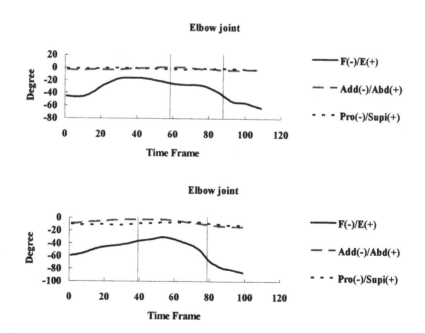

Fig 3. The angular movement of elbow joint for Group A (upper) and Group B (lower)

$$\beta = A\tan 2(-\Gamma_{23}\sqrt{\Gamma_{13}^2 + \Gamma_{33}^2})$$
$$\alpha = A\tan 2(-\Gamma_{13} / C\beta, \Gamma_{33} / C\beta)$$
$$\gamma = A\tan 2(-\Gamma_{21} / C\beta, \Gamma_{23} / C\beta)$$

where s is sine and c is cosine. Atan2[1] is used to extract these rotation angles.

3. EXPERIMENT

Three Taiwan national soft-tennis representatives (Group A) and three university tennis team members (Group B) were recruited in this study. Mean age is 23.5 years for Group A and 26 years for Group B. Mean height and weight are 172.6 cm, and 69 kgw for group A; 166.7 cm and 60 kgw for Group B. The Expert Vision video-based Motion Analysis System (Motion Analysis Corp., Santa Rosa, CA, USA) was used to collect marker trajectories with six cameras. Ten trials were sampled for each subjects at 60 Hz. Each trial lasted for 5 seconds with 3 minutes rest between trials. Then the trajectories of each reflective markers in the laboratory coordinate system were transferred to a personal computer to calculate the kinematic parameters of each joint.

4. RESULTS AND DISCUSSION

The experiment results revealed that the national athletes and college athletes had similar angular movement patterns. The patterns were described as follows:

Angular Movement of Shoulder Joint

The first phase, back swing, shown in Figure 2, the arm rotated externally and the forearm swing downward. For the coordination of the shoulder and the arm, the shoulder joint of subjects in Group A (frame 1:58) was abducted and externally rotated. But the angular movement of subjects in Group B (frame 1:39) was smaller in initial stage and was also abducted and externally rotated in late stage. The second phase, from forward swing to impact (Group A frame 59:87; Group B frame 40:78), the shoulder joint was from the lateral side to the medial and anterior site of the body. Namely, movements in the shoulder joint of subjects in both groups were

---

[1] Atan2(y,x) computes $\tan^{-1}(y / x)$ but uses the signs of both x and y to determine the quadrant in which the resulting angle lies. It is sometimes called a "4-quadrant arc tangent."

mainly adduction accompanied by a little degree of internal rotation. The third phase, follow through after impact, angular movement of shoulder joint in Group A (frame 88:109) was continuously adduction and internal rotation, the same as that in Group B (frame 79:99) but accompanied by flexion to rotate the arm forward and upward.

Angular Movement of Elbow Joint

During the back swing phase shown in Figure 3, the arm swung posteriorly and inferiorly. Thus, the movements of the elbow joints in both groups were extension similarly. During the forward swing phase, the elbow kept some degree of flexion in order to contain the stability and precision when in impact. Therefore, angular movement of the elbow joint in Group A was flexion. However, the angular movement of elbow joint in Group B was extension first and then flexion. After impact, the racket followed the ball forward and then the forearm was upward and the arm was close to the trunk to make the elbow flexion. The result was the same for both groups.

Angular Movement of Wrist Joint

The wrist joint was the final segment of the swing chain in upper extremity. During the whole swing cycle, the movement of the wrist joint shown in Figure 4 was the most frequently changed. From the beginning of the cycle, the arm moved downward and swing back. Thus, the movement of the wrist joint contained partial flexion and a little degree of ulnar and radial deviation. The wrist joint extended to swing in fit with the point at which the ball touched the ground and the rebound height. Till impact, the wrist joint was in the maximum extension during the whole cycle to produce a force which each subject could maximally generate. Wrist joints in both groups had to extend so as to follow the ball forward swing after impact. Both groups obviously decreased all the forces after the ball contacted the racket. The wrist joint had little radial deviation since the racket should be continuously moved toward the head via shoulder in the end of follow through phase.

Angular Movement of Trunk

Trunks had left rotation and right side-bending for the subjects with left dominant hand since trunk could initiate the motion of the whole upper limb from pre-swing to back-swing. Trunk transferred the forces originally from lumbar and buttock area to upper extremity in the duration from back-swing to impact. Thus, the movements of trunks in both

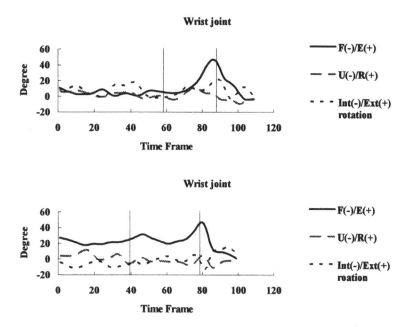

Fig 4. The angular movement of wrist joint for Group A (upper) and Group B (lower)

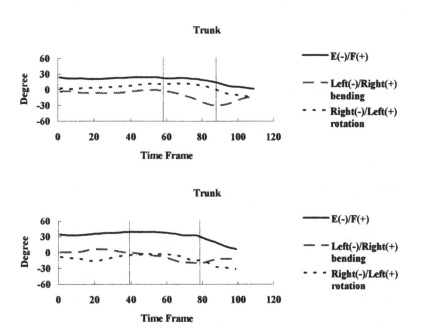

Fig5. The angular movement of trunk for Group A (upper) and Group B (lower)

groups were extension. Right side-bending and rotation of the trunk were performed due to reducing the reaction force in impact.

Range of Joint Motion

Range of joint motion was defined as the difference between the minimal and maximal angular position during the whole cycle including pre-swing, back-swing, forward-swing, impact, follow-through and finish. The average range of joint motion in both groups were shown in Table 1. The results were explicitly closed to one another among different trials for one subject. However, there was some differences among different subjects. Comparing the difference of range of joint motion between groups A and B, the ranges of abduction/adduction and internal/external rotation in group B were larger than that in group A significantly using the Mann-Whitney U-test of non-parametric statistical method ($p=0.0003$).

Table 1. Range of Joint Motion (degree)

		Group A	Group B
Trunk	F/E	29.3±3.5	33.0±10.7
	bending	25.5±7.2	30.9±9.0
	rotation	34.3±5.8	36.1±9.2
Shoulder	F/E	55.5±16.7	69.9±8.4
	*Add/Abd	53.1±4.3	72.3±4.5
	*rotation	45.3±3.1	65.0±5.3
Elbow	F/E	59.0±8.8	61.0±16.5
	Add/Abd	9.5±4.3	9.2±4.6
	Supi/Pro	7.4±4.0	8.4±3.5
Wrist	F/E	55.8±12.1	50.0±12.7
	R/U dev.	26.0±10.8	27.3±17.6

In summary, the comparison of soft-tennis forehand drive between national and university representatives were performed using three-dimensional kinematic model. The results indicated that the angular movements of flexion/extension, internal/external rotation and abduction/adduction in shoulder joint had relatively big ranges. The angular movement of flexion/extension was dominant in elbow and wrist joints. The main motion was side-bending in trunk. There were very similar angular movement for each subject. Moreover, The ranges of motion of flexion/extension and internal/external rotation in shoulder joint were smaller for national representatives compared to college representatives ($p=0.003$). In the future, a reliable kinematic model could be established completely if collecting more and more data of athletes with an experimental equipment which owns a higher sampling frequency

(more than 200Hz), in order to calculate the joint force and moment in upper extremity for the prevention and evaluation of sports injury.

ACKNOWLEDGEMENT

Supports from the National Science Council grants NSC83-0420-E-006-006 and NSC85-2331-B-006-074-M08, Taiwan are gratefully acknowledged.

REFERENCES

[1] Anderson, J. P. An Electromyographic study of Ballistic Movement in the Tennis Forehand Drive. Unpublished Doctoral Dissertation, University of Minnesota, 1970.

[2] Ariel, G. & Branden, V. "biomechanical Analysis of Ballistic V.S. Tracking Movements in Tennis Skill. In J. Groppel (Ed.), Proceedings of a national symposium on the racquet sports, pp. 125-140. Champaign, IL: University of Illinois Conferences and Institutes.

[3] Degutis, E. Simplified Anatomy and Mechanics of the Forehand Drive. Athletic Journal 46 (8), pp. 88-90, 1966.

[4] Duane V. Knudson, Intrasubject Variability of Upper Extremity angular Kinematics in the Tennis Forehand Drive. International J. of Sport Biomechanics, pp. 415-421, 1990.

[5] Duane V. Knudson, Forces on the Hand in the Tennis One-Handed Backhand. International J. of Sport Biomechanics, pp. 282-292, 1991.

[6] Giovanni Legnani, Robert Marshall, Guillermo J. Noffal. Analysis of Filmed Movements: Methodologies to Improve the Final Precision. XVth Congress of the International Society of Biomechanics. pp. 542-543. July 2-6, 1995, Jyvaskyla.

[7] Ito A., Tanabe S., Fuchimoto T. Three Dimensional kinetic Analysis of the Upper Limb Joint in Tennis Flat Serving. XVth Congress of the International Society of Biomechanics. pp. 424-425, July 2-6, 1995, Jyvaskyla.

[8] Slater-Hammel, A. Action Current Study of Contraction-Movement Relationships in the Tennis Stroke. Research Quarterly 20 (4), pp. 424-431, 1949.

[9] Xavier Balius, Carles Turro, Jorde Carles. Improving the Performance of a Top ATP Tennis Player with a Kinematical Approach, and a 3-Dimensional Interactive visualiztion of the Serve. XVth Congress of the International Society of Biomechanics. pp. 82-83, July 2-6, 1995, Jyvaskyla.

The Engineering of Sport, Haake (ed.)© 1996 Balkema, Rotterdam. ISBN 90 5410 822 3

The mechanical analysis of kicking during physical effort

Z. Waśkiewicz
Academy of Physical Education, Katowice, Poland

ABSTRACT: The technique of kicking changes according to the circumstances on the football field. Trying to understand this statement, the hypothesis, that the main factor influencing these changes is fatigue, was carried out. The experiment conducted on 8 female players of the Polish national football team showed that performed endurance effort influenced significantly the technique of the football kick.

1. INTRODUCTION

Soccer is a game with an acyclic movement structure, in which the rhythm and rate changes in relation to circumstances on the football field. The short-time efforts, characterized by high and maximal intensity are interspersed with average intensity efforts. The observation of matches showed that players during the 90 minutes of a match covered a distance from 8 to 12 km, and sometimes (English Premiership) up to 14 km. Bisanz and Gerisch (1984) determined the forms of locomotion among professional players. They established that 37% of locomotion was jogtrot, 25% marching, 20% fast running, 11% sprinting and 7% locomotion backwards. Fast runs were repeated every 30s and sprints every 90s. We should state that players have to adopt many other movement activities. During one match a football player performs approximately 1000 different activities, which change on average every 6s.

For football fans one of the most important elements of the sport is, of course, goals. One of the most frequent way of scoring goals is kicking. We can state that kicking is essentially a variation of running. The rotation of the pelvis precedes joint actions in the swinging limb. As in running hip flexion follows, accelerating the forward motion of the thigh. Knee extension comes in last adding the final speed to the kicking foot. The angular motion of the leg in space in the saggital plane is a combination of rotation of the thigh and action at the knee joint. As the non-kicking heel contacts the ground knee flexion is in opposition to thigh rotation forward so that there is little net rotation of the leg.

When knee flexion slows the leg begins to rotate due to hip flexion. When knee extension starts and accelerates, the leg gains speed. Meanwhile the thigh begins to slow and almost stop. Knee extension does not start until the thigh is past the perpendicular and is the chief contributor to speed at and through contact. The foot follows the leg rather closely since only slight ankle adjustment occurs.

The most important element of dynamic analysis of the soccer kick is to consider the pattern of muscle activity about the knee joint. Miller (1980), on the basis of quasistatic analysis, stated that the motion about the knee joint is supposedly controlled by the knee-flexor muscles. The obvious shortcoming of this analysis is the assumption that the movement is slow. But we should remember that inertial forces which are the effect of motion of the object or segment, become quite substantial and drastically alter the pattern of muscle activity that is necessary to control the movement. Indeed for such an activity as a strong kick, it even becomes necessary to consider the way in which the motion of the shank will affect the thigh and, conversely, how thigh motion will affect the shank (Putnam 1983). Enoka (1988) states that except for the first part of the kick, the resultant muscle torque is due to extensors, most of the movement is controlled by the eccentric extensor activity, and at the end of movement - concentric extensor effect. The peak resultant muscle torque occurs at the change from eccentric to concentric activity.

While looking at soccer players it is easy to establish that they have different techniques of kicking. Furthermore this technique changes according to the circumstances on the football field.

This experiment was carried out to try to understand the different techniques. The hypothesis is that the main factor influencing the technique of kicking is fatigue.

2. METHODS.

The experiment was conducted on 8 female players of the Polish national football team (weight=57.2 kg, SD=4.23, height=168.3, SD=6.11). Before the experiment the level of VO_{2max} was determined (x=52.4 ml/kg/min^{-1}, SD=12.3). All data were taken from two 100 Hz, Panasonic video cameras, and using the VIDANA computer system for 3-D movement analysis. Subjects ran on a mechanical track for 90 minutes at three speeds - 2, 5 and 7 m/s - changing according to protocol (every 30s the speed was accelerated to 5 m/s and every 90s to 7 m/s). Every 5 minutes the subject was transferred to the Department of Theory of Human Motor Behavior Laboratory to perform two football kicks.

3. RESULTS

The analysis of space flows of 22 markers allows us to determine that the foot speed (15 msec before contact with the ball) decreased from 20.9 m/s (SD - 2.98) at the start of the protocol to 16.0 m/s (SD - 4.14) at the end (Fig. 1). It shows that physical effort decreased the speed and influenced also the dispersion of results in the analysed group. This difference is statistically significant (t=2.367, p=0.05). The rate of knee extension decreased from 1754.0 degrees per second (SD - 96.6) to 1223.0 degrees per second (SD - 152.5) at the end of the intensive physical activity. These differences, shown in Fig. 2, were also statistically significant (t=4.684, p=0.01). The mechanical and physiological means by which the body achieves these final speeds are not well understood. The thigh slows or stops before contact so that it is contributing little, in a kinematic sense, to foot speed at contact. Yet it seems reasonable to assume that it does make some active contribution to the speed of leg rotation. A similar phenomenon is seen in the overarm throw. The ball in the hand does not really begin to accelerate in the lateral direction until the latter phase of spinal rotation after the shift of weight and pelvic rotation are largely completed.

The third parameter analysed in this research was the space amplitude of kicking. One of the determinants during the elimination of subjects for the experiment was body height. This variable did not vary strongly (SD=6.11cm) and allowed the

analysis of averaged traces of three markers. The dispersion of the spatial trace of the marker placed on the foot changed from 7.22% to 44.13%, on the knee from 8.13% to 37.11% and on the hip from 6.12% before effort to 31.12% in the last performed kicks.

4. CONCLUSION

It is possible to conclude that the performed endurance effort influenced significantly the technique of the football kick. It caused greater dispersion in the movement flows and decreased the analysed speed parameters.

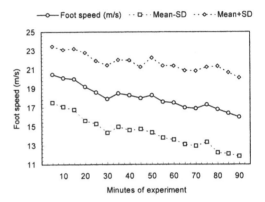

Fig. 1 The changes of foot speed (m/s) during a soccer kick for consecutive measurements.

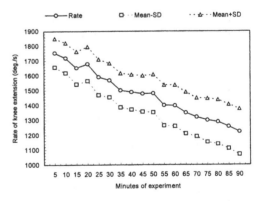

Fig. 2 The changes of knee extension rate (deg/s) during a soccer kick for consecutive measurements.

5. REFERENCES

Bisanz, G, Gerisch, G 1984. *Fusball Training, Technik. Taktit*, Reinbek.

Enoka, M R 1988. *Neuromechanical basis of kinesiology*. Human Kinetics, Champaign.

Miller, D I 1980. Body segment contributions to sport skill performance: Two contrasting approaches. *Res. Quart. J. Exerc. and Sport.*

Putnam, C A 1983. Interaction between segments during a kicking motion. In H Matsui & K Kobayashi (eds), *Biomechanics* VIII-B: 688-694. Champaign: Human Kinetics.

5 Materials in sport

The Engineering of Sport, Haake (ed.) © 1996 Balkema, Rotterdam. ISBN 90 5410 822 3

Developments of manufacturing of metal matrix composites for applications in the sports and leisure industries

M. B. Barker & A. M. Davidson
Napier University, Edinburgh, UK

D. Regener
Otto-von-Guericke University, Magdeburg, Germany

ABSTRACT: This paper presents the early results of an investigation into three major aspects of the development of metal matrix composites: processing, properties and machining characteristics. The objective of the programme is to produce cost-effective manufacturing processes leading to the successful production of components for the sports and leisure industries. The characteristics of a powder-processed metal matrix composite, comprising a matrix of 6061 aluminium alloy reinforced with particulate silicon carbide, were investigated in terms of microstructure, mechanical properties, fracture and machining performance. The properties of the material were then compared with a composite of the same aluminium matrix but which contained copper-coated SiC particles. The investigation revealed very different deformation and fracture behaviour between the coated and uncoated particle composites, with the coated particle product exhibiting superior tensile strength and ductility. The machining trials demonstrated that silicon carbide reinforced aluminium matrix composites, loaded up to 25% by weight of reinforcement, can be successfully machined into complex three dimensional shapes using electro-discharge equipment. In fact, there appeared little difference in machining characteristics between a conventional aluminium alloy and the metal matrix composite tested.

1. INTRODUCTION

In order to achieve enhanced performance of sports equipment, there is a major drive to utilise structural and functional components which are stronger, lighter, and stiffer than existing materials. Allied to this are the needs to reduce the complexity and cost of manufacturing. A family of materials rapidly moving towards these targets is Metal Matrix Composites (MMCs) and examples of commercially successful applications (Pillai 1993) are given below:

- bicycle frames such as those manufactured by Specialised Bicycle Components in the USA whose Stumpjumper mountain bike utilises an aluminium-based MMC, resulting in a 50% weight reduction compared with a similar steel-framed model.
- bicycle gears, golf club heads and arrow shafts made from aluminium reinforced with particulate silicon carbide.

A major reason for the slow commercial exploitation of MMCs is their relatively high cost of processing and the work outlined here investigates modifications to existing solid state (powder processing) and liquid routes for the production of 6061 aluminium alloy matrix reinforced with particulate silicon carbide.

Discontinuous reinforcements, such as particle-reinforced MMCs, can provide a reasonable improvement in specific strength, stiffness and wear resistance, compared with monolithic alloys. Additional advantages include isotropic properties and ease of manufacture by of conventional metallurgical and mechanical processes (Clyne 1993; Davidson 1993; Kapoor 1995; Somerday 1995). Solid state processing also has the important advantage over liquid state methods in that temperatures are lower, reducing the likelihood of the matrix reacting unfavourably with the reinforcement. It is proposed that these sintering temperatures can be lowered further, by plating a suitable metallic coating onto the surface of the reinforcing particles. In addition, it is anticipated that the presence of the coated particles will improve their wettability and/or mechanical properties and/or decrease the instances and sizes of voids and precipitates in the finished product.

Advanced materials such as MMCs are difficult to machine using conventional equipment and a possible alternative is the use of electro-discharge machining This shape-copying process, achieved by applying a succession of randomly distributed discharges between two electrodes (tool and workpiece), is described elsewhere (Metcut 1980) and is represented schematically in Figure 1.

2. EXPERIMENTAL WORK

2.1 *Materials*

All studies were carried out using 6061 aluminium alloy matrix reinforced with silicon carbide (SiC). The matrix was manufactured from atomised powder of the age-hardening Al-Mg-Si system, with a maximum particle size of 45 μm. The powder was blended with particulate SiC of nominal mean size 23 and 7 μm, in concentrations of 10 and 25 % by weight. The SiC particles were irregular blocks with a length to thickness ratio of 2 to 3.

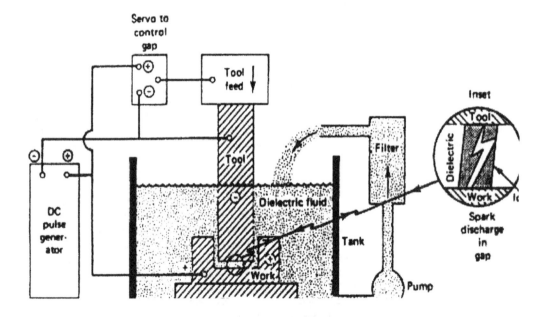

Figure 1 Electro-discharge machining equipment

The EDM of the 6000 series aluminium alloy MMC used in this programme was seen as an important step in the production of *finished components* for sports applications since manufacturing techniques of this type will enable complex three-dimensional shapes to be produced by die sinking.

A proportion of the MMC test pieces included copper-coated SiC reinforcements. The coating procedure is described by Mothersole (1994) and is currently undergoing further development at Napier University.

The manufacture of the MMCs involved degassing of the powder at 550 °C for 2 hours under vacuum, cold compaction using a pressure of 385 MPa for 5 minutes and sintering at 615 °C for 5 hours under vacuum.

Most of the specimens were then solution heat treated at 530 °C for 40 minutes, cold water quenched and artificially aged at 160 °C for 16 hours (T6 ageing treatment).

For labelling the different composite samples, the following code was used: weight per cent of SiC / uncoated (u) or coated (c) SiC particles / particle size / heat treatment (HT); for example 10(u)23HT.

2.2 Test Methods

The strength and deformation behaviour of the MMC materials were assessed with the aid of tensile test equipment contained within the specimen chamber of a scanning electron microscope (SEM). The gauge length of the specimens was 5 mm and the cross-sectional area was between 3 and 5 mm^2. During the tensile test, which was carried out at a strain rate of $0.3\mu m\ s^{-1}$, the loading was arrested several times in order to detect modifications induced in the microstructure and to select new zones for further analysis.

The microstructure of the as-received and heat treated specimens was characterised in a polished and etched state by optical microscopy and SEM. The identification of specific phases was via backscattered electron (BSE) imaging and energy dispersive X-ray analysis (EDXA), while the fracture surfaces of the specimens were studied using an SEM and confocal scanning laser microscope (CSLM).

For all EDM tests, a Transtec (150V, 90A) EDM machine was used with a paraffin dielectric. The tool material was copper and was given a positive polarity. The process parameters of interest were pulse duration; the time current flows during the pulse (time-on); the time current is not flowing during the pulse (time-off); the duty cycle i.e. the ratio of the time-on to the total pulse duration; and the peak current itself. The responses measured during die sinking were the material removal rate (MRR) and surface roughness.

3. RESULTS AND DISCUSSION

3.1 Microstructure

Light microscopy and backscattered electron imaging, revealed the existence of intermetallic compounds in all investigated composites. These are shown as light grey on the micrograph if the microstructure is unetched (Figure 2a).

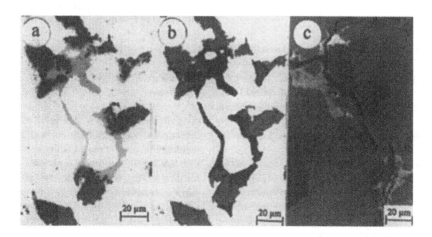

Figure 2 Microstructures including intermetallic compounds:
(a) 10µ23HT (unetched)   (b) 10µ23HT (HF/H₂O)   (c) 10µ23 (BSE)

As a result of etching with 2 ml hydrofluoric acid in 50 ml H_2O, they appear as a black constituent (Figure 2b). To distinguish between intermetallic compounds and defects like pores and microcracks, the backscattered electron imaging is particularly effective. The defects show up black on the BSE photograph and the intermetallic compounds appear as white parts depending on their composition (Figure 2c).

An EDX analysis showed that the intermetallic compounds of the matrix alloy consisted mainly of the elements aluminium, silicon, iron and chromium. Up to now, a quantitative determination of the content of the intermetallics has not been carried out. However, Liaw (1992) showed that the quantity of intermetallics decreased with increased concentration of SiC particles. Whether or not the intermetallic compounds develop depends on the composition of the matrix material and the processing methods chosen in manufacture (Davidson 1993; Liaw 1992).

After etching the specimens with Kellers reagent, fine precipitated particles were visible in the matrix of the age-hardened material. (Figure 3). It can be seen that there are precipitate-free zones along the grain boundaries and near the interfaces between SiC particles and matrix. According to literature (Voituriez 1991), these fine particles are probably Mg_2Si precipitates. Furthermore, the micrograph shows the changes experienced by the intermetallic compound due to the chemical attack.

3.2 Mechanical Behaviour

The in-situ tensile tests demonstrated considerably different deformation behaviour for materials with different thermo-mechanical histories. Figure 4 compares two of these materials, namely 10u23 (as-received condition) and 10u23HT (age-hardened condition). The results clearly indicate that tensile strength and stiffness of the age-hardened composite increase at the expense of its ductility.

The further small increase in strength, coupled with retention of ductility (10c23HT and 10c7HT) in the MMC's containing the copper coated reinforcing particles, is believed to be attributed to improved bonding conditions between the reinforcement and the matrix. However, further investigations, now in progress, have yet to confirm the mechanisms giving rise to these experimental results.

3.3 Fracture Phenomena

The in-situ tensile tests showed that the fracture behaviour of the different samples depended on several factors. In the uncoated, non-heat-treated sample (10u23), a major crack was seen to link with the many microcracks at its tip. This is evident in Figure 5a. Potential sites for crack nucleation are the intermetallics - as can be seen in Figure 2c. A micrograph of another specimen taken at a higher magnification (Figure 5b) shows debonding at the interface between the ductile matrix and brittle reinforcing particle as the dominant failure mechanism.

With the age-hardened, coated material, it was evident that crack propagation *through* the SiC particles prevailed over debonding. Evidence of the improved bonding between matrix and reinforcement is clearly presented in Figure 6b, since it shows cracks passing through a number of the reinforcing particles with little indication of separation at the matrix/particle interfaces.

However, the study of the material reinforced with 7 μm SiC particles showed that the particle-matrix decohesion is the predominant failure mechanism for both the coated and uncoated particles. High levels of particulate clustering, which are found in the material reinforced with 25 wt% SiC, decrease the resistance to damage initiation and provides a favourable path for the linkage of microcracks.

3.4 Electro-discharge Machining (EDM) of 6061 Aluminium Alloy-SiC MMC

As indicated previously, the machining parameters investigated were peak current and pulse duration, including pulse-on time and duty cycle (i.e. the ratio of pulse-on time to pulse duration). The responses measured were material removal rate (MRR) and surface roughness. Surface quality was determined using a Talysurf 4 with a cut-off distance of 0.3 mm. Ten readings were taken on each machined surface and the mean calculated.

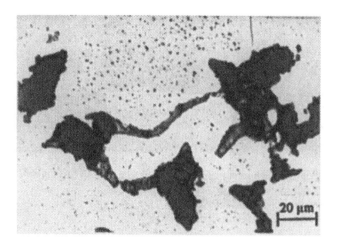

Figure 3 Fine precipitates in an age-hardened composite (10μ23HT, Kellers reagent)

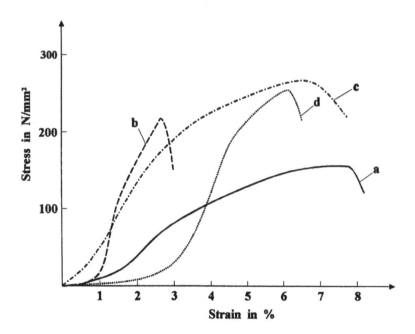

Figure 4 Representative stress vs strain plots obtained from in-situ tensile tests:
(a) 10μ23   (b) 10μ23HT (c) 10c23HT (d) 10c7HT

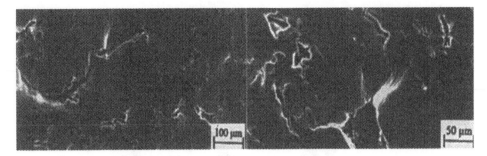

Figure 5   Deformation and fracture phenomena observed by use of an SEM during the in-situ tensile tests
of as-received specimens 10μ23:
(a)   deformation ahead of crack tip
(b)   prevailing particle-matrix decoheshion

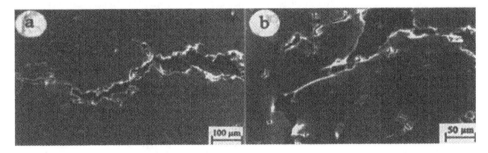

Figure 6   Deformation and fracture phenomena observed by use of an SEM during the in-situ tensile tests
of age-hardened samples 10c23HT
(a)   crack path
(b)   multiple particle cracking

Figures 7 and 8 show the effect of increasing peak current on material removal rate and surface roughness. The level of MRR increases rapidly with the rise in current as indicated in Figure 7. This is expected since the material removal is directly related to the pulse energy (the energy of the electric discharge) and thus to the current. An increase in current also resulted in an increased surface roughness.

The machined surfaces were made up of microscopic craters associated with discrete discharges. The size of these craters, and hence the surface roughness, was found to be proportional to the pulse energy and the peak current. However, as the peak current rose to 30 amps and above, Figure 8 shows the rate of increase in surface roughness slowing. This suggests a limit to the level of surface roughness as the peak energy input from the electro-discharge equipment is reached for a given set of machining conditions.

Typically only 10-20% of the material was found to be expelled from the molten area at the end of the pulse, while the remainder solidified around the crater. Due to the random distribution of the electric discharges, the surfaces generated were complex. Nevertheless, surface roughness measurements provided useful information regarding machining conditions and material response.

In Figure 9 the effect of both pulse-on time and duty cycle are shown. In these trials, the MRR was found to decrease with increasing pulse-on time, although many materials exhibit MRR versus pulse-on time curves which increase to a maximum then gradually decrease (Benhaddad 1991). The shape of the curve in this case is caused by long pulse-on times having reduced power density in each pulse. Thus less material is melted per pulse. Energy is presumably being dissipated in the dielectric while the discharge is occurring.

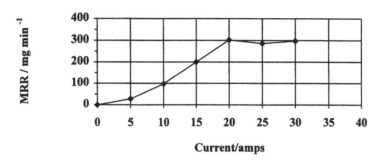

Figure 7. Material removal rate as a function of peak current
for EDM (die sinking) of Al-SiC MMC.

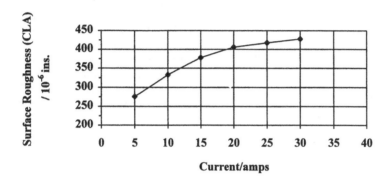

Figure 8. Surface roughness as a function of peak current
for EDM of AL-SiC MMC.

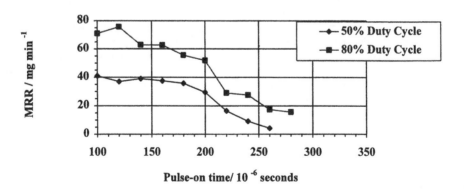

Figure 9. Material removal rate as a function of pulse-on time
for EDM (die sinking) of Al-SiC MMC.

It was also demonstrated that the MRR was higher when a duty cycle of 80% was employed compared with a 50% duty cycle. Clearly the higher the duty cycle the more material is removed, the difference being more pronounced at lower pulse-on times. This is because the period during which no material is removed is reduced to a minimum. However, stable machining can only occur if there is a reasonable pulse-off time (mainly to allow the molten material to be expelled from the discharge area and provide adequate time for the inter-electrode gap to de-ionise). Duty cycles greater than 80% resulted in short circuits and unstable machining.

4. CONCLUSIONS

(i)   The evidence presented indicates that the inclusion of copper-coated, particulate SiC reinforcements, within the aluminium alloy matrix, clearly enhances the tensile strength and ductility of the metal matrix composite.

(ii)   With age-hardened material containing coated reinforcements, it was clear that crack propagation through the SiC particles was the dominant mechanism, as opposed to debonding of the uncoated reinforcement from the matrix.

(iii)   The effects detailed in (i) and (ii) are believed to be caused by the development of a stronger bond between the coated reinforcement and the matrix, compared with uncoated reinforcements. Additional work is being undertaken to confirm the validity of this proposition.

(iv)   Microstructural observations revealed the presence of intermetallic compounds in all investigated composites.

(v)   6061 aluminium alloy matrix reinforced with 25 wt% silicon carbide particles was found to machine successfully using a die sinking electro-discharge system.

(vi)   Greater machining rates resulted in increased surface roughness up to a limiting level, depending on the machining conditions. Surface roughness for the MMC's varied between 275 and 425 μ ins (CLA) for the test conditions investigated.

(vii)   EDM has been demonstrated to be a viable technique for manufacturing complex three-dimensional shapes in aluminium-based MMC's for sports and leisure applications.

REFERENCES

Pillai, R.M., B.C.Pai & K.G. Satyanarayana 1993. *Tool & Alloy Steels*. February: 17-22.
Clyne, T.W. & P.J.Withers 1993. *An Introduction to Metal Matrix Composites*. Cambridge University Press.
Davidson, D.L. 1993. *Composites* 24/3: 248-252.
Jiang, J., C. Collado, D. Keeley & B. Dodd, 1995. *Composites* 26/11: 785-789.
Kapoor, R. & K.S. Vecchio, 1995. *Materials Science and Engineering* A202: 63-75.
Somerday, B.P., Yang Leng & R.P. Gangloff, 1995 *Fatigue and Fracture of Engineering Materials and Structures* 18/5: 565-582.
Metcut Research Associates, Machinability Data Centre, 1980. *Machining Data Handbook*, 2, 3rd Edition.
Mothersole, K.I. 1994. *An Investigation into the Electroless Copper Plating of Silicon Carbide Particles for use as Reinforcements in Metal Matrix Composites*. Project Dissertation, Napier University, Edinburgh.
Liaw, P.K., R.E. Shannon, W.G. Clark Jr. & W. C. Harrigan Jr. 1992, In: M. R. Mitchell & O. Buck (eds), *Cyclic Deformation, Fracture and Non-destructive Evaluation of Advanced Materials*, 251-277. ASTM STP 1157.
Voituriez, C. & I.W.Hall, 1991. *J. Mater.Sci.* 26: 4241-4249.
Benhaddad, M.O., J.A. McGeough & M.B. Barker, 1991. *Processing of Advanced Materials*. 1: 123-128.

The Engineering of Sport, Haake (ed.) © 1996 Balkema, Rotterdam. ISBN 90 5410 822 3

Materials in mountaineering equipment: A look at how processing and heat treatment influences the structure and properties of aluminium alloy karibiners

J.R. Blackford

Department of Engineering Materials, The University of Sheffield, UK

ABSTRACT Karabiners are clip fasteners which are used extensively in rock climbing and mountaineering. The majority of karabiners are made from 7000 series aluminium alloys (Al-Cu-Mg-Zn). The 7000 series alloys were originally developed for applications in the aerospace industry, being wrought heat treatable alloys which are strengthened by precipitation (or age) hardening. Karabiners are produced from extruded bar, and formed by either bending or forging. Changes occur to the structure of the material during the processing and heat treatment, which influence its properties. A brief review of testing and failures is given. This paper aims to give an insight into how an understanding of the interrelationships between processing structure of the material affect its behaviour in service. It deals with a specific case but the principles are generally applicable to any metallic product.

1. INTRODUCTION

Karabiners are self closing connectors which are used extensively by climbers. A karabiner's primary task is to link the rope to protection, but they are also used to clip in to belays, to hold equipment and for abseiling. Originally karabiners were designed for use by firemen, and first reports of their use by climbers was in 1910 by continental climbers. In Britain their use became common just after the 2nd World War when cheap ex-War Department equipment became available. These first karabiners were made of steel, while the majority in use today are made from high strength:weight ratio aluminium alloys. Climbers are always looking for ways of reducing weight; since on a typical route 20 to 40 karabiners may be carried. A steel karabiner may weigh 150g, while an aluminium one may weigh 50g, so a weight saving of 2 to 4kg could be made by using aluminium karabiners rather than steel ones.

The basic design is simple: consisting of a bent rod (the karabiner "body"), and hinged gate, closed by a spring (Fig 1). A variety of shapes exist, with the most common being the offset-D, as shown in Fig 1, which gives sufficient strength and also allow the gate to be opened to an extent such that clipping is made easier.

The aluminium alloys used for karabiners are from the 7000 series, the most common being an alloy designated 7075, which is based on the Al-Zn-Mg-Cu system (Polmear 1981). The composition of 7075 is shown in Table 1. These alloys were originally developed during the 1940s and 1950s for applications in the aerospace industry, they have been subject to a considerable amount of development, both in terms of quantity and type of alloying additions and heat treatment schedules. These factors will be reviewed in this paper.

Table 1 Composition of 7075

Element	amount (weight %)
aluminium (Al)	balance
copper (Cu)	1.2 - 2.0
magnesium (Mg)	2.1 - 2.9
zinc (Zn)	5.1 - 6.1
manganese (Mn)	0.3
chromium (Cr)	0.18 - 0.28
iron (Fe)	0.5
silicon (Si)	0.4
zirconium (Zr) + titanium (Ti)	0.25

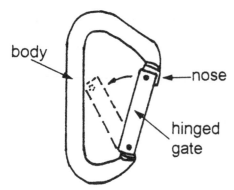

Fig 1 An *offset-D* karabiner.

2. MANUFACTURE OF KARABINERS

Karabiners are made from extruded bar generally of 10, 11 or 12mm diameter, extruded from an alloyed aluminium ingot. The bar is cut to the correct length, and shaped into the required form by one of three processes:

• bending
• cold forging
• hot forging

The nose profile (Fig 1) and gate riveting point are formed by flattening and then punching to the appropriate shape, and a hole is drilled for the gate pivoting rivet. The karabiner gates are generally machined from round bar. The bent wire process was the most common, however forging processes are becoming more widely used as they allow greater flexibility in design and shape of karabiners. The cold forging process is based on the bent wire process but allows karabiners of more complicated cross sections to be made, and in addition the forging process increases the strength of the material. The principle behind using non-circular cross sections is that the material can be "moved" to regions of higher stress thereby making the karabiner stronger (cf. use of steel I- beams in construction). Hot forging processes enable the production of more complex shapes, because the material is at a higher temperature (~400°C) and its flow stress is considerably lower, allowing more material to be "moved", the forging often being carried out in two operations to create the final shape. An example of a hot forged karabiner is DMMs Mamba, which will be considered further in section 5.

After forming, the karabiners are heat treated by a "solution treatment" (~ 40min at 475°C), followed by quenching in to water, and then "aged" (~8h at 115°C), which is a second heat treatment to produce the required strength in the material. The changes which occur to the microstructure of the karabiner during heat treatment are primarily responsible for the high strengths of these materials, and these will be considered in some detail in sections 3.1 and 3.2. After heat treatment the karabiners are then tumbled to remove rough edges before assembly.

3. ALLOY DEVELOPMENT AND MICROSTRUCTURE

The microstructure of a material influences its properties, an understanding of the microstructural features and how they may be controlled offers an opportunity to control the properties of materials. Pure metals are rarely, if ever, used in structural applications because of their inherent weakness. Additional elements (alloying additions) are added to metals and when combined with processing and heat treatments can produce beneficial effects. By suitable alloying and heat treatment it is possible to increase the yield stress of pure aluminium by as much as 40 times (Polmear 1981). A range of elements have been added to aluminium which produce a variety of effects. The major additions are those with a high solid solubility e.g. copper, zinc and magnesium. *(NB The concept of solid solution is central to metallurgy, and to the phenomenon of precipitation hardening, it is the capability of a metal to dissolve, or accommodate, another metal in its [solid] crystal lattice. In most systems that show solid solubility it increases with increasing temperature).* However, several other elements with solubilities below 1 atomic % confer important improvements to alloy properties, e.g. chromium and zirconium, which are used primarily to form compounds that control grain (or metal crystal) structure.

Aluminium alloys can be classified as non-heat treatable or heat treatable depending on their response to precipitation (or age) hardening. The precipitation hardened alloys show the highest strengths, of these the 7000 series alloys, which are based on Al-Zn-Mg and Al-Zn-Mg-Cu being the best example (Staley 1989). The addition of copper was found to be beneficial in terms of increasing the alloys strength and resistance to stress corrosion cracking (SCC) (Polmear 1981). SCC is a phenomenon which results in brittle fracture in alloys, normally considered ductile, when they are exposed to the simultaneous action of surface tension (e.g. caused by residual stresses from processing) and a corrosive environment (e.g. water vapour). The grain structure, developed during working procedures, needs to be carefully controlled in order to prevent the formation of re crystallized grains.

Although pure aluminium is a soft metal, high strength:weight ratios can be achieved in certain alloys because they show a marked response to precipitation or age hardening. A brief review of the principles of precipitation hardening will now be presented.

3.1 Precipitation hardening

The basic requirement for an alloy to be amenable to precipitation hardening is a decrease in solid solubility of one or more of the alloying elements with decreasing temperature, as shown in Fig 2(a). Heat treatment normally involves:

(i) Solution treatment at a relatively high temperature within a single phase region, e.g. α in Fig 2(a), to dissolve the alloying elements.

(ii) Rapid cooling or quenching into water, usually to room temperature, to obtain a super-saturated solid solution (SSSS) of these elements in aluminium.

(iii) Controlled decomposition of the SSSS to form a finely dispersed precipitate, usually by ageing for convenient times at one and sometimes two intermediate temperatures.
The precipitation occurs by a nucleation and growth process, the size and dispersion of the precipitate formed depend on the time and temperature of the ageing treatment, the variation of precipitate size with ageing temperature is shown schematically in Fig 2(b).

The complete decomposition of the SSSS (the ageing process) is usually a complex process, which may involve several stages and different precipitates. A more thorough description of the processes involved is given by Polmear (1982), and Starke (1989). Maximum precipitation hardening in

an alloy occurs when there is a critical dispersion, and a certain size (or range of sizes) of precipitate present. The reasons for this are considered below.

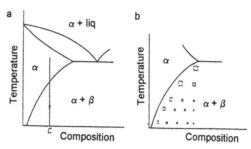

Fig 2 (a) Possible precipitation hardening system; (b) variation of precipitate size with ageing temperature.

3.2 Strengthening mechanisms in precipitation hardened alloys

The strength of an age hardening alloy is governed by the interaction of moving dislocations and precipitates (Polmear 1981). (NB dislocations are defects in crystalline materials, and the ease or difficulty with which they move has significant implications for the properties of a material). There are two fundamental ideas behind the strengthening mechanisms. The first is the result of the interference to slip by particles precipitating on crystallographic planes, as shown schematically in Fig 3. This cutting, or shearing, of particles increases the number of solute-solvent bonds and additional work must be done by the applied stress to achieve this effect; consequently this strengthens the alloy. This mechanism operates when the particles are relatively small.

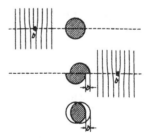

Fig 3 Representation of the cutting of a fine particle (strengthening precipitate), by a moving dislocation. (From Polmear 1982).

However, if the particles are large and widely spaced they can be readily by-passed by moving dislocations which bow out between them and rejoin (Fig 4). The yield strength of the alloy is low but the rate of work hardening is high, and plastic deformation tend to be spread more uniformly through the grains. This is the situation with over-aged alloys and the typical age hardening curve in which strength increases then decreases, with a transition from shearing to by-passing precipitates, as shown schematically in Fig 5.

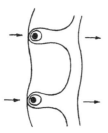

Fig 4 Representation of a dislocation by-passing widely spaced particles (larger strengthening precipitates cf. Fig 4). (From Polmear 1982).

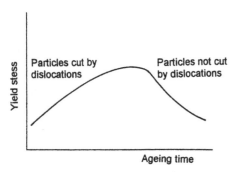

Fig 5 Variation of yield stress as a function of ageing time for a precipitation hardened alloy.

Maximum strengthening results from inhibiting the movement of dislocations. This may be achieved by having precipitates which are too large to shear easily, but also sufficiently closely-spaced to inhibit by-passing by dislocations. Recently, this has been achieved by duplex ageing treatments (for times at more than one ageing temperature), either by forming a critical dispersion of precipitates or, by forming duplex dispersions of precipitates,

consisting of small, closely-spaced particles to raise the yield strength and larger particles which will cause increased rates of work hardening and distribute plastic deformation more uniformly.

4. MECHANICAL PROPERTIES OF ALUMINIUM ALLOYS

Karabiners need to have a high tensile strength and high toughness, these properties are influenced by the microstructure of the material. This section will review the principle microstructural features that control the properties of aluminium alloys:

- coarse intermetallics
- dispersoids
- fine precipitates
- grain size and shape

4.1 Coarse intermetallics

Coarse intermetallic compounds, usually in the range 0.5 to $10\mu m$ (the largest particles found in this class of alloys), which form during ingot solidification, and contain the impurity elements iron and silicon (e.g. $FeAl_3$, Mg_2Si and Al_7Cu_2Fe), as shown in Fig 6. These particles are detrimental to the alloy properties, causing a significant decrease in fracture toughness. Significant improvements have been made by controlling the levels of these impurity elements. They have relatively little effect on yield or tensile strength, but they can cause a marked loss of ductility, as the particles may crack at small plastic strains forming internal voids which, under the action of further plastic strain, may coalesce leading to premature fracture.

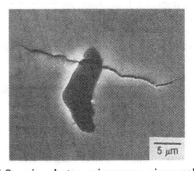

Fig 6 Scanning electron microscope micrograph of a coarse intermetallic particle (Mg_2Si), with an associated crack.

4.2 Dispersoids

Smaller, submicron particles or dispersoids, 0.05 to $0.5\mu m$ in size, which are intermetallic compounds containing the transition metals e.g. chromium, or zirconium, which form $Al_{12}Mg_2Cr$ or $ZrAl_3$ (Fig 7). These particles are primarily to retard recrystallization and grain growth (i.e. control the shape and size of the grains) in the alloys and so, indirectly, are beneficial in terms of tensile strength and toughness, see section 4.4.

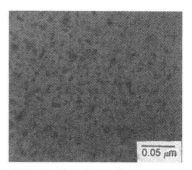

Fig 8 TEM micrograph of a strengthening precipitate ($MgZn_2$).

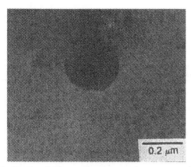

Fig 7 Transmission electron microscope (TEM) micrograph of a dispersoid particle ($ZrAl_3$).

4.3 Strengthening precipitates

Fine precipitates, up to $0.01\mu m$ in size, which form during age hardening e.g. $MgZn_2$, as shown in Fig 8. These particles are responsible for the high strengths of this class of alloys, via the mechanism discussed in section 3.2. These precipitates have at least two effects with regard to the toughness of aluminium alloys. To the extent that they reduce deformation, toughness is enhanced and it has been observed that, for equal dispersions of particles, an alloy with a higher yield stress has a greater toughness. At the same time these fine particles tend to cause localisation of slip during plastic deformation leading to bands of slip ahead of an advancing crack. Strain is concentrated within these bands and may cause premature cracking at the sites of intermetallic compounds ahead of an advancing crack.

4.4 Grain size and shape

Grain size and shape, are affected by the processing and subsequent heat treatments, as well as by the submicron particles (section 4.2). The shape and internal structure of grains often changes during heat treatment following deformation. In general, as with virtually all alloys, fine grains promote higher strengths and toughness. The effect of grain size and shape in a range of alloys based on the Al-Zn-Mg-Cu system is shown in Fig 9.

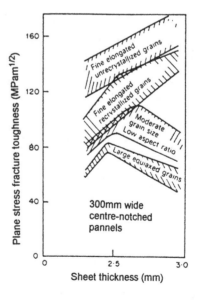

Fig 9 Effect of recrystallization and grain size and shape of various alloys based on the Al-Zn-Mg-Cu system.

The main effects of the various alloying additions are summarised in Table 2, below:

Table 2 Effects of alloying additions

Element	Effect
Mg, Zn	form strengthening precipitates
Cr, Zr	form dispersoids which retard recrystallization and grain growth
Cu	increases resistance to SCC
Fe, Si	unwanted impurities which form coarse intermetallics that reduce toughness

5. USE, TESTING AND FAILURES OF KARABINERS

Climbing has changed considerably over the last 10 years and recently has seen a substantial increase in popularity. Changes in ethics and improvements in equipment have meant that the old adage that the *leader must not fall*, is no longer the case. In fact the hardest technical routes of the present day are climbed by practising the individual moves, or short sequences of moves, and falling off, with the aim being to string all the moves together.

Climbing and mountaineering encompass a diverse spectrum of activities and potential fall situations, which causes difficulties in defining test methods. Karabiners are tested in several ways. The standard methods have been set down by the UIAA (Union Internationale des Association d'Alpinisme), and more recently have been revised as a draft CEN standard (CEN 1996). The CEN standard specifies the following tests for "basic connectors" as general purpose karabiners are now defined: static tensile tests, using 12mm diameter steel bars to load the karabiner (a) along its major axis, (b) along its minor axis and (c) along its major axis with the gate open (see Table 3). A further test is specified for the force required to open the gate, which should be a minimum of 5N.

At first sight the reasons for the different test methods are not obvious. If the karabiner is loaded in an ideal manner (i.e. along its major axis with its gate closed), test (a) as described above would be sufficient. Work has shown that forces generated in falls are normally <7kN, though a bad fall could produce a force of say 10kN, but it is virtually impossible that any karabiner loaded in an ideal

fashion, with its gate closed, will break. However, karabiner gates are not always closed during a fall, and also a karabiner is not always loaded along its major axis. There are several reason why a gate may be open during a fall: the shape of the rock may press the gate open. The karabiner may hit against the rock and flick the gate open. A rope passing through the karabiner at high speed can set up oscillations in the tape sling which are the same as the natural frequency as the gate and cause it to resonate. So, the need for high gate open strengths is clear.

Table 3 CEN Standard for karabiners

Test location	load (kN)
major axis	20
minor axis	7
gate open	7

A karabiner is not always loaded along its major axis as it may swivel or twist during use (this may occur with the rope running through it). Fortunately the design of karabiners is such that non-major axis loading rarely occurs. However related problems have been experienced in the past with low strength karabiners (often those made from 9mm diameter bar) when using two ropes, or wide tape/web, clipped into a single karabiner, as this tends to load the karabiner away from its main axis which leads to an increased likelihood of failure. These situations can be avoided by clipping only one rope into a karabiner, and by using narrower web.

A design which virtually eliminates the problem of non-major axis loading and also has a considerably higher gate open strength is a captive eye design. One example of this is DMMs Mamba (Fig 10), which is a hot forged karabiner, with a slot in one end to take a tape sling and a gate open strength of 10kN. There are other factors which should be considered when using karabiners, and other designs and types of karabiners which may be more appropriate for use in certain circumstances, but these are beyond the scope of this paper, however, reviews of the use, abuse and choice of karabiners have been compiled (Raleigh 1992, and Reid 1992).

Fig 10 DMMs Mamba karabiner.

6. CONCLUDING REMARKS

The comments in this final section represent a personal view, of which matters are significant and what deserves further attention.

It is rare that karabiners break. In many ways this is not surprising as the design of karabiners has evolved over a period of time, and the materials have been developed in substantial research programmes primarily for the aerospace industry.

However, karabiners can and do break. Factors that can be addressed in order to limit failures are:

- education of users
- design
- material

Changes made to the design and material and the way they are processed are interrelated, so changing one factor may influence another.

In terms of materials for instance, it may well be relatively easy to improve a particular property but this may be at the expense of another, for example increasing the tensile strength of an alloy often leads to a decrease in its toughness [although the two properties are not linked by a simple inverse relationship], and resistance to stress corrosion cracking.

An appreciation of how one factor may adversely affect another is particularly important when one considers that the majority of testing of karabiners (and all the testing that is specified by the international standards) is by static tensile testing, so no measure of a karabiners toughness or resistance to corrosion *has* to be carried out, though it would seem prudent to do so.

Incremental improvements will be made in materials and design (possibly by stress analysis), by using established ideas and principles from materials science and engineering, from this it seems that an interdisciplinary approach is likely to produce the greatest gains. Finally, there will always be a place for improvement from inspired design.

ACKNOWLEDGEMENTS

Financial support for this conference from the British Mountaineering Council, and DMM International Ltd., and copies of micrographs from Dave Hanlon (Department of Engineering Materials, University of Sheffield), are gratefully acknowledged.

REFERENCES

CEN 1996. No. prEN12275 *Draft Standard for Mountaineering Equipment - Connectors - Safety Requirements and Test Methods*. Copies may be obtained from the Head Office of The BSI.

Polmear, I.J., 1981.*Light Alloys, Metallurgy of the Light Metals*. London: Edward Arnold.

Raleigh, D, 1992. *Smooth Operators, Karabiners are not as simple as open and shut*. Climbing. August/September 101-111.

Reid, S, 1992. *Lightweight Karabiners*. On The Edge. 31: 24-29 and 68-70.

Smallman, R.E., 1963. *Modern Physical Metallurgy*, 293. London: Butterworth.

Staley, J.T., 1989, *History of Wrought-Aluminium-Alloy Development*. Treatise on Materials Science and Technology, 31:3-31. London: Academic Press.

Starke, E.A., 1989, *Heat Treatable Aluminium Alloys*. Treatise on Materials Science and Technology, 31:3-31. London: Academic Press.

Thompson, D.S., 1975, *Met Trans*. 6A:671-683.

The Engineering of Sport, Haake (ed.) © 1996 Balkema, Rotterdam. ISBN 90 5410 822 3

The effect of microstructure on the impact dynamics of a cricket bat

C. Grant & S. A. Nixon

Department of Mechanical, Materials and Manufacturing Engineering, University of Newcastle upon Tyne, UK

ABSTRACT: The construction of a reliable computer model lies at the heart of research aimed at improving the dynamic performance of a cricket bat through design. Bats are invariably made of English Willow which is relatively light and durable. The face of the blade is heavily rolled during manufacture to produce a inhomogenous structure that significantly affects the dynamic behaviour of the bat. Willow has a cellular microstructure with a maximum theoretical compression of 3 6. Based on an average compression factor of 3, density measurements of samples taken from a rolled blade imply a compressed layer about 1.4mm thick. The compressed layer can also increase modal frequencies by up to 10%. Micrographs of the compressed layer provide qualitative graphical support for the results.

1 INTRODUCTION

The construction of a reliable computer model lies at the heart of research aimed at improving the dynamic performance of a cricket bat through design (see Grant and Theti, 1994; Grant and Baird, 1995). The laws of the game (see Lewis, 1994) insist that the blade of the bat shall be made of wood, a natural material with a complex cellular micro-structure. Bats are invariably made of English Willow which is relatively light and durable and like all woods has anisotropic, visco-elastic mechanical properties. However, impact takes place over such a short period that time dependent effects are of little significance.

The face of the blade is heavily rolled during manufacture to produce a surface better able to withstand severe impact without incurring excessive damage. This results in a non-homogenous structure that significantly affects the dynamic behaviour of the bat.

Grant and Nixon (1996) produced a modal model of bats based on a uniform elastic material. To correlate the model with actual behaviour, it was necessary to significantly increase the density of Willow used in the model. The principal aim of this work is to investigate the effect of rolling on the microstructure and its influence on flexural vibrations that determine the dynamic performance of the bat.

A layer of the blade close to the surface is likely to be most affected by the rolling process. For this reason, small square section beam samples to BS 373 (1957) were used in preference to tests on a whole bat. The effect of a surface layer on the samples would be enhanced, and uncluttered by the relatively complex structure of the bat as a whole.

2 MICRO-STRUCTURE OF WILLOW

Wood is an orthotropic material in which the stiffness and strength parallel to the axis of the tree are much greater than in the radial or tangential direction (Gibson and Ashby, 1988). Blades of cricket bats are always cut with the long axis parallel to the tree axis, and with the face parallel to a radius as shown in Figure 1.

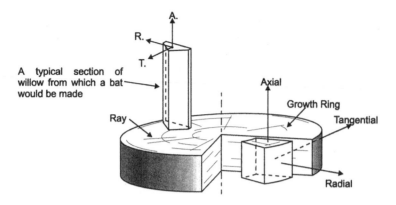

Figure 1: Orientation of a cricket bat blade relative to the tree

Concentric circular *growth rings* and radial *rays* are visible on the axial section. Growth rings comprise regions of cells that vary in size with the annual growth cycle. Rays are radial arrays of smaller cells that add significantly to the structural properties of the section.

On a macroscopic scale, wood appears as a porous structure with densely spaced sap channels that conduct fluid up the tree. Figure 2 shows the distribution of sap channels in a sample of bat Willow. Although considered to be structurally unimportant, the sap channels in Willow are about 100 μm diameter and contribute to the compressibility of the section in the rolling process.

The underlying micro-structure of the section consists of long hollow fibres giving the appearance of a cellular section. The fibres are typically 15-20 μm diameter with a wall thickness of 1.5 to 2 μm as shown in Figure 3.

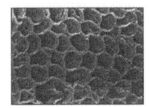

Figure 3: Cell structure of bat Willow(X1000)

Cell walls consist principally of cellulose, and these give the section its strength, weight and stiffness. The physical properties of the cell walls have been measured and are similar for all types of wood. Table 1 shows the bulk properties of Willow compared to the properties of cell walls.

The ratios shown in Table 1 give an indication of the theoretical limit to the amount that the section can be compressed. The slightly differing figures for density and stiffness correspond to differences in composition of cell walls in different timbers.

Table 1: Physical properties of English Willow

	density kg/m^3	Young's Modulus GPa
Willow*	417	6.6
Cell wall+	1500	35
Ratio	3.6	5.3

* after Lavers, 1983
+ after Gibson and Ashby, 1988

Figure 2: Sap channels in Willow (X20)

Table 2: Physical tests on beam samples

	density ρ kg/m^3	stiffness EI Nm^2
Sample A	466	80.9
Sample B	483	100.2
Standard	417	88.7

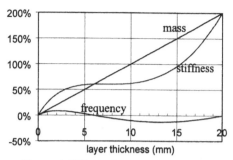

Figure 4: Effect of a dense layer on mass, flexural stiffness and frequency

3 BEAM TESTS

Beams 20mm x 20mm x 300mm were cut from a new blade as supplied by the manufacturer in the rolled condition. Two samples were taken from the central region of the blade, either side of the centre line. Each sample had one long face that formed a part of the rolled face of the blade. The samples were first weighed and then tested in three-point bending with the rolled face uppermost.

The bend tests were conducted in accordance with BS 373:(1957). The beams were strained at 6.6mm/min up to a maximum load of 500N. Table 2 shows the results of the tests in terms of the mean density and flexural stiffness of the two samples compared to calculated values based on the standard properties of Willow given in Table 1.

The samples show a mean density 14% greater than the standard properties with a 4% difference between samples. Flexural stiffness are only 2% greater on average but with a 24% difference between samples.

Time dependent effects were very much in evidence in the bend tests, so the results quoted in Table 2 are the average of loading and unloading tests extrapolated to zero load.

4 COMPUTER MODEL OF A ROLLED SURFACE LAYER

If the density and elastic modulus of a beam could be uniformly increased by a common factor, then the weight and flexural stiffness of the beam would increase in proportion. However, the frequencies of all flexural modes of vibration in the beam would remain unchanged. This is because the square of the frequency of each of the different vibration modes in the beam is pro-portional to the flexural stiffness divided by the mass per unit length of the beam.

If a thin layer of the top surface of the beam were to have its density and elastic modulus increased, then the mass and flexural stiffness would increase, but not necessarily in proportion. It follows that the frequencies of natural vibration would also change.

Figure 4 shows the percentage increases in mass, flexural stiffness and natural frequency calculated for a beam of similar section to those used in the bend tests described above. The surface layer has its density and elastic modulus increased by a factor 3, similar to what might be achieved by the rolling process used in the manufacture of cricket bats.

When the thickness of the dense layer is either zero or covers the whole depth of the section, there is no effect on frequency as predicted above. Mass increases linearly with thickness as might be expected, but the flexural stiffness changes in a more complex manner.

According to Figure 4 the weights of the two beam samples tested above imply layers of 1.2mm and 1.6mm thick respectively. Table 3 gives the implied layer thickness and corre-

Table 3: Implied increase in mass, stiffness and frequency of beam samples

increase	Sample A	Sample B
thickness (mm)	1.2	1.6
mass %	12.0	16.0
stiffness %	28.4	35.1
frequency %	7.1	7.9

sponding increases in mass, stiffness and frequency for both samples.

5 MICRO-STRUCTURE OF A ROLLED SURFACE LAYER

Beam samples used in the tests described above were prepared for examination in a Scanning Electron Microscope. Figure 5 to Figure 8 show the cell structure at the face, and at depths of 1mm, 2mm and 3mm from the face respectively.

The micrographs clearly show the crushed structure at the face decreasing in severity with depth. Measurements of cell aspect ratio suggest that the degree of compression varies from about 4 at the face reducing to unity at a depth of about 3 mm. These qualitative results compare well with the results derived from density measurements.

6 CONCLUSIONS

The theoretical upper limit for compression of English Willow is 3 6. With an assumed mean compression factor of 3 in a surface layer, density measurements indicate a layer thickness of 1.4mm ± 0.2mm. Static flexure tests were unreliable due to time dependent effects.

A dense, stiff surface layer affects mass and flexural stiffness by disproportionate amounts, resulting in a net effect on the frequencies of flexural vibrations. Depending on the density factor and depth of beam, flexural stiffness can be increased by about 50% whilst corresponding increases in frequency are up to 10%. For the beam samples, the increases were around 30% and 7 to 8% respectively.

Figure 5: Cell structure at the bat face (X1000)

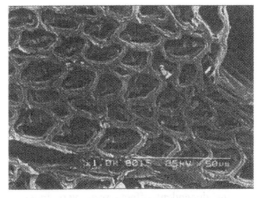

Figure 7: Cell structure 2mm from the bat face (X1000)

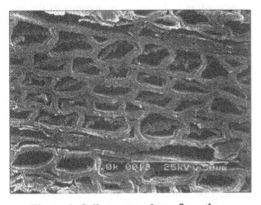

Figure 6: Cell structure 1mm from the bat face (X1000)

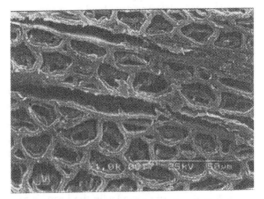

Figure 8: Cell structure 3mm from the bat face (X1000)

Photo-micrographs of the section close to the rolled face provide qualitative support for the results based on density measurements.

7 REFERENCES

BS 373: Methods for testing small clear specimens of timber, British Standards Institute, London, 1957.

Gibson L.J. and Ashby M.F., *Cellular solids: Structure and properties*, Pergamon Press, 1988.

Grant C. and Baird A.D., *Modelling for an improved cricket bat*, Institute of Physics Annual Congress, Physics of Sport, Telford, 1995.

Grant C. And Nixon S.A., *Parametric modelling of the dynamic performance of a cricket bat*, Proceedings of the International Conference on The Engineering of Sport, Sheffield, U.K., July 1996.

Grant C. and Thethi P., *Recent Advances in Experimental Mechanics*, Vol. 1, Silva Gomes J.F. et al (eds), Balkema, Rotterdam, 1994, 669-674.

Lavers G.M., *The strength properties of timber*, 3rd Edn.,HMSO, London, 1983.

Lewis T., *MCC coaching manual*, Weidenfield & Nicholson, London, 1994.

The Engineering of Sport, Haake (ed.) © 1996 Balkema, Rotterdam. ISBN 90 5410 822 3

Materials selection for sports equipment

U.G.K.Wegst & M.F.Ashby
Cambridge University Engineering Department, UK

ABSTRACT: Sports impose extreme demands on the materials on which they depend. Success depends not only on the skills, strength or stamina of the athlete, but on the ultimate efficiency of the materials of which his or her equipment is made. In this paper, the mechanical performance of materials is evaluated using the Cambridge Materials Selector (CMS) software package, which is based on two key ideas: that of material selection charts and that of material indices. The high structural efficiency of natural materials, in comparison with engineering materials such as metals, polymers and composites, is illustrated using CMS. Case studies are presented to demonstrate how CMS may be used to select materials for sports equipment such as oars, archery bows and vaulting-poles, both from an historical and a present-day perspective. Particular attention is given to natural materials such as woods and bamboo, which, historically, have been used extensively for sports equipment.

1 MATERIAL SELECTION CHARTS AND MATERIAL INDICES

Material selection charts and material indices provide an objective method of comparing the performance of different materials for a given application (Ashby 1989, 1992). A 'material index' is a grouping of material properties which characterises performance. The familiar 'specific stiffness', E/ρ, is one such index; so too is the 'specific strength', σ_f/ρ (here E is Young's modulus, σ_f the failure strength and ρ the density).

Figure 1 illustrates, schematically, the idea of a 'material property chart'. It shows one material property, in this case Young's modulus, E, plotted against another, the density ρ. The scales are logarithmic to accommodate the vast range of materials available to us, from low density polymeric foams and materials such as cork which have very low moduli, to dense metals with high moduli, such as steel and tungsten, or ceramics such as silicon carbide. The chart illustrates that materials of a given class cluster together; each envelope on the figure encloses all members of the material class it represents.

In the case studies presented later, we use material charts at a higher resolution, produced using the Cambridge Materials Selector software (CMS 1995). In those, each individual material is represented by a bubble. The size of the bubble indicates the range of the property. The width of this range depends on the composition of the material, its purity, texture and structure.

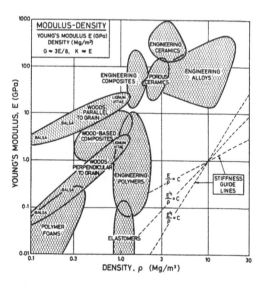

Figure 1: An example of a material property chart for engineering materials showing Young's modulus plotted against density. Guidelines show the slopes of three material indices. Their use is explained in the text.

In the case of engineering materials, these features are influenced and controlled by the manufacturing process. In the case of natural materials, it is the growth conditions which determine the structure and properties.

The modulus-density chart is just one example of a material property chart. Other charts relate other mechanical, thermal and electrical properties. In this paper, we concentrate on mechanical performance and therefore on mechanical property charts.

So far the material property charts have proved to be a clear way of plotting and comparing data. They become even more useful in conjunction with material indices, which provide a powerful tool for the selection of materials for given applications. One index has been mentioned already: the performance of materials as light, stiff ties (tensile members) is measured by the index E/ρ — the larger the value of E/ρ, the lighter is the tie for the same stiffness. The form of the index depends on the mode of loading: axial loading, bending and twisting lead to different indices. As an example, the performance of a light, stiff beam (a component loaded in bending) is measured not by E/ρ, but by the index $E^{1/2}/\rho$; that for flat plates in bending is $E^{1/3}/\rho$. The logarithmic scales allow all three to be plotted on the same figure; each appears as a set of straight, parallel lines. One member of each is shown on figure 1, labelled 'stiffness guide lines'; the required set can be constructed from these.

There are many material indices, each measuring some aspect of efficiency in a given mode of loading. The way in which these indices are derived is illustrated in the following three case studies.

2 MATERIALS FOR OARS

Rowing was the means of propelling vessels in ancient times until the boats became too large and heavy and sails were introduced. The advantage of oars was well understood by the Romans, the Vikings and the Venetians; in 54 BC Caesar depended largely on oars when crossing the Channel, Vikings rowed and sailed across to enter the estuaries along the coast of England and the Venetians ruled the waves of the Mediterranean Sea.

The first recorded amateur oars-men are probably the islanders who entertained Odysseus on his return to Ithaca. In England, the first recorded rowing competition dates back to 1716, when watermen, whose profession it was to row ferries with people and goods across the Thames, raced for the first time. The popularity of rowing increased gradually and both rules and equipment became increasingly sophisticated. The establishment of rowing as an Olympic sport in 1900 was a strong stimulus for the development of high performance boats and oars. Since then, both have exploited to the full craftsmanship and material of the day.

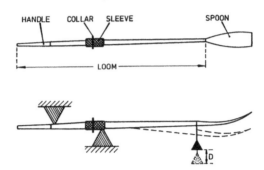

Figure 2: Oars are designed on stiffness, measured in the way shown in the lower figure, and they must be light.

Table 1: The design requirements for oars.

FUNCTION:	Oar = Light, stiff beam
OBJECTIVE:	Minimise mass

CONSTRAINTS:
(a) Length specified
(b) Bending Stiffness, S, specified
(c) Toughness, $G_c > 1$ kJ/m^2
(d) Cost, $C_m < 75$ £/kg

2.1 The Model

Mechanically speaking, the oar is a beam loaded in bending. It must be strong enough to carry the bending moment exerted by the oarsman without breaking, it must have just the right stiffness to match the rower's own characteristics to give the right 'feel' and — very important — it must be as light as possible. The strength constraint is easily met. Oars are designed on stiffness, that is, to give a specified elastic deflection under a given load. The upper part of figure 2 shows an oar: a blade or 'spoon' is bonded to a shaft or 'loom' which carries a sleeve and collar to give positive location in the rowlock. The lower part of the figure shows how the oar stiffness is measured: a 10 kg weight is hung on the oar 2.05 m from the collar and the deflection at this point is measured. A soft oar will deflect nearly 50 mm; a hard one only 30 mm. A rower, ordering an oar specifies how hard it should be.

The oar must also be light; extra weight increases the wetted area of the hull and the drag that goes with it. So there we have it: an oar is a beam of specified stiffness and minimum weight.

The mass, m, of the oar — treated as a solid cylinder of radius, r, and length, L — is

$$m = AL\rho = \pi r^2 L \rho \qquad (1)$$

where A is the cross-sectional area and ρ the density of the material of which it is made. The stiffness of the beam is

$$S = \frac{cEI}{L^3} \qquad (2)$$

where E is the modulus, I the second moment of area of the beam and c is a constant (roughly 24 — the material selection is independent of this value). For a solid cylinder

$$I = \frac{\pi r^4}{4} \qquad (3)$$

The length, L, and stiffness, S, are specified; they are constraints. The free variable is the radius, r. We use equations 2 and 3 to eliminate r in equation 1, giving

$$m = 2 \left(\frac{\pi S L^5}{c} \right)^{1/2} \left(\frac{\rho}{E^{1/2}} \right) \qquad (4)$$

The mass, m, of the oar is minimised by choosing materials with large values of the material index

$$M_1 = \frac{E^{1/2}}{\rho} \qquad (5)$$

The design requirement lists two further constraints — on toughness, G_c, and cost, C_m. These are frequently taken for granted — the designer subconsciously rejects materials which are too brittle (ceramics, for instance) or too costly (platinum). It is better to make them explicit. We therefore require that the limits set out in table 1 are met.

2.2 The Selection

The selection requires two stages. Figure 3 shows the first, a chart of Young's modulus, E, plotted against density, ρ, using the generic database. A selection line for the index is shown on it. It identifies three classes of materials: woods, bamboo, palm, carbon-fibre and glass-fibre reinforced polymers and certain ceramics (table 2). Ceramics meet the first set of design requirements, but are brittle; a ceramic oar, if dropped, might shatter.

Shock-resistance requires adequate toughness, G_c (not just fracture toughness, K_{Ic}). A useful rule-of-thumb for this is to choose materials with a toughness, G_c, such that

$$G_c = \frac{K_{Ic}^2}{E} > 1 \text{ kJ/m}^2 \qquad (6)$$

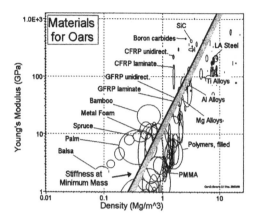

Figure 3: A chart showing Young's modulus, E, plotted against density, ρ, using a generic-materials database. The materials index, M_1, is shown by a diagonal line of slope 2.

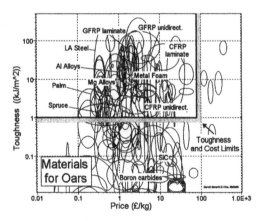

Figure 4: A chart showing toughness, G_c, plotted against cost per kg, C_m. The box isolates materials with $G_c > 1$ kJ/m^2 and $C_m < 75$ £/kg.

We require a second stage, a chart of toughness, G_c, against cost, C_m, shown in figure 4. The toughness axis is created by generating a user-defined property combination with K_{Ic} and E, as per equation 6. A box selection specifies materials with $G_c > 1$ kJ/m^2 and $C_m < 75$ £/kg.

Ceramics are eliminated because they are brittle. The recommendation is clear: make your oars out of wood or — better — out of glass-fibre reinforced polymers (GFRP) or carbon-fibre reinforced polymers (CFRP).

Table 2: Materials for oars.

MATERIAL	COMMENT
Woods, bamboo, palm	Properties not easily controlled, low G_c and heavy
CFRP	More control of properties and cheaper than wood
GFRP	Cheaper than CFRP, but lower G_c
Ceramics	Good M_1, but brittle — eliminated by low G_c

If we want to make wooden oars, we need to know which wood to chose. To determine this, we create a modulus-density selection chart, this time with the CMS wood database (figure 5): it identifies the softwoods spruce, fir and cedar, bamboo, palm and the tropical woods Balsa and Cuangare as good materials for oars.

There are, of course, additional non-numerical requirement in oar-making. The wood must, for instance, be available in long enough beams with straight grain, that it is dimensionally stable and must not warp due to changes in moisture content.

2.3 Postscript

This selection indicates which materials the oars should be made of, but what are they made of in reality? Many racing oars and sculls are still made of wood, but more and more the high performance composite CFRP replaces this traditional material.

Wooden oars are still manufactured, in the true sense of the word: craftsmen shape them by hand as 100 years ago. Shaft and blade are made of Sitka spruce from the northern US or Canada, since the short growing season in this part of the world favours a fine grain. The oar is a composite of four wooden strips which are laminated together for more uniform properties. The compression-side of the oar is stiffened by a hardwood strip, typically ash; the blade is glued to the shaft. The semi-finished oar is shelved for some weeks to settle down before it is finished by cutting, polishing and varnishing. The final spruce oar weighs between 4 and 4.3 kg and costs about £200.

The advantage of the composite oar over the wooden oar is that it can be made into a more efficient shape, a thin-walled tube, and therefore is significantly lighter for the same stiffness (typical weight: 3 kg). The component parts are fabricated from a mixture of carbon and glass fibres in an epoxy matrix, assembled and glued. This allows for a greater control of performance: the shaft is moulded to give the stiffness specified by the purchaser. And that for a price which is lower than that of a wooden oar: a CFRP oar costs about £170.

Is there scope to do better? For normal construction, we read from the charts that wood and CFRP offer the lightest oars for a given stiffness. Novel composites, such as metal foams and others not at present shown on the chart, might permit further weight saving and functional-grading (a thin, very stiff outer shell with a low density core such as a metal tube filled with a metal foam) might do it.

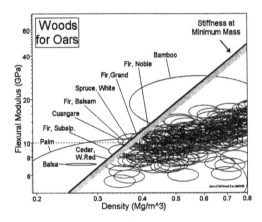

Figure 5: A chart showing Young's modulus, E, plotted against density, ρ, using a database of wood properties. The materials index, M_1, is shown as a diagonal line of slope 2.

3 MATERIALS FOR ARCHERY BOWS

Archery bows have been used for many tens of thousands of years; the earliest arrowheads found in North Africa, already rather sophisticated, date back to about 50,000 BC. Originally designed for hunting and warfare, the archery bow today has developed into a high-tech piece of sports equipment. Bows may be divided into two classes, simple or 'self' bows made singly of wood or wood like materials such as bamboo or palm, and composite bows with layers of various kinds of material, traditionally such as wood, sinew, bone or horn, nowadays such as wood, carbon-fibre and glass-fibre reinforced polymers.

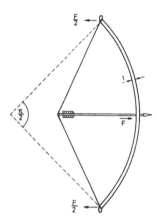

Figure 6: Archery bows are designed on energy storage per unit mass. They must not break when fully drawn.

Regardless of the material of which it is made, the bow has the prime mechanical function of storing elastic energy. For reasons of simplicity we concentrate on 'self' bows, which are believed to have been used widely in Europe from about 10,000 BC; the English longbow, famous from the English victory in the battle of Agincourt, is one example of this kind.

Table 3: The design requirements for archery bows.

FUNCTION:	Archery bow = Leaf spring
OBJECTIVE:	Maximise stored elastic energy per unit mass
CONSTRAINT:	May not break when fully drawn

3.1 The Model

Mechanically speaking, the archery bow is a leaf spring. It stores elastic energy when drawn which is transformed into the kinetic energy of the arrow, the bow and the string when the bow is released. Since the kinetic energy is distributed to the individual components in proportion to their mass, the most efficient bow is the one which stores the most elastic energy per unit mass.

The elastic energy stored in a unit mass of a material loaded in pure bending is

$$U_m = \frac{1}{6}\frac{\sigma^2}{E\rho} \tag{7}$$

The material loaded in bending will be damaged when the stress σ exceeds the modulus of rupture, σ_{MOR}, of the material of which it is made, giving

$$U_m \leq \frac{1}{6}\frac{\sigma_{MOR}^2}{E\rho} \tag{8}$$

The best choice of material to meet this criterion is the one with the highest value of the material index M_1

$$M_1 = \frac{\sigma_{MOR}^2}{E\rho} \tag{9}$$

There are two more constraints. The first is that the archer must be able to draw the bow fully; the second that it must not break when he does so. The first constraint imposes restrictions on the modulus, E, of the material of the bow; and the second imposes a lower limit on the quantity σ_{MOR}/E. They are defined as follows. We assume that the bow, when fully drawn, describes a quarter-circle of radius $R = 2L/\pi$.

The force, F, required to bend an elastic beam to a quarter-circle, that is to bend each arm through an angle of $\pi/4$, is (from elementary beam theory)

$$\theta = \frac{\pi}{4} \approx \frac{FL^2}{16EI} \tag{10}$$

Assuming, that the cross-section of the beam is square and the second moment of area I is

$$I = \frac{t^4}{12} \tag{11}$$

We find that a bow of thickness, t, and length, L, will be bent to a semi-circle by the drawing force, F, if its modulus has the value

$$E = \frac{3F}{\pi}\frac{L^2}{t^4} \tag{12}$$

Typically, the draw-force, F, for a 'self' bow is about 450 N and the length, L, is about 1.9 m. There are practical limits for t, which we assume to be in the range of 15 to 30 mm. Inserting those values we find the practical limits for E

$$2 < E < 30 \text{ GPa} \tag{13}$$

The requirement that the bow must not break is treated as follows. The strain in the surface of a beam of thickness, t, bent to a radius, R, is

$$\epsilon = \frac{t}{2R} \tag{14}$$

179

and the stress in its outer ligaments is

$$\sigma = \frac{Et}{2R} \qquad (15)$$

The stress, σ, may not exceed the modulus of rupture, σ_{MOR}, if fracture of the bow is to be avoided when the bow is bent to the minimum radius

$$R_{min} = \frac{t}{2}\frac{E}{\sigma_{MOR}} \qquad (16)$$

We require that the bow does not snap when bent to a quarter-circle with the radius $R_{min} = 2L/\pi$. Rearranging equation 16 with this in mind we find

$$\frac{\sigma_{MOR}}{E} > \frac{\pi}{4}\frac{t}{L} \qquad (17)$$

3.2 The Selection

Figure 7 allows selection of materials for bows. It shows the quantity σ_{MOR}^2/ρ plotted against Young's modulus, E. These axes were chosen because they allow both the index M_1 of equation 9 and the limits on E of equation 13 to be shown on a single diagram (broken lines). The additional constant of equation 17 has no influence on the selection, and is not shown. This figure draws on the generic database, containing data for a wide range of material classes.

Table 4: Materials for archery bows.

MATERIAL	COMMENT
Woods, bamboo, palm	Cheap, but properties not easily controlled
GFRP	Good control over properties
Epoxies	Matrix material for GFRP

The selection, identified in the figure, is given in table 4: the best choices are woods, bamboo, palm and GFRP. This suggests exploring the properties of wood more closely. The same selection, using a database of wood properties, is shown in figure 8. It identifies the softwood Yew, the temperate hardwoods Birch, Hornbeam and Walnut, the tropical woods Azobe, Benge, Degame, Greenheart, Rosewood, bamboo and palm. Besides Hornbeam, a wood used for arrow-making, Azobe and Benge, which may be used at least locally, all of these are known as speciality woods for bow-making.

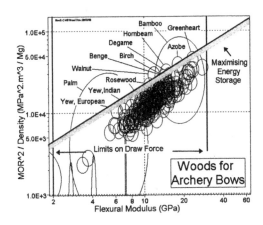

Figure 8: A chart showing the ratio σ_{MOR}^2/ρ plotted against Young's modulus, E, using a database of wood properties. The materials index, M_1, is shown by a diagonal line of slope 1; the limits on E imposed by the draw force, F, are indicated.

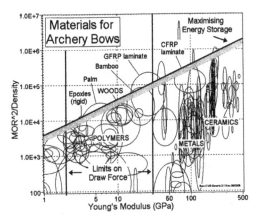

Figure 7: A chart showing the ratio σ_{MOR}^2/ρ plotted against Young's modulus, E, using a generic-materials database. The materials index, M_1, is shown by a diagonal line of slope 1; the limits on E imposed by the draw force, F, are indicated.

3.3 Postscript

The bow-wood of historical fame is, of course, Yew. One feature, which contributed to its success, is that it resembles a composite: the sapwood has a higher tensile strength and a lower modulus than the heartwood, which performs better in compression. The mediaeval bow-makers were aware of that fact; they made bows with an elongated D-shaped cross-section with sapwood at the straight edge of the back of the bow. This way the lower modulus leads to a shift of the neutral axis towards the geometrical centre of the section and the tensile stress is carried by the Yew sapwood.

The mechanical properties of the bulk material described above are not sufficient to select the best material for archery bows. Other properties and features of the woods need to be considered. The wood must be available in sufficient length, straight grained, without defects such as knots and worm holes, since these weaken the wood considerably.

4 MATERIALS FOR VAULTING-POLES

Pole-vaulting dates back to the times when poles were used to cross natural obstacles such as ditches, creeks and fences. From this necessity developed the sport of pole-vaulting for distance which was succeeded by pole-vaulting for height, an Olympic discipline of modern times. One of the last male bastions in athletics, pole-vaulting is also one of the few disciplines in which the make-up of the equipment is not regulated. According to IAAF rule, the pole may be 'made of any material or combination of materials and of any length and diameter'. This leaves immense scope for experimentation and innovative design.

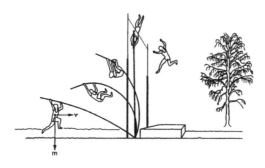

Figure 9: Vaulting-poles are designed on energy storage per unit mass. They must neither break nor buckle when bent to the minimum radius and their stiffness must be matched to the vaulters weight and vaulting technique.

To approach the problem, we need to know, how pole-vaulting is carried out today. The pole vaulter accelerates down a 45 m long runway, carrying the pole with both of his hands and roughly parallel to the ground. He then thrusts the pole into a slot which is set into the ground below ground level beneath the crossbar, pulls himself upwards until he is almost doing a handstand, and twists as he approaches the crossbar to arch over, feet first and face down. The athlete releases the pole just before the fly-over, giving it a push backwards to prevent it from knocking off the crossbar. Finally, the vaulter lands on his back in a cushioned pit. The very best pole vaulters can clear heights of about 6.2 m.

The pole itself is flexible, typically has a diameter of 40 mm and a length of 4.5 - 5.5 m. The latter means that the athlete needs to be catapulted more than half a meter if he is to vault more than 6 m. This pole length range is prescribed indirectly by the mechanics of the sport: it must be long to be of advantage to the vaulter by merely acting as a pivoting column, but not too long, since the athlete has to carry it while running at high speed. For the mechanical analysis of the pole below, we assume that the length of the pole is fixed.

4.1 The Model

From this description we see that, mechanically speaking, the vaulter's pole has the same function as the archery bow: it acts as a leaf spring. The kinetic energy of the vaulter at take-off is stored as strain energy in the pole. It is the release of this strain energy which catapults the athlete across the crossbar. This means that, to optimise the efficiency of the pole, we must maximise the energy it stores and search for the material which stores the most energy per unit mass (energy losses play a secondary role and will not be considered here). The requirement 'per unit mass' is important for two reasons: the first is that the kinetic energy of the vaulter ($U_{kin} = mv^2/2$, where m is the body-weight and v is the take-off speed) increases with a decrease in pole mass; the pole is 'dead' mass, which needs to be accelerated, but does not contribute to the kinetic energy of the athlete's body. The second is that, as with the archery bow, the energy stored in the pole is released directly proportionally to the mass of the pole and that of the vaulter, i.e. the smaller the mass of the pole, the more energy is transferred to the athlete.

The constraint in pole-vaulting is that the bending stiffness of the pole must be tuned to the mass of the athlete; too stiff, and the vaulter's mass is insufficient to bend it; too flexible, and it will bend far too much. Another, if indirect, effect of the stiffness of the pole is that on the time dependent performance of both the pole and the athlete: too stiff a pole leads to too rapid and straight a flight path leaving little time for the vaulter to perform the complex vaulting manoeuvre; a more athletic vaulter may prefer a more flexible pole, since he might take better advantage of its catapult effect than a less agile vaulter who benefits more from the pivoting column effect. The aim is therefore to optimise the stiffness for a given vaulter according to his weight and vaulting technique, to allow for a smooth flight path with minimal energy loss in transfer.

Table 5: The design requirements for vaulting-poles.

FUNCTION:	Vaulting-pole = Leaf spring
OBJECTIVE:	Maximise stored elastic energy per unit mass

CONSTRAINTS:

(a) May not break when bent to minimum radius of curvature
(b) Optimise pole stiffness for individual vaulter
(c) Poles with tubular cross-section may not buckle locally under applied moment

We treat the pole as a hollow tube of length, $L = 5$ m, outer radius, r, typically 18 - 24 mm, and wall thickness, t, which we are free to choose to optimise the stiffness and maximise the energy storage.

We deal with energy storage first. To do so, we derive an objective function describing the maximum elastic energy per unit mass, that the pole can store. The vaulter exerts a moment, M, to the tube, causing it to bend through an angle θ. The energy stored in it is

$$U = \frac{1}{2}M\theta \qquad (18)$$

The moment is related to the angle by

$$\theta = \frac{ML}{EI} \qquad (19)$$

giving

$$U = \frac{1}{2}\frac{M^2 L}{EI} \qquad (20)$$

The second moment, I, of a thin-walled tube is

$$I \approx \pi r^3 t = \frac{\pi r^4}{\phi} \qquad (21)$$

where $\phi = r/t$ measures the slenderness of the cross section (it is the 'shape factor' defined by Ashby (1992)).

The bending, and thus the energy stored in the pole, is limited by failure. The failure moment of the pole is (using equation 21)

$$M_f = \frac{I}{r}\sigma_f = \frac{\pi r^3}{\phi}\sigma_f \qquad (22)$$

where σ_f is the stress at which the tube wall fails. The maximum energy that the pole can store is therefore (using equations 20, 21, 22) given by

$$U = \frac{\pi r^2 L}{2}\frac{\sigma_f^2}{\phi E} \qquad (23)$$

The mass of the pole is

$$m \approx 2\pi r t L \rho = \frac{2\pi r^2 L \rho}{\phi} \qquad (24)$$

where ρ is the density of the material of which it is made. Dividing the energy by the mass gives the energy per unit mass:

$$\frac{U}{m} = \frac{1}{4}\frac{\sigma_f^2}{E\rho} \qquad (25)$$

There are two mechanisms by which the tube wall could fail. The first is irreversible deformation when the stress in the tube wall exceeds its modulus of rupture, σ_{MOR}. The second is local buckling: a catastrophic loss of stiffness when the tube wall stress exceeds the local buckling stress, σ_b (Calladine 1983), given by

$$\sigma_b = 0.3\frac{E}{\phi} \qquad (26)$$

The optimum shape, ϕ^*, for the tube-wall for a given material is that at which failure and local buckling occur at the same stress

$$\phi^* = 0.3\frac{E}{\sigma_f} \qquad (27)$$

Local buckling is a defect-sensitive, catastrophic failure mode. It is normal design-practice to ensure that this mode is not the dominant one. Tubes are, therefore, designed with $\phi \leq \phi^*$, ensuring that failure — should it occur — is by irreversible deformation, not by buckling. This allows σ_f in equations 25 and 27 to be identified with the modulus of rupture, σ_{MOR}. We thus find the index for selecting materials for the pole from equation 25; it is

$$M_1 = \frac{\sigma_{MOR}^2}{E\rho} \qquad (28)$$

Materials with a high value of M_1 have exceptionally good energy-storing capacity per unit mass.

The stiffness constraint is approached by noting that the moment, M, exerted by the vaulter will bend the pole through an angle

$$\theta = \frac{ML}{EI} = \frac{ML\phi}{E\pi r^4} \qquad (29)$$

(where use has been made of equation 21). Thus for the vaulter to bend the pole through an angle, θ, requires that the modulus, E, of the material satisfies the equation

$$\frac{E}{\phi} = \frac{ML}{\theta\pi r^4} \qquad (30)$$

The best choice of material for the pole is thus that which maximises M_1 and which has a value of E/ϕ which lies between limits given by inserting typical values for M, L, r and θ into this equation.

4.2 The Selection

The selection, using the generic database, is made using the diagram of figure 10. Here the axes are $\sigma^2_{MOR}/\phi\rho$ and E/ϕ; contours of constant index M_1 are lines of slope 1, since

$$M_1 = \frac{\sigma^2_{MOR}}{E\rho} = \frac{\sigma^2_{MOR}}{\phi\rho} \frac{1}{E/\phi} \qquad (31)$$

These axes are chosen because they allow both the index and the limits to be plotted on the same diagram. Limits on E/ϕ are found by inserting the typical values $500 < M < 1000$ Nm, $L = 5$ m, $18 < r < 24$ mm and $\theta = \pi/2$ into equation 30, giving

$$1.5 < \frac{E}{\phi} < 10 \text{ GPa} \qquad (32)$$

The selection is shown in table 6: bamboo, GFRP and CFRP are the optimum choices.

4.3 Postscript

It is not surprising that the vaulting technique changed in parallel to the development of pole material and design. In the early times of pole-vaulting the pole was made of solid ash, hickory or cedar and had a relatively high stiffness. It was equipped with a spike at one end, which was dug into the ground as the vaulter planted his pole. In the beginning of the 20th century a slot-like box was sunk into the ground below ground level at the base of the uprights.

From the materials point of view, major improvement in vaulting performance was achieved by the introduction of the bamboo pole in 1904, which, as we saw on the material property charts, is significantly more efficient than wood: lighter for a given stiffness and strength, and with a natural taper which leads to a further weight saving. This taper, of course, is not symmetrical from the centre to the ends of the pole, but from one end of the pole to the other, like one limb of the archers

bow, a less efficient shape than the 'two-limbed' pole.

The bamboo pole was very successful. A short interval of aluminium and steel poles in the fifties and beginning sixties did not improve the performance; the selection chart in figure 10 shows that these metals are less efficient than bamboo. The high stiffness of metal poles favoured speed and power, which, as we know now, are rather disadvantageous.

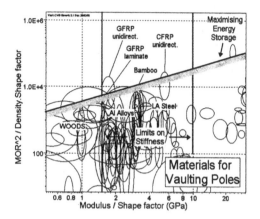

Figure 10: A chart showing $\sigma^2_{MOR}/\phi\rho$ plotted against E/ϕ using a generic-materials database. The material index, M_1, is shown by a diagonal line of slope 1; the limits on E/ϕ imposed by the stiffness requirement are indicated.

Table 6: Materials for vaulting-poles.

MATERIAL	COMMENT
Bamboo	Efficient due to tubular shape and taper, cheap, but properties not easily controlled — the traditional material
CFRP	Good control over properties and easily shaped
GFRP	Good control over properties, easily shaped and cheaper than CFRP

It was the introduction of glass fibre composites in the sixties which brought revolutionary changes in both style and performance. The pole of lower stiffness with a longer relaxation time put more emphasis on gymnastic agility than the more rigid bamboo and metal poles. The new material allowed the vaulter for the first time in pole-vaulting history to perform a rather complex manoeuvre of swinging past the pole whereby he turns upside down and awaits the right instance when the bowed pole recoils to give him his vital lift to cross the bar face down and feet first. The efficiency of the pole was further improved by the introduction of carbon fibre composites in the eighties, which led to a further reduction in pole weight.

What the sporting world did not realise when the tubular bamboo poles replaced poles of solid cross-section, was that they switched to a fibre composite of high structural optimisation. Materials scientist are only beginning to explore the principles of its efficiency now, more than 90 years later, even though, unknowingly, they partly 'reinvented' bamboo's structure and efficiency, first in the form of glass-fibre then in carbon-fibre reinforced polymer poles. How well bamboo performed as both a material and a structure is best shown by the success of those athletes who used bamboo: their records were broken as late as 1957.

5 CONCLUSION

Many types of sports equipment perform mechanical functions: oars, bows and vaulting-poles are examples. In designing and choosing materials to make them, two classes of selection criteria emerge. The first is driven by the wish to maximise some aspect of performance (such as stiffness per unit mass, or energy-storage per unit mass). The second is set by the necessity to tune the response of the equipment to some characteristic of the athlete: the bodyweight (the vaulting-pole), the maximum draw-force of which the athlete is capable (the bow); or simply the 'feel' of the equipment ('the softness' of an oar — probably, again, a matching of muscle stiffness to oar stiffness). In this paper, we have shown how materials selection charts, material indices and material limits can be used to identify candidate-materials which optimally satisfy both classes of criteria. It is part of a general framework for materials selection to meet engineering design goals.

ACKNOWLEDGEMENTS

The authors wish to thank Dr.S.C.Burgess for stimulating discussions on the peculiarities of pole-vaulting and Mrs.S.Mason for skillfully drawing the sports equipment.

REFERENCES

Ashby, M.F. 1989. On the engineering properties of materials. *Acta Metall.* 37: 1273-1293.

Ashby, M.F. 1992. *Materials selection in mechanical design.* Oxford: Pergamon Press.

Brockhaus Enzyklopädie 1986. 19th edition. Mannheim: Brockhaus.

Burgess, S.C. 1996. The design optimisation of poles for pole vaulting. To be presented at The Engineering of Sport, Sheffield, United Kingdom, 2-4 July 1996.

Calladine, C.R. 1983. *Theory of shell structures.* Cambridge: Cambridge University Press.

CMS software 1995. Granta Design Limited, 20 Trumpington Street, Cambridge, U.K.

The new Encyclopaedia Britannica 1974. 15th edition. Philip W. Goetz (editor-in-chief). Chicago: Encyclopaedia Britannica, Inc.

Hardy, R. 1992. *Longbow: a social and military history.* 3rd edition. Sparkford: P. Stephens.

Kooi, B.W. 1994. The design of the bow. *Proc. Kon. Ned. Akad. v. Wetensch.* 97 (3): 283-309.

Redgrave, S. 1992. *Complete book of rowing.* London: Partridge Press.

Rees, G. 1993. The longbow's deadly secrets. *New Scientist.* 1876, 5 June 1993: 24-25.

6 Mechanics

The Engineering of Sport, Haake (ed.) © 1996 Balkema, Rotterdam. ISBN 90 5410 822 3

Mechanics and design of a windsurfer mast

A.J. Barker & J.L. Wearing
The University of Sheffield, UK

ABSTRACT: This paper is focused towards improving the communication barrier between windsurfing mast designers and the recreational sailor. To begin with the idea of an "Integrated Rig Concept" for a windsurfing rig is introduced. The components and the workings of a slalom windsurfing rig are then discussed briefly with special attention given to the mast. A new test rig is then used to provide a more comprehensive alternative to the already available IMCS test. The results from the test rig successfully identified the mast's curve and flex characteristics.

1. INTRODUCTION

Since 1967, when Jim Drake claims to have invented sailsurfing, windsurfing equipment has gone through a fast evolution. The early developments in board design came from the sport's surfing cousin, while sail design borrowed ideas from other wind driven craft such as the catamaran. Modern developments in this exciting new sport, however, rely heavily on the technical disciplines of mechanical and materials engineering. This paper is directed towards the design of a windsurfing rig with special attention given to the mast.

The mast forms the backbone of the rig and has, not surprisingly, the greatest influence on the sail's dynamics. The combination between sail and mast is what the industry calls the "Integrated Rig Concept". The "Integrated Rig Concept" focuses on the creation of a perfectly harmonic functional unit for the fastest and most responsive rig. Selecting a mast that is either too stiff or too soft is a pitfall of many novice windsurfers. With a mind to helping those troubled by their own inexperience, this paper aims to improve the communication barrier between designer and sailor.

The present study takes two different but necessary approaches towards understanding the mechanics of the windsurfing rig. The first stage is to discuss briefly the basic theory behind a windsurfing rig. During this stage a slalom windsurfing rig is to be fully assembled with its main features then analysed and discussed. In order to put numerical values to the theory observations of controlled laboratory experiments are recorded. The Xcess1 test rig, which has been specifically designed by one of the authors (A.J. Barker) of this paper, has been used to investigate the load deflection relationship and the curvature of a slalom-race windsurfing mast.

Personal interest of A.J. Barker and the lack of published work related to windsurfing mast mechanics have provided the motivation for the present study.

2. THE WINDSURFING RIG

Before discussing the mast which is the main focus of this paper, it is important to have an understanding of the workings of a windsurfing rig. The standard windsurfing rig consists of four basic components: the sail, the mast, the boom and the mast foot. Other components such as battens and camber inducers may be added to improve aerodynamic entry to reduce drag. Figure 1 indicates a slalom rigs main features.

The rig forms the engine for the board which is fuelled by the wind. The sail breathes in at the power point and exhausts at the head. The rig is designed to convert any available wind energy into kinetic energy of the whole arrangement by pressure concentration through the sailors legs and the mast foot. The way the sail works is much the same as the aerofoil of an aircraft wing. As the air passes over the sail a driving force is generated.

The sail should deliver an even and constant pulling force, as shown in figure 2, against the sailors centre of gravity for effortless sailing. By taking the power from the sail and driving it against

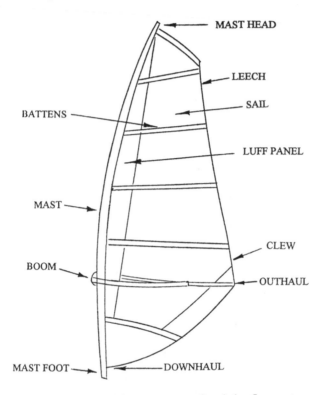

Fig. 1. The components of a windsurfing mast.

The sail works on the basic fluid mechanics theory that if fluid velocity increases, local fluid pressure must fall and vice versa. Because the wind on the leeward side of the sail has further to travel than that on the windward side it has to travel faster. The result it low pressure on the leeward side and the rig being driven forward.

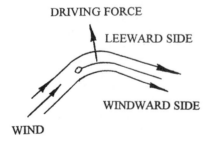

Fig. 2. Wind Velocity and Driving Force.

the boards centre of lateral resistance, the board should begin to move.

The foil is shaped with the greatest camber rotation at the base and boom area and this rotation reduces with the distance from the bottom of the rig until rotation is zero at the top of the rig. At the base the aim is to make a smooth curve with a fine entry into the sail. The tension in the luff panel should be as tight as a drum avoiding any bumps and creases that may cause air detachment. The point of maximum fullness should be at about 35% behind the boom and in front of the surfer. Here, the power is easiest to control and most effectively converted into board velocity.

The top-half of the sail provides a flat exhaust area for getting rid of this power. The air is allowed to escape by the twisting open of the leech of a powered up sail. This twisting action begins just above the boom to a maximum at the head. Modern race sails, for example, are designed for a large amount of twist, because modern racers want to stay powered up over varying wind conditions.

The rig can be trimmed by adjusting boom height and fine tuning the sail tension. Individual trimming is essential to achieve the best shaped foil in order to give easy handling and above all maximum velocity. The trim of all components depends on the sailor's body weight, sailing conditions and the individual's sailing style. A rig trimmed for a high lift and a low drag foil shape will result in good draft stability. A lack of tension will cause the draft and centre of effort to move during gusts making the sail unstable. The general rule is to give a large amount of downhaul tension for a flatter sail profile and more leech twist. This improves control in high winds and increases the top speed. Reduce downhaul slightly for a fuller sail. Profile with less dynamic twist but a tighter leech. This increases power, acceleration and improves overall light wind performance. By adjusting the outhaul and downhaul tension the power in the sail can be changed.

The geometry, materials and the manufacturing process determine the flex characteristics of a mast and hence the way a sail is to handle. For easy handling the mast is engineered for the tip to flex when hit by a gust while maintaining the balance of tension in the sail. The faster the mast can return to its original trimmed position, the longer the rig is able to hold the ideal profile shape, the more propulsion and thus velocity can be produced. A mast with well engineered reflex dynamics allows the rig to convert gusts directly into forward thrust by a quick transmission of driving force. While preserving the mechanical properties of the mast designers are continually trying to build masts with reduced diameter. This is to even the airflow over the sail and reduce the drag.

Modern masts are made from the three main materials of glass, aluminium and carbon. Carbon is becoming increasingly more popular because of its light weight and good flex characteristics. Some masts even contain kevlar for improved impact resistance. However, masts are in general made from materials by using the best compromise of flex, low centre of gravity, strength, weight and also cost. The way the masts are manufactured also varies enormously. Filament winding and cross carbon construction technology are to name but two.

The following section is devoted exclusively to the design of this very important component. The section uses the Xcess1 test rig to investigate, and then attempts to explain, the mast's design features.

3. TESTING THE MAST'S MECHANICS

The new IMCS (International mast check system) provides a general guide to the mast's rigidity. Simply the higher the IMCS number the stiffer the mast. Modern masts range from an IMCS number of about 20 to about 32. The IMCS value of a mast can be found by attaching a 30kg load to the midpoint. The measured deflection at the centre is then divided by the mast's effective length to give the IMCS value. The IMCS also gives the flex characteristics as a percentage of central deflection at the ¼ and ¾ points. Unfortunately the IMCS test lacks realism and numerical quality for a study of mast mechanics.

The mast is to serve a fast changing dynamic environment with varying loads. The Xcess1 test rig has been developed to provide a more comprehensive alternative to the IMCS. A schematic diagram of the test rig can be seen in figure 3.

By adjusting the tension in the ropes it is possible to account for the changes in sail pressure that occur over a wide range of sailing conditions. The tension-key arrangement has been carefully located so as to simulate, under the controlled conditions of the laboratory, changes in pressure at the sails centre of effort. In order to represent low wind pressure the test rig is set at a low tension mark. As wind pressure on the windward side increases the tension in the ropes is increased.

The Xcess1 test rig makes it possible to simulate different wind loading conditions and to observe, measure and record numerical data regarding the mast's load deflection relationship for these varying operating conditions. This allows the operator to determine the best bend curve for the different types of sail. The test rig is easily adapted to the simulation of different sized sails and masts by adjusting the clew and rope lengths.

A Gaastra racing mast from a 6.2m^2 sail was used for the test. The test rig was set up for a boom length of 186cm and a luff panel of 487cm. The mast length is 460cm, weighing 19.8N with a carbon content of 70%. The IMCS value was determined to be about 27.5. This indicates that the mast is relatively stiff, an expected characteristic of racing masts. In order to capture the working characteristics of the mast, a simulated load of 300N to 450N, is imposed at the artificial centre of effort. The bend curve of the mast can be adequately defined by taking angle readings at each of the contact points. The deflection at the tip was also measured. The results of this simulation are

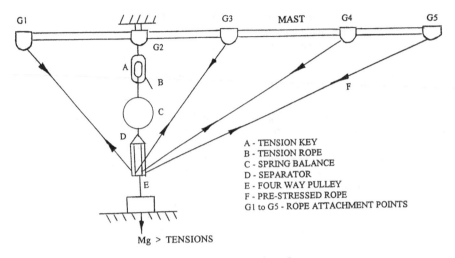

A - TENSION KEY
B - TENSION ROPE
C - SPRING BALANCE
D - SEPARATOR
E - FOUR WAY PULLEY
F - PRE-STRESSED ROPE
G1 to G5 - ROPE ATTACHMENT POINTS

Mg > TENSIONS

Fig. 3. The Xcess1 test rig and components.

shown in Table 1 and figure 4 in the following section.

4. THE RESULTS TO THE TESTS

Table 1 shows the resulting tip deflection with corresponding flex characteristics at the rope attachment points (to the mast) as a percentage of the flex at point G2. Figure 4 shows the results in graphical form.

LOAD N	Tip Deflection (mm)	CURVE %				
		A	B	C	D	E
0	20	100	100	100	99	97.5
300	170	77.5	100	91	89.5	88
350	225	75	100	89	87	85.5
400	250	73	100	87.5	85	84
450	270	71	100	86.5	83.5	82.5

Table 1. Tip deflection and curve percentage at the nodes.

5. DISCUSSION OF THE RESULTS AND THE MAST'S OBSERVED BEHAVIOUR

Under zero induced loading the mast tip was visibly deflected by its own weight to about 20mm. Table 1 shows that the mast seems to deflect relatively freely to begin with under the light loading. As this loading increases the mast becomes progressively more resistant to deflection. This particular design feature is so that the mast can be easily flexed to match the curve of the sails luff. After rigging, this behaviour quickly decays so that the mast can support the foil shape and balance of tension in the sail.

Above the boom the mast gently curves through attachment points G3 to G5. This helps to achieve a flat exhaust area for the air exiting from the foil. The upper end of racing masts are designed to be especially stiff to improve the rig's response to gusts and to help the rig stay powered up over varying wind strengths. For some recreational sails this mast is probably too stiff in this upper section. When rigging up, the head battens in the exhaust

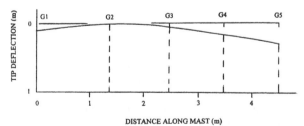

Fig. 4. Mast's deflected shape at a load of 450N.

190

area will tend to stick more easily. A large amount of downhaul may reduce this sticking sufficiently. If, however, the battens lower down are sticking, the best remedy is to try a softer mast.

From the results in figure 4 it can be seen that the mast is shaped with a significantly greater bend at the base. The reason for this large curvature at the base of the mast is to help the foil to adapt a deep powerful profile at the sails centre of effort. A difference of about 10% between top and bottom ends seems to be maintained throughout. For example, based on a maximum centre bend of 100% the mast bottom bend is 71% and the mast top bend is 81%. The almost constant difference could provide a useful way, in addition to the IMCS value of comparing, masts compatibility with sails.

When powered up the mast is required to bend to keep the leech tight with a flat head. It is this elasticity and/or stiffness in the mast and sail which enables the board to be controlled in chop and gusty winds. A mast with a correctly selected stiffness helps the sail to breathe and also eliminates upper leech stall. There are many different types of mast available so as to cover the whole performance spectrum. For example, race sails require a stiff mast for optimum racing efficiency where the wind may vary two or three wind forces in strength. However, by using a stiff mast in the smaller wave sails power could easily be lost during lulls in the wind with an overall unstable feeling. Additional design aims include an extremely high breaking strength, reduced overall mast deflection in high wind load and chop, while still maintaining the necessary flex to match the curve of the sails luff. The mast must also be able to withstand the loading created in the high stress around the tack, foot batten and clew areas of the sail.

6. CONCLUSION

The workings of a windsurfing rig and the idea of the "Integrated Rig Concept" to provide a perfectly functional harmonic unit has been discussed. The Xcess1 test rig was successfully used to simulate the loading on a windsurfing mast. The bottom and top ends have different flex characteristics. These must be able to maintain the balance of tension in the sail and to hold the rig's profile. In general the mast is designed for a responsive bottom end with a firm stable top end. The testing showed that the larger the sail and the heavier the sailor the stiffer the mast needs to be. The message is not only should the bend curve of the mast suit the sail's luff,

but the stiffness should suit the intended type of windsurfing (e.g. Slalom, race and wave).

Based on the work discussed in this paper, a technique has been developed which deals with all the necessary variables to enable the most effective mast to be selected to suit the requirements of the board's operating conditions.

The Engineering of Sport, Haake (ed.) © 1996 Balkema, Rotterdam. ISBN 90 5410 822 3

Friction coefficient of golf balls

W.Gobush

Titleist and Foot-Joy Worldwide, Fairhaven, Ma, USA

ABSTRACT: Tangential and normal forces of golf balls were measured at highly lofted impact angles using piezoelectric force transducers. An air cannon was used to eject golf balls at the force transducer surface mounted almost parallel to the incoming golf ball direction. Two types of golf ball constructions used in the playing of golf were measured to determine the dynamic coefficient of friction on a grooved surface. These measurements were compared with the photographic analysis method for obtaining the dynamic coefficient of friction. A one-camera system was used to image capture the golf ball at two positions before and after impact with the force transducer. From the position measurements of markers on the golf ball, spin rate and ball velocity were measured before and after impact. Using the rigid body theory of impact, the resulting dynamic friction is calculated from these photographic measurements. A comparison of the two methods for measuring the friction coefficient and the effect of velocity on the resulting spin rate and friction coefficients of the golf balls are discussed in this study.

INTRODUCTION

The measurement of friction between golf club and golf ball is important in assessing the spin characteristics of a golf ball for a high lofted iron. In this range of golf swings, the ball is sliding throughout the period of impact with the club. In this study, we compare two methods for measuring the coefficient of friction between ball and grooved clubface and how it is affected by the relative speed of the collision.

1 INDIRECT METHOD OF MAW

As discussed in a paper by N. Maw et al (1981), the coefficient of friction can be calculated by measuring the ball velocity parallel and normal to the clubface surface just before and just after impact. Using this model, one obtains the following equation for the coefficient of friction, μ:

$$\mu = \frac{V_x^A - V_x^B}{\left(V_z^A - V_z^B\right)\left(1 + 1/K^2\right)} \quad (1)$$

in which $K^2 = I/MR^2$ where

I is the moment of inertia of the golf ball

M is the golf ball mass and

R is the radius of the golf ball impacting the plate.

V_x^B, V_x^A = tangential speed of ball before and after impact.

V_z^B, V_z^A = normal speed of ball before and after impact.

In the experimental setup shown in Figure 1, balls are fired from the air gun through a small aperture in the box and rebound from the steel block which is set at a 20° slant to the flight direction of the ejected golf ball. This gives a grazing collision between the ball and club insert which is mounted to a force transducer.

Using stroboscopic methods, an electronic CCD camera captures the view shown in Figure 2 and the position of seven markers on the golf ball are automatically acquired by computerized video analysis. From the known time interval between strobe firings and the positional information on the golf ball at locations before and after impact, the velocity components of the ball can be calculated and the friction coefficient calculated using equation 1. A mathematical description of a one-camera photogrammetric analysis of spin and ball velocity is described in U.S. Patent 5,471,383, Gobush et al.

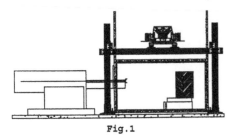

Fig.1

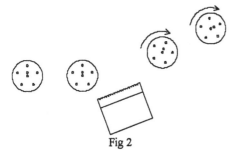

Fig 2

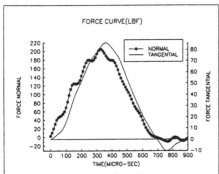

Fig 3

In this study, a wound balata-covered liquid center golf ball was compared with a hard covered two piece ball. These represent the two extreme construction types in golf ball design. Two different incoming velocities were used in this study to see the dependence on initial surface speed.

2   DIRECT FORCE MEASUREMENT

As described in an earlier paper (W. Gobush, 1995), a piezoelectric force transducer can be used to make high speed measurements of forces acting normal and tangential to a sur-

face. By time averaging the normal and tangential forces and taking the ratio of these timed average forces, the coefficient of friction can be calculated. In Figure 3, we see the shape of these curves as sampled for the balata wound ball at 42 feet per second. In Table I we list measurements of the coefficient of friction by this method along with the indirect method described in section 1.

TABLE I

Ball Type	Incoming Tangential Speed (FPS)	Friction Coefficient (Photo-graph)	Friction Coefficient (Trans-ducer)	Spin Rate (RPS)
Wound Balata	88	0.28 ± 0.02	0.30 ± 0.02	99.5
Two Piece	85	0.073 ± 0.002	0.077 ± 0.002	25.2
Wound Balata	42	0.37 ± 0.03	0.38 ± 0.02	66.2
Two Piece	42	0.16 ± 0.02	0.16 ± 0.02	25.5

3   DISCUSSION OF RESULTS

The utilization of piezoelectric transducers for measuring friction coefficients has been found to compare well with the photographic method of measurement. The coefficient of friction is much higher in wound balata golf balls due to the softness of the cover and is dependent on impact speed..

ACKNOWLEDGEMENTS

The author thanks Bill Morgan of the Acushnet Company for his support and encouragement on this project. Computer support by programmer Diane Pelletier and camera system design by Charles A. Days is greatly appreciated.

REFERENCES

Maw, N., Barber, J.R., and Fawcett, J.N., The Role Of Elastic Tangential Compliance In Oblique Impact, Journal of Lubrication Technology, Vol. 103, 1981, pp. 74-80

Gobush, W., "Spin and the Inner Workings of a Golf Ball", Golf The Scientific Way, edited by A.J. Cochran, Aston Publishing Group, 1995, pp. 141-145.

The Engineering of Sport, Haake (ed.)© 1996 Balkema, Rotterdam. ISBN 90 5410 822 3

A motion based virtual reality training simulator for bobsled drivers

R.K. Huffman
NASA Johnson Space Flight Center, LinCom Corporation, Houston, Tex., USA

M. Hubbard
University of California, Davis, Calif., USA

ABSTRACT: A bobsled training simulator developed at UC Davis during 1991-1995 and used to train the US bobsled team for the 1992 and 1994 Olympics is described. It is similar to flight simulators used to train commercial aviation pilots, and addresses four human sensory systems; visual, vestibular, tactile and auditory. The synchronization of the various engineering subsystems which furnish these four sensory inputs provides much of the design challenge, since time and other mismatches between them can result in a non-realistic experience, ineffective training and even motion sickness. The fact that everything must be computed in real time means that virtually any variable of interest can be chosen for feedback to the driver to be used in improving driving strategies. Because of the large expense of training at real tracks, the simulator can be a cost effective means of providing large amounts of training, even during the summer, and of practicing risky driving strategies in absolute safety.

1 INTRODUCTION AND MOTIVATION

Over the past two decades, the sport of bobsledding has seen many technological improvements. Most of these have been related to increases in the aerodynamic and frictional efficiencies of the sled and crew. In addition to technical efficiency, however, two distinct contributions of the team members are important to a winning time. These are labeled the push phase and driving phase, and are shown schematically in Fig. 1.

1.1 The bobsled competition

The push phase begins before the clock starts with the driver, pushers (one or three in the two- and four-man events, respectively) and sled 15 meters behind the start line as shown in Fig. 1. After receiving clearance from the start house the team accelerates the sled to give it as much initial velocity as possible. Official timing of the run begins when the sled nose crosses the start line but

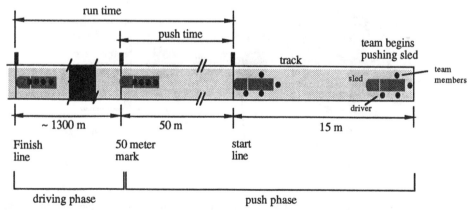

Figure 1 - Two phases of team performance. The push phase begins 15 meters behind the start line and ends at the 50 meter mark, where teams must be in the sled. Push time is defined as the time from the start line to the 50 meter mark. The driving phase extends from the 50 meter mark to the finish line, during which time the only active participant is the driver. Total run time is measured from start line to finish line.

the team continues to push the sled until no longer able to accelerate it. According to the rules, the entire team must be in the sled before it has traveled 50 meters past the start. The brakeman for elite teams usually enters the sled at about the 40 meter mark

The standard measure of push phase performance is called the *push time*, defined as the time required to travel from the start line to the 50 meter mark. Representative push times are around 5 or 6 seconds, but the difference between good teams and bad ones can be 1 second or more.

Although a good push is vital to a world class run, what happens during the driving phase is equally important. During this phase only the driver, who steers, is active. It is essential that the driver steer a "fast line" to take advantage of a good push.

A typical sled, shown in Fig. 2, has four runners; two fixed in the rear, and two in the front whose steerable yaw rotation causes them to experience a lateral component of the ice friction force. Even though this is used to control sled heading and lateral position, bobsleds are relatively uncontrollable. Ullman and Cross (1979) have estimated that the lateral friction coefficient for a bobsled runner is only approximately 0.07.

1.2 Motivation for a simulator

Although inexpensive and safe methods for power and speed training for the push phase have allowed the performance of the US team to increase steadily over the years, driver training is more problematic. Indeed, the main motivation for a bobsled training

simulator is financial. Because the track at Lake Placid has been forbidden for international competitions by the FIBT, the US team must travel to the track in Calgary, Canada, at prohibitive cost. It has been estimated (Roy, 1995) that each training run by the US Bobsled team at the Calgary track costs roughly $700. Even when there is a chance to train on real ice, logistics limit the number of actual training runs to 3 or 4 per day. A second important motivation is safety. Although a simulator cannot replace on-ice training, it can provide as many practice runs as desired without the expense and danger of an actual track.

Another motivation of simulator training is that quantitative feedback data can be made available to the driver both during and after the run, allowing the driver to better understand and correct his or her mistakes. A simulator can also, in theory, re-create any course in the world.

Of course these benefits can be realized only if the simulator is realistic enough. The driver in the simulator must feel, insofar as possible, as if he were in a real sled going down a real track. And when a driver steers the simulator, it must respond as a real sled does.

1.3 Simulator overview

The simulator explained in this paper is analogous to flight simulators which have been used successfully by both the military and commercial airlines for decades. In those, and in ours, an attempt is made to "convince" as many of the pilot's pertinent sensory organs as possible that he is indeed flying an airplane or, in our case, driving a bobsled. It is generally

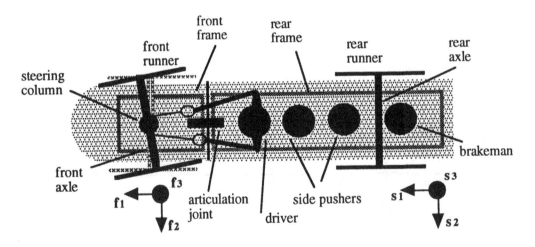

Figure 2 - Steered four-man sled. The sled body is made up of a rear frame in which the team sits and which is supported by four runners which ride on the ice. The two rear runners are attached to, and are free to rotate about the long axis of, the rear axle. The rear axle is fixed to the rear frame of the sled. The two front runners are attached to, and are free rotate about the long axis of, the front axle. The front axle can be rotated by the driver through steering ropes about f_3 thereby steering the sled.

accepted that the two most important sensory systems for the simulator to satisfy are the pilot's visual and vestibular (motion sensing) systems (Meiry, 1966; Reid & Nahon, 1985). In fact many useful simulators are completely visual, one example of which is the fixed-based bobsled simulator described by Huffman, Hubbard, and Reus (1993), which was used in preparation for the 1992 Winter Olympic Games.

Tactile input can also be critical if the pilot, or driver, is expected to perform tasks that require "feel," such as steering a car or bobsled. Tactile input is also significant as it relates to vehicle motion. It is important that the forces which move the driver in the simulator are closely related to the forces that would create actual driver motion in the real vehicle. In the bobsled, forces on the driver are applied mostly in the seat; a simulator that moved a driver by pulling on his arms would be ineffective.

Where sound contains significant information, such as the noise of an car engine, it can be very important. However, sound only for the sake of realism is thought to be of little benefit in feedback, but not detrimental (Young, Kenyon, & Hollister, 1988).

The simulator described below attempts to satisfy the drivers visual, vestibular, tactile, and auditory systems. The visual system is addressed through computer generated graphic scenes which would be seen driving down the real course. The vestibular inputs, or "motion cues," are provided by motion of the simulator cockpit in pure roll which, while different from that of the real sled, is designed to make the driver's vestibular system react as much as possible as if it were experiencing the real motion.

Tactile inputs are simulated in two ways. First with regard to motion, the cockpit of the simulator was designed to be ergonomically as much like a real bobsled as possible. This places the contact points between the driver and the sled, and therefore the forces that move him, in locations that are consistent with the real sled. Second, the tactile sensation of the steering mechanism of the simulator was made to "feel" as much the like the real sled as possible through the inclusion of a force feedback system simulating the ice-runner interaction forces that would be perceived through the steering ropes.

Finally, sound as recorded in an actual bobsled run is played in the cockpit. Although the benefit from the audio input is probably small, the expense of time and money to provide it was minimal.

The remainder of this paper presents some details of these efforts. First we discuss dynamic modeling of three dimensional bobsled motion. Next, simulator motion theory, which has been developed mostly in relation to aircraft simulators, is related to bobsled motion. Justification for limiting the motion of the bobsled simulator to one degree-of-freedom (roll) is presented. Hardware design and implementation of the simulator are discussed, including the motion base, simulator cockpit, and force feedback system hardware. Finally an evaluation of the simulator's usefulness, including driver evaluations, is presented.

2 BOBSLED DYNAMICS

In this section we discuss briefly the dynamic equations of motion which are integrated in real time in the simulator to describe the motion of the sled. Of particular interest are translational and rotational trajectories of a sled reference frame labeled S, which represents the passenger compartment of the sled, and is characterized by three body fixed unit vectors, s_1, s_2, and s_3 as shown in Fig. 2 It is assumed that the motion of the sled reference frame is synonymous with motion of the driver. Thus S determines the driver's eye point location and view direction (assumed to be along the s_1 axis), and the angular and translational motion that the driver's vestibular system would experience.

Until recently, the only models of bobsled dynamics were one-dimensional (1D) in nature. Baumann (1973) developed a single differential equation describing sled position and velocity. Ullman and Cross (1979) developed another 1D model based on a curve-fit of measured velocities at the 1978 World Championships in Lake Placid. These models were used to evaluate the effect of various factors on sled performance and, while insightful, concerned themselves only with along-track position and speed.

Hubbard, Kallay, and Rowhani (1989) presented the first 3D model of bobsled motion, in which the sled was modeled as a single particle, and used it to calculate details of the motion on a single turn at the Calgary track. Huffman, Hubbard, & Reus (1992) used an enhanced version of the particle model in a fixed based version of the simulator used by the US Bobsled Team during the 1992 Albertville Olympic Games. Although four models of increased sophistication have been developed subsequently, in this paper we will focus on only the particle model. First however we review the track description which is common to all sled dynamic models.

2.1 Track description and parameterization

In order to locate only those points which are actually on the track we characterize the track surface using two independent dimensionless parameters a and b; a increasing from zero at the start in the along-track direction and b varying in the cross-track direction. In terms of a and b, r can be written as

$$r = r(a,b) \qquad (2.1)$$

where every vector $r = [x(a,b), y(a,b), z(a,b)]$ within the parameter ranges $0 < a < a_{max}$ and $0 < b < b_{max}$ corresponds to a point on the track surface.

Next a grid of points known to be on the surface is generated by the intersections of lines of constant alpha (a_1, a_2, ... , a_n) and constant beta (b_1, b_2, ... , b_m). Although older "natural" tracks (e.g. St. Moritz) are built anew from snow each year and have little if any documentation of their shapes, newer refrigerated tracks are constructed from drawings which specify the shape and location of many cross-sections along the track, lines of constant alpha. Similar information could be obtained for natural tracks by surveying. From the shape and location information of the known cross-sections, points of constant b at that value of a and their inertial coordinates can be determined. The grid is complete when every b_j on each a_i is associated with its inertial coordinates $x(b_j, a_j)$, $y(b_j, a_j)$, $z(b_j, a_j)$.

Standard bicubic spline interpolation routines can then be used to approximate the inertial coordinates of r at any and every alpha and beta in the ranges $0<a<a_{max}$ and $0<b<b_{max}$. The spline function is a product of cubic polynomials in a and b which matches the position, slope, and curvature at each grid point. The coefficients of these polynomials change in each patch, a rectangular region in a,b space bounded by adjacent lines of constant a and b. Thus the spatial derivatives, r_a, r_b, r_{aa}, r_{bb}, and r_{ab}, (here r_{ab} denotes the mixed partial derivative of r with respect to a and b, $\partial^2 r /\partial a\partial b$) at any point on the surface, can be determined easily by differentiation of the polynomial with respect to a and b. These partial derivatives play a large role in the dynamic equations. An interpolation accuracy analysis shows that, by including the side walls in the surface to be approximated, the track approximation error using bicubic splines can be reduced to less than 0.1 mm.

2.2 Particle dynamic model

The particle model of Huffman, Hubbard, and Reus (1992) is a two degree-of-freedom, four state (a, b, da/dt, db/dt) dynamic model for the sled based on a particle which is constrained to remain in contact with the track surface. This implies that the mass center of the simulated sled is on the track surface rather than at a point above the surface as occurs on the real sled. The error induced by this assumption has been shown to be insignificant.

By writing the translational velocity of the particle computed by differentiating the position vector with respect to time as

$$v = dr/dt = u_1 r_a + u_2 r_b \qquad (2.2)$$

where $u_1 = da/dt$ and $u_2 = db/dt$, the acceleration of the particle, a, can then be obtained by direct differentiation as

$$a = dv/dt = du_1/dt\, r_a + du_2/dt\, r_b$$
$$+ u_1{}^2 r_{aa} + u_2{}^2 r_{bb} + 2u_1 u_2 r_{ab}. \qquad (2.3)$$

Six forces are assumed to act on the particle (Hubbard, Kallay, and Rowhani, 1989). These are W = weight, L = aerodynamic lift, D= aerodynamic drag, N = normal force, C = Coulomb ice friction, S = steering force, exact expressions for which are omitted here for brevity. Finally, using Kane's method (Kane & Levinson, 1985) the equations of motion can be shown (Hubbard, Kallay, & Rowhani, 1989) to be

$$du_1/dt = \frac{(F_1 + A)r_b{\cdot}r_b - (F_2 + B)r_a{\cdot}r_b}{m((r_a{\cdot}r_b)^2 - (r_a{\cdot}r_a)(r_b{\cdot}r_b))} \qquad (2.4)$$

$$du_2/dt = \frac{(F_1 + A)r_a{\cdot}r_b - (F_2 + B)r_a{\cdot}r_a}{m((r_a{\cdot}r_b)^2 - (r_a{\cdot}r_a)(r_b{\cdot}r_b))} \qquad (2.5)$$

where:

$$F_1 = (W + L + D + C + S){\cdot}r_a \qquad (2.6)$$

$$F_2 = (W + L + D + C + S){\cdot}r_b \qquad (2.7)$$

$$A = -m\left(u_1{}^2 r_{aa} + u_2{}^2 r_{bb} + 2u_1 u_2 r_{ab}\right){\cdot}r_a \qquad (2.8)$$

$$B = -m\left(u_1{}^2 r_{aa} + u_2{}^2 r_{bb} + 2u_1 u_2 r_{ab}\right){\cdot}r_b \qquad (2.9)$$

The two additional kinematical differential equations are

$$da/dt = u_1 \qquad (2.10)$$

and

$$db/dt = u_2. \qquad (2.11)$$

Equations 2.4, 5, 10, 11 are the four first-order differential equations which are numerically integrated in real time in the simulator to describe the translational trajectory of the particle and the rotational motion of the sled reference frame. The angular position, velocity, and acceleration of the sled reference frame will also be required in the simulator's motion base calculation. Each can be shown to be calculable given the four states a, b, da/dt, db/dt. Although we have here, for brevity, focussed exclusively on the particle model, four additional dynamic models of increasing sophistication are available in the simulator, each with its advantages and limitations. The most complex of these has two rigid bodies which rotate about a roughly horizontal articulation axis as does the real sled.

2.3 Model verification using actual data

As a measure of the accuracy of the model we compare recently acquired data from actual runs

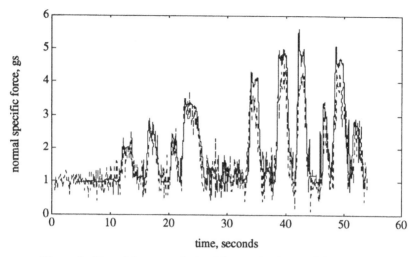

Figure 3 - Normal force as calculated from a simulator training run (solid),
and data from an actual run (dashed) at Lillehammer.

taken on the Lillehammer Track. This data included three-axis accelerations, speed, steering angle, and articulation angle, in addition to other variables, such as runner shoe deflection, that are not included in any of the simulator models. For brevity only normal force is presented here. Figure 3 shows the normal specific force calculated in a simulator training run (solid), and the actual normal specific force (dashed) recorded using an onboard accelerometer. With some slight adjustment to the initial time of the recorded data, which was not exactly specified, the plots can be seen to be well synchronized.

Note that both the simulated data and the real data show short spikes of normal specific force below 1 g. In the simulated data this is caused by imperfections in the track approximation which create small areas of negative curvature. What causes these spikes in the real data is open to question, the possibilities being random noise, or normal accelerations of the sled bouncing down the track. In either case, it is important that the normal specific force never reaches zero, indicating that the sled remained in contact with the ice at all times. This lends credibility to the fundamental on-track constraint common to all of the dynamic models.

Also of note is that in the lower g turns, the plots are not only time synchronized but also match very closely in magnitude. However in the higher g turns towards the bottom of the course, the recorded specific force is slightly lower. Without knowing the specifics of the data collection system and its limitations, it is difficult to say which is correct. Because the normal force calculation is relatively straightforward we are tempted to suspect errors in the data collection.

While the verification of the model presented in this section is the first of its kind, it is far from

complete. It is hoped that in the future, more extensive, higher quality data can be obtained from actual runs. Ultimately, a method of measuring the actual position of the sled on the track, and its orientation, is needed. This might be accomplished by somehow measuring the distance of the sled from the side wall of the track.

3 SIMULATOR DESIGN

The simulator is a self-contained motion-based training device which allows US Bobsled Team members to practice driving skills. The intended users range from novice drivers trying to learn the basics, to members of the US Olympic Team fine tuning their elite techniques. US Team members have used the simulator to prepare for both 1992 and 1994 Winter Olympic Games, and it will soon be installed permanently at the Olympic Training Centers in Lake Placid, New York and Salt Lake City, Utah. Figure 4 shows the simulator in schematic form.

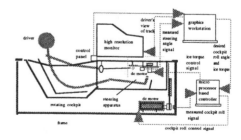

Figure 4. Bobsled simulator schematic

199

3.1 Steering mechanism

The driver steers the simulator through ropes and "D-ring" handles just as in a real sled. Pulling on the ropes in the real sled rotates the front runners in yaw thereby turning the sled. In the simulator pulling on the ropes turns a gear connected to an optical encoder. The steering signal from the optical encoder is fed into the computer and used to compute the steering force in the simualtion.

The steering mechanism is also connected to a dc motor designed to generate the torque (called the "ice torque") which would be imposed on the real steering column by the ice-runner interaction. The ice torque is calculated in the computer and relayed to a microprocessor based controller which controls the current to the motor.

3.2 Visual interface

The driver interacts visually with the simulator through the computer monitor mounted in the cockpit. This monitor presents the driver with many types of information, the most important of which is the graphical representation of the bobsled track and surroundings. The scene corresponds to what would be seen by the driver from the moving sled and is generated based on the position of the sled calculated by the dynamic simulation. The drivers eyepoint is calculated as a fixed point in the sled reference frame. Since this visual scene is driven by the dynamic simulation, it therefore responds to steering commands and track dynamics. The graphic software used to render the images was based on the computer manufacturers proprietary graphics library, and was optimized to allow the scene to be generated in real time (Huffman, Hubbard, & Reus, 1992).

In addition to graphical images of the track, the computer monitor in the cockpit gives the driver performance feedback. Using a menu system controlled by a button control panel mounted in the cockpit, the driver can access simulator performance data. This includes graphical portrayal of steering, speed, and position during the run, and how those variables compare to those on the best simulated runs. Thus the simulator serves as a powerful learning tool. The button panel and menu system can also be used to modify simulation options such as changing tracks and dynamic models, and of course to start the simulation. At present four tracks are supported; Calgary, la Plagne, Lillehammer and Salt Lake City.

3.3 Motion platform design

In addition to the visual scene presented to the driver, the simulator cockpit was designed to roll in response to the roll calculated by the equations of motion. Roll motion was chosen because it is the dominant angular motion of the sled as can be seen in Fig. 5. Because roll only motion will give false tilt cues, causing a misalignment of the apparent g vector (always felt through the seat in a bobsled), a "washout" filter was used to limit the motion to onset ques at the entrances and exits of turns (Meiry, 1966).

3.4 Cockpit design

The welded aluminum cockpit frame has the same shape as a typical sled frame. This is important to the realism of the simulator because the drivers sit directly on the frame in the real sled. Additionally, the entire cockpit was covered with a fiberglass shell (not shown in Fig. 4) which was created using a mold for actual bobsled shells.

The shell serves two purposes. First it enhances the realism of the environment in the same manner as the shape of the cockpit frame. Its importance to the realism of the motion simulation is much greater, however. Recall that simulator motion is not the full six degree-of-freedom motion of the real

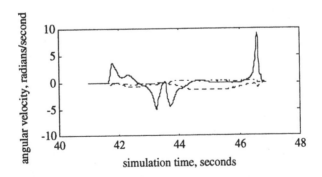

Figure 5. Simulated roll (solid), pitch (dashed), and yaw (dot-dashed) angular velocities of sled on turns 14 and 15 at Lillehammer.

sled, but rather a filtered roll-only motion designed to make the driver's vestibular system perceive full bobsled motion. Paramount to the effectiveness of this illusion is that the driver's visual sense be presented with a scene consistent with the actual motion. If the cockpit were not surrounded by an opaque shell, the driver would see the fixed inertial environment, and the realism would be greatly diminished because the visual scene would take precedence.

Cockpit roll motion is accomplished using a dc motor attached to the steel frame supporting the cockpit. The motor drives the cockpit to the desired roll angle through a belt drive, in closed loop. The computer calculates the roll angle from the equations of motion detailed in Section 2, processes this signal with the washout algorithm discussed above, and communicates the processed roll signal to the microprocessor based controller.

4 SIMULATOR EVALUATION

Ideally, the simulator should be judged by a demonstrated improvement in performance on the real track that is directly attributable to training on the simulator. Unfortunately, due to many factors, this is a nearly impossible task.

4.1 *Judging performance in an actual sled*

The inability to distinguish the driver's improved performance is a result of a number of largely uncontrollable influences in the actual event. These are unaffected by the driver's steering and therefore obscure his contribution. While it is true that the driver must steer well to have a fast time, one can steer perfectly and still have a poor run due to other factors such as the push time, the friction and aerodynamic inefficiencies of the sled, and even the condition of the ice at the time of the run. All of these influences are analogous to noise on a measured signal. In this case it is nearly impossible to separate the noise from the signal.

Another issue which makes driver performance evaluation difficult is the lack of data available. As discussed previously, the quantitative data available on a real bobsled run is typically limited to the push time, the final time, a speed at one point on the track (usually somewhere near the bottom of the course), and a small number of split times. From this limited data it is very difficult to pinpoint quantitatively where driving mistakes were made.

All of this seems to indicate that there is little understanding of whether or not a driver has driven well. However, the indications of a bad run are fairly obvious. These include wall collisions, large oscillations in turns, and oversteering to avoid trouble.

If the run is clean, however, i.e. free of the above mentioned pitfalls, then judging the difference between a good run and a great run is much more difficult. In fact with a clean run but a poor finish time, there is a tendency, with some rationalization, to blame the equipment or the condition of the ice for the poor performance. While there exist volumes of anecdotal speculation on fast lines down a track and other driving strategies, there is very little quantitative evidence to support any of them.

4.2 *Judging performance in the simulator*

Evaluation of a driver's performance on the simulator is quite a different story. The simulator has only one controllable influence on performance, the driver's steering. All other factors (i.e., friction, aerodynamic efficiency, etc.) are held constant. Furthermore, virtually every variable of concern can be recorded and used for post-run analysis. This is a monumental advantage of training drivers on the simulator compared with the real track. When a driver has a subpar time on the simulator it is certain that this was due to driving and that adjustments are required to improve the time. This is in marked contrast to the real sled where after a bad time one can only speculate on the cause.

Because the effect of simulator steering has been completely decoupled from the other uncontrollable factors mentioned above, the success rate for drivers is universally high in being able to learn quickly how to drive the simulator to a fast time. Figure 6 shows driver performance during a typical training session and illustrates significantly improved performance after only a fairly small number of runs. Recent research has shown that an optimal steering profile exists for the particle model (Zhang, Hubbard, and Huffman, 1995). Using this profile, drivers could theoretically learn to drive the fastest

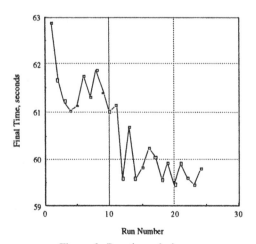

Figure 6. Run times during
a typical training session

possible line down a particular course. Although it is clear that drivers can quickly learn how to drive the simulator, the lack of a quantifiable connection with increased performance on the real track makes a completely quantitative evaluation of the benefit of the simulator impossible.

5 DRIVER FEEDBACK

Even though a quantitative measure of simulator effectiveness is highly desirable, using driver feedback we can obtain a meaningful qualitative assessment of its realism and usefulness. The limitations of this type of an analysis are obvious. Standards can vary substantially because what seems realistic to one driver may not to another. Additionally, drivers may be unaware of the subtle benefits gained during simulator training. Again, this is attributable to the difficulty of evaluating a driver's contribution to performance on a real track. As discussed previously, the driver may have difficulty recognizing differences in his steering performance, and even if he does, it is almost impossible to attribute definitively the differences to simulator training. Even with these limitations, driver feedback is still the best evaluation tool available, and can provide meaningful results.

Members of the US Bobsled Team have used the simulator extensively in training on two separate occasions. The first was at UC Davis, the second was in Calgary in preparation for the 1994 Olympic Games.

5.1 *Motion testing At U.C. Davis*

While at U.C. Davis two drivers were asked to judge the realism of the motion simulation. Tests which varied motion algorithm parameters were conducted, and drivers were asked to score the motion simulation between 1 and 10. A score of 10 corresponded to "indistinguishable from the real sled motion", a score of 5 indicated that it was no improvement over the simulation with no motion at all, and a score less than 5 meant that the motion was determental to the simulation. These tests showed that the drivers viewed the motion as beneficial, with scores above 8 and as high as 9 for the preferred motion configurations.

Table 1 shows the results of the test which varied the "washout" gain, K with the filter break frequency held constant at zero. These results showed that once the gain was about 0.125, only 12.5% of the actual roll motion, the drivers felt that the simulation was greatly enhanced over a fixed base simulation. Similar results were found while testing other motion parameters.

Table 1. Driver evaluations of the simulator motion at different gains. Each of the numbers to the right of the gain are scores on a specific run. A session average for each driver is shown below.

Washout Gain, K	Driver 1	Driver 2
0.0625	5	6
0.1250	8	8, 8,8
0.1875	8	8, 8, 9
0.2500	8, 9, 8, 9	8, 7, 7
Average	7.85	7.70

5.2 *General evaluation at Calgary*

The team also used the simulator in Calgary for five weeks in January-February 1994. During this time they were in rigorous training using both the physical training facilities at the University of Calgary, and the bobsled track at Canada Olympic Park, and drivers as well as other members of the team took hundreds of simulator runs. Near the end of the period, we distributed an evaluation questionnaire, with questions ranging from the very general, "How realistic is the simulator? Explain", to quite specific, "Comment on the realism of the graphics." The following is a summary of important comments.

There was general consensus that the simulator was very real visually. In addition, the drivers were uniformly impressed with the motion simulation during their training exercises. One driver commented that it was "easier to drive with the motion" and all drivers thought the simulation was enhanced by motion. Drivers were also asked to comment on the realism of the dynamic model, i.e., does the simulated sled take the same trajectory as a real sled, and is the steering response of the simulated sled consistent with that of the real sled? The answers were almost universally, "it's close." Several drivers felt that it was very realistic most of the time but that a few turns required steering slightly more or less than on the actual course.

6 SUMMARY

The design, construction, and use of an interactive virtual reality bobsled simulator has been described. A brief introduction to the sport of bobsledding, with specific emphasis on the methods currently used for training drivers, showed a strong financial motivation for a training simulator. A simple particle dynamic model was reviewed which serves as a basis for the computer simulation of three dimensional bobsled motion. Simulated normal force from this model compares well with data taken from actual runs at the Lillehammer track. The rotational and translational motion that the human vestibular system experiences during an actual run were investigated and roll was shown to be the dominant angular motion perceived by the driver. It is possible to develop roll motion algorithms which add realism to the fixed base version of the simulator. The most important issue related to the simulator is the fidelity of the dynamic simulation. Roll-only motion was universally judged by drivers to increase the simulator's usefulness and to be an improvement over the fixed base version. The drivers opinion of the simulator as a training tool was uniformly positive, and many now consider it to be part of their training routine. We believe the simulator s present benefit is outweighed by its potential, especially the prospect of faster computers running more complex and accurate dynamic models.

ACKNOWLEDGMENTS

Without the financial research grant assistance of the US Olympic Committee and the US Bobsled & Skeleton Federation this research would not have been possible. Also computer equipment loans from NASA, SGI, and IBM were invaluable. The data in Fig. 3 were provided by Chassis Dynamics, Inc. Finally, the help of Jim Reus with graphics programming and Brian Richardson in cockpit fabrication is gratefully acknowledged.

REFERENCES

Baumann, W. 1973. The influence of mechanical factors on speed in tobogganing. Medicine and Sport 8:453-459. Biomechanics III, Basel: Karger.
Hubbard, M., M.Kallay & P.Rowhani 1989. Three dimensional bobsled turning dynamics, *Int. J. Sport Biomech.* 5:222-237.
Hubbard, M., M.Kallay, K.Joy, J.Reus & P.Rowhani 1989 Simulation of vehicle and track performance in the bobsled, Third Joint ASCE/ASME Mechanics Conference: San Diego.
Huffman, R.K., M.Hubbard & J.Reus 1993. Use of an interactive visual simulator in driver training, 1993 ASME Winter Annual Meeting.
Kane, T.R. & D.A. Levinson 1985. *Dynamics: Theory and Applications,* New York: McGraw-Hill.
Meiry, J.L. 1966. The vestibular system and human dynamic space orientation, NASA Report No. CR-628.
Reid, L.D. & M.A. Nahon 1985. Flight simulation motion-base drive algorithms: part 1 - developing and testing the equations, UTIAS Report No. 296.
Roy, M. 1995. Personal communication.
Ullman, D.G. & C.Cross 1979. Engineering a new-generation bobsled, ASME paper 79-DE-E-5.
Young , L.R., R.V. Kenyon & W.M. Hollister 1988. Fundamentals of flight simulation. Man-Vehicle Laboratory Report, Dept. Aeronautics & Astronautics, Cambridge, Ma.: MIT.
Zhang, Y., M.Hubbard & R.K.Huffman 1995. Optimum control of bobsled steering. *J. Opt. Theory &Appl..* 85(1):1-19.

The Engineering of Sport, Haake (ed.) © 1996 Balkema, Rotterdam. ISBN 90 5410 822 3

Mechanics of the modern target archery bow and arrow

S.H.Joseph
Department of Mechanical and Process Engineeering, University of Sheffield, UK

S.Stewart
School of Manufacturing, Materials and Mechanical Engineering, University of Plymouth, UK

ABSTRACT: Archery is an olympic sport with an international following at local club level. The equipment combines features recognisable from their asian, european and american roots with modern composite materials and lightweight alloys. The quality and setting up of the equipment is important enough to be a significant factor in the archer's performance for all except the very beginner. The understanding of the mechanics of equipment and set up is still at the pre-scientific level amongst archers in general, and amongst manufacturers and suppliers of such equipment in the UK. This paper outlines the basic mechanics of the shot. The bow as an energy storage and conversion system is described, and factors determining efficiency and stability are identified. Methods for studying the static and dynamic aspects of the variable geometry of the bow limb in draw and shot are proposed.

1 INTRODUCTION

Archery competitions take many forms around the world, depending on local history and custom. In the great majority, the aim is to hit a mark as closely as possible. In Olympic target archery the equipment used is restricted to what is effectively a traditional bow and arrows, made in any materials, with an adjustable plain front sight and an attachment to the bow just above the bow hand to rest the front of the arrow on. Distances to the target vary from 30m to 90m for outdoor shooting, and are known in advance. The accuracy obtained by a good club level male archer is to hit within a 500mm diameter circle at 70m range; world record accuracy is within a 120mm circle at that range The scoring system is cumulative, so that winning is a matter of not shooting bad shots, rather than of obtaining a certain number of excellent shots.

Accuracy is mainly determined by the archer's technique. Strength plays a part, in that a bow of a given efficiency drawn with a greater force will deliver a greater speed to the arrow. Higher speed produces a flight that is less subject to variations in wind, and permits a stance in shooting that varies less with target distance. Accuracy is also affected by the precision of the equipment, and its correct adjustment, or *tuning*. With careful tuning the arrow can be launched with minimal contact with the bow handle,

or *riser*, and with nearly pure translational motion along the arrow direction. With poor tuning the arrow may strike the riser, and leaves the bow rotating about an axis transverse to its flight.

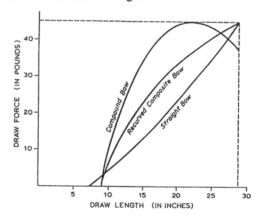

Fig. 1 Typical plots of force vs draw for three types of bow

A plot of draw force against distance for a number of different bow types is shown in Fig.1. A typical equipment setup is shown in Fig.2. The characteristic *recurved* limb shape is explained by looking at the dra curves: a bow without recurve produces a concave plot, up to the holding force at full draw; the recurved limb produces a convex plot, and thus stores the same energy at a lower holding force. The lower holding force allows an archer of a given strength better control of the aim and release of the shot.

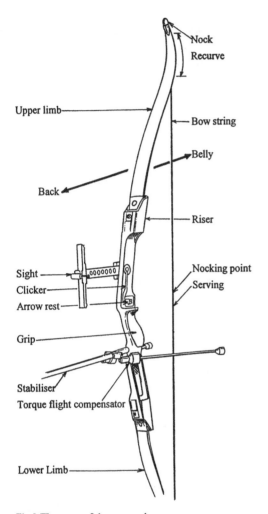

Fig.2 The parts of the recurve bow

Techniques vary, but an archer might spend a second or two aligning the bow and herself with the target before drawing, a second in drawing the bow almost fully, and two to five seconds aiming and completing the draw before the loose.

Limbs are made of GRP, which stores elastic energy with low loss, and are of low mass so that their residual motion when the string has straightened has minimal kinetic energy. The string is made of oriented PET or PE, so as to be light and inextensible and again minimise lost energy. Arrows are made of carbon fibre composite, to give maximum stiffness at minimum weight. The riser is NC machined in stretched aluminium alloy plate, to give accuracy and

rigidity. A variety of *stabilisers* may be attached to the riser, consisting of weights of 50 to 150 gms mounted on light, rigid rods of length 200 to 900 mm. The stabiliser system may contain compliant and/or damping elements amongst the weights and rods.

The release of the arrow (the *loose*) and its acceleration by the string is a dynamic event lasting about 20 ms. Elastic strain energy from the bow *limbs* is converted into kinetic energy of the arrow with an efficiency of about 75% The arrow, string, limbs, riser, stabilisers, and the archer's body form a mechanical system that should deliver the same force along the same path in every shot. The complete system is only partly understood We identify two important episodes in the shot sequence to study in more detail: the loose, and the moment of string straightening. At the loose the tensions in the archer's muscles from holding the draw change suddenly, with the load coming off the string hand, and the rapid acceleration of the limbs interacting with the force from the bow hand produce accelerations in the riser. Thus the disposition of the archer's body at the loose, combined with the arrangement of riser and stabilisers, governs the delivery of force to the arrow in the early part of its acceleration. At the time that the string straightens the limbs are sharply decelerated, and their energy largely transferred to the arrow. The remainder of the energy emerges as vibrations in the limbs, and movement in riser and stabilisers. Thus the arrangement of limbs and stabilisers controls the delivery of force to the arrow in the later part of its acceleration.

In the following sections we look at the anchor and loose, and the straightening string episodes in more detail, report experiments on measuring the disposition of the archer's body and the accelerations of the riser and stabiliser system, and begin a basic theoretical analysis of the launch process.

2 THE ANCHOR AND THE LOOSE

Anatomical considerations have been paramount to coaches of archery ever since the middle ages. Roger Aschaum, an archery advisor to Henry VIII wrote a treatise on the subject which is just as relevant to archers today as it was then. From a highly simplified mechanical point of view the archer can be considered as a system of linkages made up by the arm and shoulder bones as shown in plan view in Figure 3a

The left hand acts as a prop, pushing the bow out with the minimum necessary muscle activity. The fingers of the right hand hook the string on the distal finger crease. Then, with all the long bones of the right arm roughly horizontal (or with the elbow slightly elevated depending on style) the elbow is levered back under the activity of the shoulder and back muscles. The arm muscles as such are only minimally involved, except for contraction of the finger flexors to prevent the string from inadvertently being prematurely released.

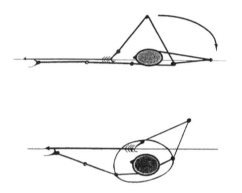

Fig.3 (a) Ideal and (b) real alignment of the archer's anatomy

Ideally, once the bow is drawn, all the links are aligned with a line of force directly through the arrow to the target. Because of the mechanical advantage of the position it should require very little force to maintain the bow at full draw. The little force that is involved is supplied by the muscles controlling the scapula and the posterior shoulder complex This is the basis of the instruction echoed down the ages from Roger Aschaum to the modern day "get it on your back"

Unfortunately this position is an ideal: the nearer to it we can get, the more efficient the draw and the less energy required from the archer However due to anatomical differences, such as large protruding abdomens or breasts, short limbs, short tendons and limited mobility in the shoulders, this is extremely difficult for some archers to even approximate, and for them considerable muscle force is required both to draw the bow and to hold the bow steady while aiming.This condition is sketched in Fig.3b

Having achieved the nearest approximation to the optimally efficient position the archer strives to exactly regain this position every time he shoots an arrow. It is very tempting to anatomically "lock" all the joints. i.e. to put all the joints at the extreme of their articular range. However apart from the possibility of slicing the inside of the elbow with the string, all tension due to straining muscles is considered undesirable, as this can be transmitted to

the bow, and then to the arrow, throwing it off course. Joint locking is encouraged only in the shoulder where it is recommended that the archer uses all available muscle effort to extend the drawing shoulder as far as possible This means that the bones are at the extreme edges of the articular surface, and only in contact over a small area.

This puts a lot of strain on the joint and the articular cuff of supporting ligaments. It is not uncommon in archers with flexible and mobile joints, that the humeral head actually dislocates posteriorly as much as 20mm, and the head of the humerus transmits the drawing load directly to the soft tissue posterior to the joint, rather than to the load bearing surfaces of the joint. However it does give the archer a feeling of stability, but potentially can cause long term damage if the archer has a bow which is too heavy or too powerful for them to draw. The skeletons of bowmen recovered from the Mary Rose warship showed evidence of damage to bone structures which had distorted to compensate This included flats on the surface of the articular surface of the proximal humerus which indicates that the joint was repeatedly partially dislocated under load.

What is at first glance quite simple therefor, is in reality a complex action and takes many years to master fully, and can potentially cause significant damage to ligaments and bones and eventually arthritis if it is not taught in an informed and responsible way.

3 THE STRAIGHTENING STRING

3.1 *The limbs*

Measurements of the bow as the arrow is being shot have been made using high speed film, and using accelerometers attached to the bow and limbs at various points. These show the forward movement of the limbs at the loose, and the rapid deceleration of the limb tips at the moment that the string straightens. This causes a considerable shock pulse throughout the system, and oscillations which decay in a classical pattern. The decay may take from 0.25 to 3 seconds depending on the tune of the bow and the skill of the archer.

If the movements of the limbs are plotted from a slow motion video we can see that after the string has reached vertical the forward momentum of the limbs causes them to continue to travel toward the target. The almost rigid constraint of the straight string means that they can only do so by buckling. The subsequent mode of oscillation is sketched in Fig 4.

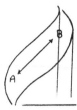

Fig.4 Diagram of the dominant mode of vibration in the limb after the string straightens

different makes of stabiliser varies considerably. Some examples are shown in Fig.6.

Fig. 6 Accelerometer traces from stabiliser systems (l to r): Arten, Hoyt + AGF, with washer.

Clearly, the extent of this oscillation, and hence the magnitude of the energy lost to it, depends on the stiffness and mass distribution of the limbs. These parameters also determine the draw curve of the bow, and hence the important holding force and total energy storage figures. The best design compromise between the static and dynamic requirements is at present found by trial and error.

Fig.5 Accelerometer traces from (l to r) Border, Hoyt, PSE and Stylist limbs.

It is noticeable that limbs made by different manufacturers vary considerably in their dynamic characteristics. Some typical accelerometer traces are sketched in Fig.5.

3.2 Stabilisers

The self evident role of stabilisers is to reduce unwanted movement as the bow is being aimed by increasing the smaller moments of inertia of the bow, and to move the static balance of the bow forward so that the nock is not deflected downwards during the launch. It is also claimed by manufacturers that stabilisers absorb some of the energy generated at loose.

Accelerometer readings indicate that the addition of a stabiliser can indeed reduce the time taken for the oscillations to decay, and that the performance of

The most rapid damping in a conventional system was obtained using a *long rod* of length 53cm, 3cm diameter, parallel walled hollow tube made of carbon composite filled with polystyrene foam and rubber washers. This is in contrast to commercial systems that use lengths of over 70 cm. The best damping of all was achieved by the simple addition of a rubber tap washer between the bow and the stabiliser assembly. Clearly the subject of damping and stabilisation remains little understood by manufacturers and archers.

3.3 The riser

The acceleration of the bow handle itself measured at the centre point in the direction of the target can be as much as 680ms-2. The riser is designed to be rigid, but high speed video of a CNC machined aluminium riser shows the ends flexing as much as 12mm, as the two ends oscillate forwards and backwards together relative to the centre. If the bow is badly tuned other modes are strongly present, to the extent that no single mode can be seen to dominate.

3.4 The archer

If the archer grips the bow tightly on loose then they too become part of the system and some of the energy is dissipated through the bones and soft tissues of the archers arm and shoulder. Apart from the possibility of long term injury this is considered to be a bad fault in performance terms, as tensions in the muscles on loose can cause unwanted movements in the bow and the arrow to fly off course.

4 AN ANALYSIS OF THE LAUNCH

4.1 *An energy audit*

One reason for the success of the bow and arrow as a device for launching projectiles is the extreme lightness of the final link in the launching chain: the string. Fibres have been the lightest and strongest materials available for many years, and remain so with the advent of modern materials. Another reason is the elegance of the mechanical system whereby the comparatively heavy limbs of the bow pull the string outwards from its ends, so that they move the arrow at an ever increasing velocity ratio as the string straightens. This means that the kinetic energy in the limbs is largely extracted and delivered to the arrow by the time it is launched In a typical system approximately 50 joules of energy are stored under the draw curve of Fig.1, and the final arrow speed is about 50 ms^{-1} The energy delivered to the arrow is about 38 Joules: we will look at how the 12 Joules is lost, and attempt to quantify the time in which it happens.

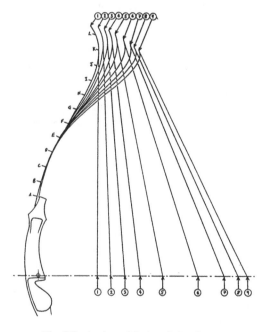

Fig. 7 Static view of the bow being drawn.

A diagram illustrating the bow in various states of draw is shown in Fig.7. For the sake of an initial dynamic analysis we will assume that the deflection of the limb along its length is purely transverse and proportional to the movement of the limb tip, and that the movement of the limb in launching is quasi static, and that the string is inextensible. The model of the limb then becomes one of a light rotating arm (Fig.8) with an effective mass at its end and a torsion spring driving its rotation. With this model, we see that at the loose the arrow and limb tip move at a similar speed, and when the string straightens the tip is brought to rest as the arrow is launched. Knowing the final arrow speed v relative to the limb tip, we can estimate the acceleration f of the limb tip along the string direction at the moment of launch as $f = v^2 / l$, where l is the distance from the arrow nock to the tip. This gives a figure of 3600 ms^{-2}, or 360g, sufficient to generate substantial tensile forces in the string and compressive forces in the limb The importance of these forces is recognised, in that modern high stiffness string materials and light weight arrows are only recommended for use with bows of sufficient strength to resist them.

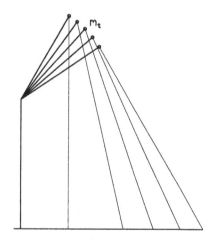

Fig 8 Diagram of the dynamic model used

We can further quantify the forces by estimating the effective mass m_t for movement along the string direction. This has been done by measuring the natural frequency for small oscillations of the limbs ω as 13 Hz, and the static stiffness k as 5.3 Nmm^{-1}, in that direction. The effective total mass is then given by $k / \omega^2 = 800$gm, or $m_t = 400$ gm for each limb. Using the above figure for limb tip acceleration, the force in the string is estimated at 1400 N, some seven times the static force at full draw.

The extension of the string under this large transient force is sufficient to absorb significant energy. The static extension C is measured as approximately 2.5

mm.m^{-1} kN^{-1}, resulting in an estimated extension of 5 mm at launch, and a storage of about 3.5 Joules at that point. This is an over estimate, as will be explained below, but clearly the assumption of inextensibility in the string needs to be abandoned. To move forward, we need to estimate the dynamic characteristic of the string / tip mass system on the model, and to compare it with the time interval in which the final straightening of the string takes place. The characteristic response time of the tip in this mode is estimated from the effective tip mass and extensibility of the string as 0 66 ms.

The timing of the string straightening is best illustrated by an estimate of how the fraction of the total kinetic energy that resides in the limbs varies with time on the simplified model under consideration.

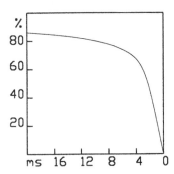

Fig.9 Theoretical plot of % energy remaining in limbs against time in ms

This is plotted in Fig.9, using the figure for m_t obtained above. We notice that the energy transfer from the limbs occurs in the last 4 ms, or 20%, of the launching process. At 0 66 ms from launch, about 16% of the energy remains in the limbs. The response time of the elastic string and tip mass system is thus fast enough to ensure that most of the energy is transferred, but slow enough to make a proper analysis of the dynamics of the straightening rather difficult. We can say, however, that the extension of the string means that the limbs will still be travelling forward when the string straightens, so that v, and hence the force on the string, is reduced. Thus if the residual energy in the limbs is 16% their forward tip speed at launch is about 10 ms^{-1}, the string tension is about 900 N, and the stored elastic energy in the string is about 1.6 Joules.

A further destination of lost energy is the kinetic energy of the string itself. The natural frequency of

the string in transverse vibration is 100 Hz, giving an effective mass of 1.6gm, or storing 2.5 Joules at launch speed.

As an initial theoretical energy audit, we find that about 4 Joules is expected to be lost in the string, and the remaining 8 Joules may be left in the limbs at the time of launch.

5 DISCUSSION

The archer requires an efficient bow and an accurate shot. The efficiency of the bow depends on a complex interaction between the static and dynamic characteristics, mainly of the limbs, but also of the riser and string. The analysis presented here shows that some simple experiments and calculations provide a plausible explanation for the loss of energy from the arrow to the rest of the system. We believe that further work is needed on the dynamic modelling of the events at the straightening of the string, so that criteria for good limb design can be obtained for use by UK manufacturers.

Considerable time, effort and money is employed by archers and manufacturers in the pursuit of the optimally damped bow. There is considerable doubt about its relevance. The arrow cannot be diverted from its course by any activity of the bow once it has parted company from the string. Prolonged oscillations are a symptom of a badly tuned or inefficient bow, but the problem should be cured at source rather than have its symptoms suppressed by damping. The archer and bow system should be adjusted so that the energy that causes the oscillations is diverted back into the arrow where it belongs.

The accuracy of the shot is mainly governed by the archer's technique. This is dependant on the archers ability to control the forces incurred by drawing the bow, aiming and loosing, in the most anatomically efficient manner. It is universally observed that the dynamic response of the bow depends strongly on technique: a bad shot can be felt directly by the archer. We believe that further work is required to study the relationship between the stance and muscular state of the archer, and the dynamic response of the system at launch, so that the performance of top UK archers and the enjoyment of club archers can both be improved.

ACKNOWLEDGEMENTS

Su Stewart wishes to thank the Daphne Jackson Memorial Trust for supporting her in this work.

The Engineering of Sport, Haake (ed.) © 1996 Balkema, Rotterdam. ISBN 90 5410 822 3

The fatigue life of nylon monofilament as fishing gear material

H. Kanehiro, S. Takeda, T. Kakihara & K. Satoh
Tokyo University of Fisheries, Japan

ABSTRACT:The fatigue and durability of monofilament fishing gut and multifilament fishing gut were compared in fatigue tests employing repeated elongation under a load and flection. The fatigue life of both types of gut was shorter in the case of flection than elongation.

Also, in terms of elongation, the monofilament fishing gut exhibited greater durability than the multifilament type. In the case of multifilament fishing gut, a great reduction in strength was observed as the fatigue progressed. In the case of the monofilament fishing gut, no reduction in strength was observed at all. These results suggest that the fatigue mechanism of the two types of fishing gut is different and that the twist and the existence of interaction between the filaments exert an influence on the fatigue characteristics of multifilament fishing gut.

In recent years in the fishing–as–a–sport and the tuna long–line fishing industries, nylon monofilament has replaced wire and multifilament. Compared with multifilament fishing gut,monofilament fishing gut lacks flexibility but its excellent transparency has resulted in greater catches and this, together with its low cost, has led to its widespread use. As a result of the tuna long–line operation tests conducted over a period of three years from 1988–1990 aboard the Tokyo University of Fisheries Training Boat Shinyo–maru (649 tonnes), it was also found that,by comparison with multifilament fishing gear(cremona), monofilament (nylon) fishing gear resulted in excellent catch rates. However, fishing gut is exposed to a number of complex physical effects during operations such as contact with hulls, bites from sharks, etc., intertwining with other fishing gut and bending stress. Once fishing gut becomes damaged, its transparency deteriorates and it can easily be broken. Therefore, it is usually replaced with new gut after a short period of use. An understanding of the durability of the material is important not only in terms of practical use but also in terms of material economy. For the purposes of this research, fatigue tests (elongation and flection) were carried out using hypothetical stresses that occur during operations in order to obtain information about the usage life of fishing gut.

1 TEST METHOD

Nylon fishing gut No.150 (consisting of one monofilament referred to hereafter as monofilament) and 3–ply nylon fishing gut No.50 (hereafter referred to as multifilament (No.50 × 3)) were used in this test. Also, in order to investigate the influence of the number of filaments and the twist on the fatigue–resistance characteristics of the fishing gut, a test was carried out on a multifilament consisting of a (No.50 × 3) filament and a (No.150 × 3) filament. A static break test (tension and bending tension strength) and a fatigue test (direct elongation and flexural elongation fatigue hereafter referred to as elongation fatigue and flection fatigue) were carried out employing a universal tension tester. The sample used in the test consisted of a 70–cm length of fishing gut joined at both ends with aluminum tubes and attached in a loop to a cylindrical (diameter:50mm) up–and–down

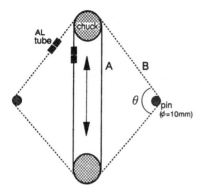

Figure 1. Methods of mounting samples in fatigue tests. Both ends of a loop sample 70 cm long were joined by aluminium tubes for the present tests The sample was fatigued under (A) tension cycling and (B) tension cycling under bending (θ=100°)

chuck as shown in Fig. 1.   Considerable care was taken to ensure that the fishing gut did not break at the point where it was joined to the tubes.   In the flection fatigue test, a 10− mm stainless steel pin with a polished surface was employed and the tension angle during   the test was maintained between 100−110deg . The fatigue test was carried out using the fixed load repeat test method. The test was carried out with   the minimum repeat load for every test set at 0 and the maximum load (S) was varied within 95−30% of the fishing gut break strength (To). With a tension speed of 500mm/min and a repeat frequency ranging from 1− 10000,the frequency (fatigue life, N)was measured until breakage occurred.   Data regarding breaks at the aluminum tube joints was excluded. Samples which did not break even in the 10000 frequency test were left standing for a   few days. Then, they were observed for any external changes and the tension and flection break tests were carried out again.

2 RESULTS AND CONSIDERATION

The results of the static break test are shown in Table 1. The values for flection strength were much lower than those for tensile strength. The decline in strength of multifilament fishing   gut   was   particularly

conspicuous and it was found to be weak in response to bending. Also, even with the same No.150 equivalent fishing gut, monofilament fishing gut has greater tensile and flection strength than multifilament fishing gut (No.50 ×   3).

Table 1. Details of the tested samples

Materials		Nylon-6	Nylon-6
Construction		mono (No. 150)	3-ply (No. 50×3)
Deniers		33,000	33,000
Diameter (mm)		2.05	2.65
Twist factor (turns/100 cm)		0	60
Breaking strength T_0 (kgf)	Tensile	198.0	173.3
	Under bending*	194.4	151.0
Breaking elongation E_0 (%)	Tensile	29.2	35.5
	Under bending*	26.5	32.1

* Tensile strength and elongation under bending at about 100° over a pin ($\phi \approx 10$ mm).

2.1 Elongation and Elongation under Bending Fatigue of Monofilament Fishing Gut

The   results   of   the   elongation   and elongation under bending fatigue tests are shown in Fig.2. In the elongation fatigue test, a

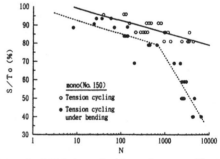

Figure 2. S-N fatigue data for nylon monofilament. S, stress (% tensile load, S/T_o); N, number of cycles to failure

virtually linear relationship was observed in the frequency (hereafter referred to as the breaking frequency) N between cyclic loading S/T_0(%) until breaking. In the elongation under bending test also, in the cyclic load range of 95− 80% (where N is below 1000 cycles), the same linear relationship was observed between the cyclic load and N. However, at the boundary where the cyclic load was 80%, the breaking frequency suddenly decreased and a pronounced deviation from the line was observed. In the comparison with the same cyclic load, it was discovered that the breaking frequency was less in the case of elongation under bending fatigue than in the case of elongation fatigue . At loads below 80%, this rate was conspicuous. In terms of elongation fatigue, monofilament fishing gut possesses durability when large loads are applied but, in terms of flection (bending) fatigue, it was found to be comparatively weak.

2.2 The Fatigue Durability of Monofilament and Multifilament Fishing Gut

The results of fatigue tests on both types of fishing gut are shown in Fig.3. The results of the elongation fatigue test indicate that the same linear relationship was observed with multifilament fishing gut as with monofilament fishing gut (Fig.3−a). The S−N relationship between the two types of fishing gut obtained through the least square method.

monofilament fishing gut:
$$S/T_0 = 1.05 - 0.68 \mathrm{Log}N$$
multifilament fishing gut:
$$S/T_0 = 1.00 - 0.112 \mathrm{Log}N$$

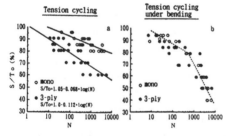

Figure 3. Comparison of monofilament with 3-ply filament in fatigue life. Fatigue test (a) tension cycling, (b) tension cycling under bending

In the case of multifilament fishing gut , the twist and the existence of interaction between the filaments exert a great influence on the fatigue characteristics. From Fig.3−a and the equation obtained through the linear method, when both types of fishing gut are subjected to the same repeat load, the fatigue life (break frequency) of multifilament fishing gut is one tenth or less that of

monofilament fishing gut.In terms of flection and elongation fatigue, as shown in Fig.3−b, virtually no difference was found between the fatigue life (break frequency)of both types of gut when subjected to the same repeat load test. In the case of multifilament fishing gut ,

it was observed that the break frequency decreased rapidly below a repeat load of 80% S/T_0.This suggests that the fatigue mechanism in terms of both flection and elongation fatigue in both types of gut differs at the 80% load boundary. In Fig . 4, a comparison

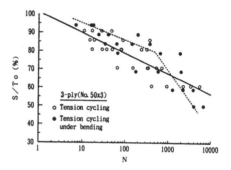

Figure 4. S-N fatigue data for nylon 3-ply filament

of the results of the flection and elongation fatigue tests on multifilament fishing gut is provided.The clear difference observed in monofilament fishing gut in both fatigue tests was not observed to the same degree in multifilament fishing gut (Fig . 3). It appeared that fatigue life was rather longer in the case of flection elongation under bending than in the case of elongation with heavy loads. From the above results , it was found that, in comparison with multifilament fishing gut, monofilament fishing gut exhibited superior durability in terms of elongation fatigue. However, in terms of flection elongation under bending fatigue, hardly any difference in durability was observed between the two types of fishing gut. In the flection elongation under bending fatigue test, breaks in the fishing gut all occurred in the bends and it appeared that the compression deformation of the fishing gut contributed greatly to fatigue breakage. Consequently, it appeared that the difference in the number of filaments had little influence on flection elongation under bending fatigue.

Table 2. Residual strength of nylon filaments after 10^4 cycles fatigue test

Filaments	Breaking strength T_0 (kgf)	Types of fatigue test	Imposed max. load S/T_0 (%)	Residual strength T (kgf (%))	Permanent elongation (%)	Breaking elongation (%)
Mono	198.0	Tension cycling	75 75 75	216.3(109%) 212.5(107%) 207.5(105%)	4.8	21.3
	194.4	Tension cycling under bending*2	30 30	205.0(105%) 205.0(105%)	—	
3-ply	173.3	Tension cycling	50	153.8(88%)	6.5	22.0
	151.0	Tension cycling under bending*2	40	110.0(73%)	—	
mono*1	182.2	Tension cycling	65	192.5(106%)	—	

*1 One filament in 3-ply filament nylon ((No. 150)×3). This filament was the same with the monofilament (No. 150), but was different only in the twist (the former twisted, the latter not).
*2 Tension fatigue test under bending at about 100° over a pin (a = 10 mm).

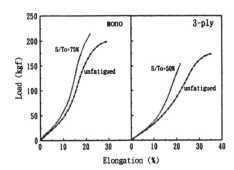

Figure 5. Load-elongation curves for unfatigued (original sample) and fatigue tested sample (remained unbroken after 10^4 cycles loading). The S/T_0 values in the figures denote the cyclic loading condition

2.3 The Residual Strength Fatigue Test

Specimens which did not break in the 10000-cycle   fatigue test were measured for permanent distortion. Subsequently, they were broken and the residual strength was measured. The load-elongation curve before and after the fatigue test is shown in Fig.5 and the dynamic characteristics are shown in Table 2. The rising inclination and decrease in elongation of the load-elongation curve before and after the fatigue test were distinctive. An increase of 5-10% over the static break strength To in terms of both elongation and flection fatigue was observed in the residual strength T of monofilament fishing gut. In particular, in the case of elongation fatigue, in spite of the fact that the repeat load strength was 75% and the stress was high, no decrease in strength was observed . In the case of multifilament fishing gut, the fatigue characteristics were greatly influenced by the interaction between the filaments and the increase in strength was explained mainly by the stabilization of the twist structure resulting from repeated stress. The increase in strength of the nylon filament fishing gut used in this research is thought to originate in the internal structural changes in the fiber resulting from the repeated load stress. This kind of structural change is thought to occur not only as a result of elongation stress but also as a result of flection elongation under bending fatigue. On the other hand, in the case of multifilament fishing gut, the residual strength decreases greatly by 12-27% in the case of both elongation fatigue and flection elongation under bending fatigue. It didn't break in the 10000-cycle test but the gradual advance of fatigue is discernible. As shown above in the case of monofilament fishing gut, it is thought that decrease in strength occurs gradually in the course of the fatigue test. This difference in fatigue characteristics is attributable to the different fatigue mechanisms of each type of gut.

2.4 The Influence Exerted by Twist on Fatigue

In order to examine the. influence exerted by twist on the fatigue of fishing gut, a fatigue test was conducted using monofilament fishing gut and a multifilament fishing gut (No.150 ×   3) composed of 3-ply of the same thickness of No.150 with one of these three filaments removed. Any difference with

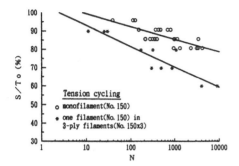

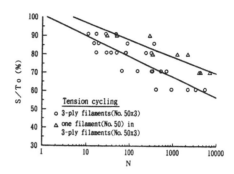

Figure 6. Effect of twist on the fatigue life of nylon monofilament. Nylon monofilament (No. 150) and one filament in nylon 3-ply filament (No. 150x3) were compared. Both samples were the same monofilament (No. 150), but were different only in the twist (the former having no twist, and the latter having twist (32 turns/100 cm).

Figure 7. Effect of filament-filament interaction on the fatigue life of nylon filaments. Nylon 3-ply filament (No. 50x3) were compared. Both samples had the same twist, but were different only in the number of filaments (the former had three, while the latter had one).

No.150 monofilament fishing gut indicate the existence of twist. The strength T_0 of the same single filament was greater in the case of the monofilament fishing gut at 198.0 kgf (Table 1) than in the case of multifilament fishing gut at 182.2 kgf. A comparison of the elongation fatigue test results for both types of gut is shown in Fig.6 .

A clear difference was observed in the ability to withstand fatigue of the two types of gut. The fatigue life of the monofilaments of which multifilament fishing gut is composed is shorter than that of monofilament gut without twist. Even with the same number and amount of filaments, one filament of multifilament fishing gut has a higher tendency to break through fatigue because of the strong effect of torsion deformation resulting from twist. When the residual strength in a 10000−cycle fatigue (S/T_0=60%) test on one filament of multifilament fishing gut was measured,a strength increase (T=192.5 kgf) of approximately 6% was observed.

2.5 The Influence Exerted on Fatigue by the Number of Filaments

In order to examine the influence of the number of filaments fishing gut is composed of on the ability of fishing gut to withstand fatigue, a comparison was made of (No.50 × 3) multifilament fishing gut and one filament removed from such gut. The results of the

elongation fatigue test are shown in Fig.7. The repeat load break frequency of 3−ply multifilament fishing gut was found to be low at approximately one tenth that of a single filament removed from the gut. The results indicated that 3−ply multifilament fishing gut had a tendency to break easily. In addition, the strength of 3−ply multifilament gut was 173 kgf (Table 1) whereas that of one filament was 62.3 kgf, indicating that the 3−ply structure is weaker per filament. 3−ply multifilament gut experiences friction between filaments whereas no friction occurs with one filament. It is thought that this is the reason for the difference in static strength and ability to withstand fatigue between the two types of fishing gut. In fact, the results of microscopic observation indicated abrasion damage on multifilament fishing gut broken by fatigue.

REFERENCES

K.SATOH , I.KASUGA & S.TAKEDA 1990. Nippon Suisan Gakkaishi.56:1605−1609.
K.KANEHIRO & H.FUJII 1992.Nippon Suisan Gakkaishi.58:2321−2327.
M.C.KENNEY ,J.F.MANDELL & F.J.McGARRY 1985.J.Mat.Sci.20:2045−2059
K.KANEHIRO 1984.Nippon Suisan Gakkaishi. 50:443−449
A.R.BUNSELL & J.W.S.HEARLE 1971.J.Mat.Sci.6:1303−1311

The Engineering of Sport, Haake (ed.)© 1996 Balkema, Rotterdam. ISBN 90 5410 822 3

Bicycle chain efficiency

Matt D. Kidd, N. E. Loch & R. L. Reuben
Department of Mechanical and Chemical Engineering, Herriot-Watt University, Edinburgh, UK

ABSTRACT : Roller chain drives are to be found in use throughout the world of engineering for the transmission of power. The chain's main advantage over other methods is that it is highly efficient. It is widely accepted that efficiencies of 98% are easily achievable and maintained; however such figures may only be possible in the favourable environment of a clean chain in an oil bath. Such states do not exist on today's bicycles. The bush roller chain, as invented for Starley's safety bicycle by Hans Renold in 1880, has undergone many modifications but little or no research has investigated actual efficiency as a power train. This paper introduces a method for the evaluation of bicycle chain efficiency as a function of power train geometry and condition. A mechanical model of the chain is also presented which will be used to analyse the results and suggest ways of improving efficiency in use.

1 BACKGROUND

The bicycle industry has been swamped by new engineering technology in recent years. Much work has been done to improve aerodynamics and to manufacture bicycles from previously considered exotic materials such as carbon fibre, titanium and metal matrix composites. Despite all these developments, the transmission of a cyclist's power is still dependent on a roller chain. Improvements have been introduced to make it simpler, lighter, stronger and to shift better but any chain manufacturer, bicycle shop or engineer will give wide ranging and conflicting answers on the best means of lubrication. The most common misconception cited by these people is that the chain is 98.6% efficient. Believing this to be true for today's application of bicycle chains is very hard. The oft-quoted figure originates from an unobtainable experimental paper produced from work carried out at the National Physical Laboratory, UK, some time in the 1930s. The test consisted of a ¼" pitch chain running between two co-planar sprockets, transferring 740W and submerged in an oil bath, conditions not relevant to current riders and their bicycles. It is believed that such high efficiencies for the bicycle chain are both unobtainable and that peak efficiency is short-lived.

2 THE CHAIN

The majority of chains on today's bicycles are of a bushingless construction, pioneered by Sedis, a French component manufacturer. A bushingless chain link pair consists of 8 parts, two pairs of inner① and outer plates②, two pins③ and two rollers④ (see Figure 1). This design allows an amount of lateral flexibility most important for the derailleur gear changing mechanism. Now widely accepted as the lightest and widest ratio-permitting of all gear mechanisms, it will be found on all competition

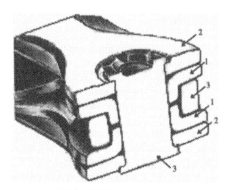

Figure 1 : Chain components

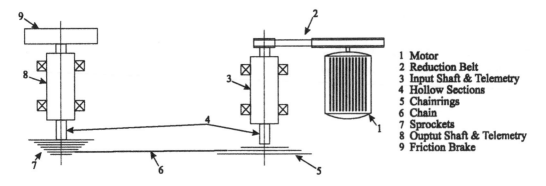

Figure 2 : Chain test rig

1	Motor
2	Reduction Belt
3	Input Shaft & Telemetry
4	Hollow Sections
5	Chainrings
6	Chain
7	Sprockets
8	Output Shaft & Telemetry
9	Friction Brake

machines and the majority of recreational bicycles. By allowing the chain to run on any of the 5,6,7 or 8 gear sprockets at the rear wheel, excess chain length is taken up by a sprung arm. The derailleur mechanism is mostly to blame for the reduced efficiency of chains. The chain path is only ever straight and ideal in one gear combination. In other gear selections the chain line can be as much as 2.5° away from ideal, creating increased friction and wasted lateral forces. The sprung arm consists of a pair of small jockey wheels around which the chain must follow. This doubles the total chain articulation from 360° to 720°. Additionally, tension is applied to the return chain which forms another loss of power. The size of the jockey wheels is 11 teeth which further reduces the efficiency because of the tight angles of rotation formed. Secondary effects of the derailleur system deny the use of a chain oil bath or case. By starving the chain of fresh or clean oil and allowing exposure to contaminants, proper lubrication is a major problem.

3 LUBRICATION

The lubrication of a bicycle chain can be as simple or complicated as the owner wants. While some are happy to add fresh oil to old, others will remove the whole transmission system, soak it in a cocktail of chemicals and then apply another cocktail of lubricants to match not only the season but the locale. Lube manufacturers supply an array of lubricants full of additives. The lubes may be wax, molybdenum disulphide, synthetic or grease-based. Recent additives include PTFE in various amounts, a previously considered ineffective lubricant for high pressure applications. Neglect of the transmission

system leads to degradation in shifting performance, and expensive replacement bills let alone to loss in efficiency.

A chain lubrication company is now about to produce an onboard chain oiling device. Their novel approach is to use a water-soluble oil for primary lubrication. When the chain becomes dirty and the oil contaminated it may be flushed out by the application of water (carried on bicycle). Test riders reported an associated boost in performance with the consequential re-application of fresh oil. Performance is not the only benefit of this system though, as the cleanliness of the chain leads to improved transmission lifespan, a cleaner bicycle and a cleaner rider.

4 RESEARCH

In researching the alleged performance benefits and the general efficiency of the bicycle chain, an experimental rig was completed in 1995. The rig (Figure 2) consists of bicycle components and an accurate torque measurement device. Modified front chainrings⑤ are driven by a 500 W variable-speed motor① between 50 and 200 r.p.m. Gears are selected on the adapted rear sprockets⑦ by a derailleur mechanism and a friction band brake⑨ provides up to 400 W of dissipation. Both the input and output torques are measured by strain gauges on the hollow drive shafts④. Simple multiplication by rotational velocity gives accurate power data. Original apparatus used a hired radio telemetry system to transmit the strain data on the rotating shafts. This system has now been replaced with a novel optical link prototype.

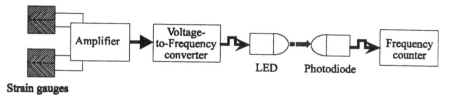

Figure 3 : Optical link for strain data

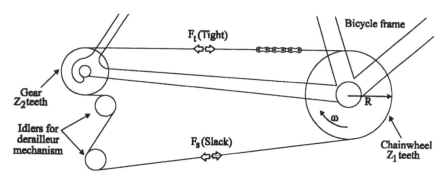

Figure 4 : Chain set-up on a bicycle

The simple and cheap optical system (Figure 3) involves conversion of the strain gauge voltage signal produced by a bridge amplifier through a voltage-to-frequency converter. This pulsing DC signal drives an infrared LED that is positioned axially along the shaft. A static photo sensor receives the data which is fed into a frequency counter. There is a direct relationship between applied torque and output frequency. A torque range of 0 to 50 Nm correlates with 10 Hz to 10 kHz frequency, thus providing torque data with a resolution of 0.005 Nm.

5 THEORY

Many researchers have investigated the performance and efficiency of chain transmission systems with emphasis on industrial applications where the driving and driven sprockets are co-planar[1][2][3]. Additionally many of these systems feature comprehensive lubrication in a largely dirt-free environment - conditions alien to most bicycle chains.

Figure 4 represents a typical bicycle chain systems for derailleur operation in which the chain follows a more tortuous path than is normally found in industrial drives.

Figure 5 shows a section of the chain wheel and the associated forces assuming a rigid body approach and steady-state operation. Referring to Figure 1, frictional losses occur at contact points between:

- pin and bush

- outer and inner plates
- inner plate bush and roller.

Coenen and Peeken [1] evaluated energy losses at these contact points for a simple chain drive (see Figure 6) and their approach, which forms the basis of this paper, is summarised here:

$$F_1 = F_t \frac{\sin(\gamma + 2\alpha - \phi)}{\sin(\gamma + 2\alpha)} \qquad (1)$$

$$N = F_t \frac{\sin\phi}{\sin(\gamma + 2\alpha)} \qquad (2)$$

where γ = sprocket tooth pressure angle

2α = articulation angle

ϕ = instantaneous articulation angle
    with $0 \leq \phi \leq 2\alpha$

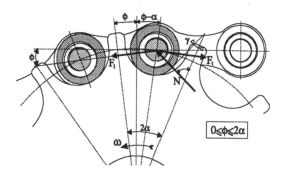

Figure 5 : Force analysis

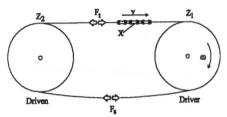

Figure 6 : Simple chain drive

Figure 7: Lateral forces

Obviously, forces F_1 and N are functions of ϕ and hence mean values are employed to obtain mean energy losses over the range 0 to 2α.

i.e.
$$\begin{aligned}(F_1)_m &= a_o F_t\\(N)_m &= a_N F_t\end{aligned} \tag{3}$$

where a_o and a_N are coefficients defined by tooth geometry over the range $0 \le \phi \le 2\alpha$.

During articulation, frictional energy is dissipated at the pin, bush and plate surfaces, i.e.

typically, $E = \left(F\mu\dfrac{d}{2}\right)(2\alpha) = F\mu\dfrac{d}{2}\left(\dfrac{2\pi}{Z}\right)$ for a single

contact point, where d is the pin diameter and μ the

coefficient of friction. Now $E_{Tot} = \displaystyle\sum_{i=1}^{n} E_i$ represents

the total energy loss for n contacting surfaces. Coenen and Peeken [1] showed that for an industrial drive system with co-planar sprockets having $Z_1 = Z_2 = Z$, the total energy loss for one cycle of operation is given by :

$$E_{Tot} = (1 + a_o)\frac{\mu d \pi X}{Z}(F_t + F_1) + a_N \frac{\mu D \pi X}{Z}(F_t + F_1) \tag{4}$$

where D = inner bush diameter
X = total number of links in the chain
Knowing the chain speed v and chain pitch p, the total power loss due to friction is given by

$$P_L = \frac{\mu \pi v}{Zp}(F_t + F_1)\left[d(1 + a_o) + Da_N\right] \tag{5}$$

Now Efficiency $\quad \eta = \dfrac{(Power)_{Output}}{(Power)_{Input}}$

$$\eta = \frac{P_I - P_L}{P_I} = 1 - \frac{P_L}{P_I} \tag{6}$$

And $P_I = (F_t - F_s)v$ at input.

Hence $\eta = 1 - \dfrac{\mu\pi(F_t + F_1)}{Z(F_t - F_s)}\left[d(1 + a_0) + Da_N\right]$ (7)

Equation (7) relates to the system studied by Coenen and Peeken [1], as shown in Figure 6.

Comparing Figures 4 and 6, it is observed that the bicycle case will incur additional losses due to the extra articulations of the derailleur change

mechanism. Also, for some gear ratios the driver and driven sprockets will not be co-planar thereby introducing small lateral forces, as shown in Figure 7.

Hence using the Coenen and Peeken [1] model as a base, equations (4) and (7) will be modified to account for the effects of:
(i) additional articulation of the derailleur mechanism
(ii) variation in gear teeth Z_1 and Z_2
(iii) effect of lateral forces in chain associated with (ii)
(iv) changes to the coefficient of friction to account for poor lubrication, worn teeth, dirty chain, etc.

6 CONCLUSIONS

This work is in its infancy at present but the aims and objectives of the research programme are:

(a) to commission a research rig in which a typical bicycle chain may be investigated experimentally under a variety of simulated conditions relating to load, lubrication and wear from which realistic estimates of transmission efficiency may be made.

(b) to extend the Coenen and Peeken theoretical model to include additional effects peculiar to the operation of bicycle chains with derailleur change mechanism.

7 ACKNOWLEDGEMENTS

The authors wish to thank Dr. G. M. Alder, Mr. F. Mill & Mr J. Muhl (University of Edinburgh) and Mr. F. Scott (Scottoilers Ltd.) for their previous and current help.

8 REFERENCES

(1) Coenen, W. and Peeken, H., 1985, "Wirkungsgrad von Rollenkettengetrieben unter besonderer berücksichtigungdes schmierungs-zustaandes.", *Antriobstechnik* Vol.24#1, pp56-61.
(2) Chen, C-K., 1989, "On the Fundamental Mechanics of Roller Chain Drives and the Dynamic Analysis of Universal Joints.", *PhD thesis*, Columbia University.
(3) Naji, M.R., 1981, "On Timing Belt & Roller Chain Load Distributions", *PhD thesis*, University of Houston.

The Engineering of Sport, Haake (ed.)© 1996 Balkema, Rotterdam. ISBN 90 5410 822 3

The dynamic response of a golf club head

J.S.B. Mather
Department of Mechanical Engineering, University of Nottingham, UK

J. Immohr
BIBA, University of Bremen, Germany

ABSTRACT: The work presented is the first stage of our experimental and theoretical investigations of the movement of the face of the golf club under the impact of a golf ball. The results suggest that the contact time of the ball is less invariant than has been previously suggested and that the variations in the shape of the force pattern between the ball and the face is perhaps due to harmonic vibrations of the club. Results for different materials show quite distinct patterns.

1. INTRODUCTION

The rules of golf quite clearly state that *the material and construction of the face* (of a golf club) *shall not be designed or manufactured to have the effect at impact of a spring*. The statement is unequivocal in its meaning but it is doubtful if the manufacturers of golf clubs heads can comply with the constraint, simply because, in the main, they are not aware of the movement of the faces on their clubs. Indeed one manufacturer actually suggests in sales material, without any scientific evidence to prove the claim, that the design of their head is such as to increase the velocity of the ball at impact. This is supposedly achieved by concentrating the head mass near to the centre of gravity and immediately behind the ball impact point. Whereas, another manufacturer suggests that by distributing the mass around the periphery of the head a similar increase in velocity can be attained.

The reason for these anomalous situations is because the golf industry is always striving to find a new story that will enable more clubs to be sold. The use of face inserts of different material which in some way increase the distance carried by the ball or the spin rate generated at impact is one fashionable story at present, and statements are often made of the advantages of a particular design over others. Indeed some manufacturers have patented designs to achieve more distance and control when even a casual observation of the proposed technology would

confirm that no such benefits are possible with their designs.

Notwithstanding the lack of evidence, the golf industry may very well soon pursue the concept of increasing the distance of ball flight by correctly harnessing the spring effect of the head. It is essential therefore that the impact process is fully understood and that some idea of the possible benefits are evaluated. This paper aims to investigate that process and perhaps add to the knowledge of the topic.

The results of such a study would be twofold. If the overall conclusion were that any benefit is marginal then the governing body of golf could allow manufacturers to pursue the application of the process without any action needed under the current rules and without fear that the game is being harmed. If, however, there are indications that some combination of head shape and material can greatly increase ball velocity and perhaps flight distance, then perhaps the governing body of the game should consider a stronger constraint and/or a specific control test.

Finally the reader should know that the achievement of distance does not necessarily make for better play. Riccio (1990), in his statistical studies of amateur golfers in the USA, shows clearly that the ability to overcome distance deficiency ranks in the top 5 variables but is well below that of being able to hit the greens in regulation figures. In addition the golfer would have to achieve an increase of about 10 yards per shot to reduce his average score for the round by one shot. However such increases and more may be possible and the golfer may be encouraged to seek it.

1.1 Current design of clubs

Clubs are traditionally divided into two groups - irons and woods, so called because of the materials originally used in their construction. The iron heads are required to be flat and attached to the shaft at one end of the face. Their shape is trapezoidal and modern designs have appendages and cavities which purport to reduce the movement of the head at impact. The wood heads, now made dominantly in steel, are altogether more complex in shape, in the form of a bulbous shell with a strong face, sometimes reinforced by webs. The rules on face geometry in woods allow for curvature across and down the face, and this is proven to affect the ball spin.

Obviously tests on the face movement of such heads could be designed but they would require very careful interpretation and more importantly would probably be specific to the head and the ball used. In the future such tests may become the norm.

The authors felt, therefore, that it might be profitable at this early stage to evaluate the possible clubhead response with suitable analytical models (confirmed by experiments). To this end simple models of impact with flat plates with various boundary conditions are in construction and will be evaluated. This paper reports on the first stage of that work.

1.2 Modal analysis and finite elements

In a parallel exercise to the one reported here, Mather (1996) and Knowles (1996) have indicated that the response of the clubhead at impact is governed by the design of the whole golf club. Their work is being pursued vigorously and developed from the quasi-static situations reported so far to the dynamic.

It was also felt that more insight could be gained by evaluating the ball impact using analytical methods rather than the numerical methods of, say, finite elements. In this way, a better understanding of the effect of individual variables used in the design of the club could be achieved.

The elasticity of the ball is a clearly a powerful factor in the calculations and experiments were done to find the appropriate ball properties which are used as input to the impact theory, for the evaluation of the deflection of both the ball and the

impact surfaces modelled as flat plates with relevant boundary conditions.

2 EXISTING EXPERIMENTAL AND NUMERICAL WORK ON IMPACT

There are many reported studies of the impact of ball and clubhead.

The Cochran and Stobbs (1968) results showed that the impact time of ball and head was about 500 usecs, and independent of head velocity. The forces between the ball and the face ranged from 6 to 10 kN. Gobush (1990) measured the normal and tangential forces of impact by firing a ball from an air gun onto a force plate set into a large rectangular block. His work confirmed Cochran's suggested impact time and presented plots of the variation of force with the type of ball used and the angle of impact. He also remarked that the deflection of the ball was about 25% lower in a dynamic impact than under the equivalent static load, implying that the deflection depends on strain rate. Scheie (1990) calculated mean forces of about 15 kN for ball impact velocities of 68 m/s by determining the ball acceleration from high speed photography. Ujihashi (1994) studied the impact of balls on rods of different masses. He found that the peak force was almost entirely dependent on the ball velocity and not the mass of the rod. Mean forces ranged up to 16 kN for ball velocities of 47 m/s with impact times of about 400 usecs. The coefficient of restitution was calculated to be between 0.7 and 0.8 and depended on the construction and velocity of the ball.

Chou, Liang, Yang and Gobush (1994) compared the experimental results of the last named author with finite element solutions. The ball was modelled as a one-piece construction with a Poisson's ratio of 0.49 and an elastic modulus of 103 MPa. At a velocity of 46 m/s, the model predicted that the ball leaves the surface after 400 μsecs and the coefficient of restitution is 0.74. Overall the FE solutions and the experimental results agreed well.

A two piece model of the ball was used in finite element calculations by Iwata, Okuto and Satoh (1990). The model uses a Poisson's ratio of 0.4 for the core of the ball and 0.45 for the cover and calculates the stresses induced in the head. A point to note is that the contact time is significantly lower than previous results at 340-370 μsecs.

Thomson, Whittaker, Wong and Adam (1990) also evaluated a two-piece model which had the same Poisson's ratio of 0.49 for the core and the cover but different elastic moduli of 50 MPa and 60 MPa respectively. The results are presented in the form of stresses and deflections in the ball which agree well with observations.

It is worth pointing out that in all of these investigations, either the face of the impacting surface is deliberately made very rigid or it is assumed to be so.

3. BALL TESTS

Although some data exists on the characteristics of golf balls, we decided to confirm that data and perhaps add to our knowledge of the effect of strain rate and of the differences between a wider range of modern golf ball than we had previous data for.

3.1 *Ball construction*

Static tests were carried out on two piece and three piece balls commercially available in golf shops. The latter are often referred to as wound balls. The wound ball has a small centre, of about 27mm diameter, and a density of 1.7 g/cm^2. It is wound over with 25m of highly stretched rubber thread to give a final core diameter of 38 mm. A cover is then moulded onto the outside of the core to give a final diameter of 42.67 mm as required by the rules of golf. The cover may be of a durable ionomer (e.g. Surlyn) or balata which is a natural polymer. Synthetic balata is now available. The two-piece ball has a solid core of cross-linked rubber composition which is again 38mm in diameter. The moulded cover is normally an ionomer and of a range of different hardness.

3.2 *Compression on the club face*

Many tests have been done on the compression of the ball at impact. It is now accepted that the radius of the contact area is about 15 mm for a hard 2 piece ball and as much as 25 mm for a wound ball with a soft cover. The radial compressions associated with these areas are between 1.5 and 4 mm (this figure has to be compared to half of the compression values given in the results of the next section because the ball is being compressed at both ends of its diameter.). Surprisingly, in some of our own tests on ball spin generation we found only small variations in the shape of the area from a circle.

3.3 *Elastic modulus tests*

The elastic modulus can be found, as usual, from the load/deflection curve. Because of the non-linear elastic behaviour of the rubber-like materials the modulus of elasticity varies with strain, and therefore we will find upper and lower bounds to the modulus for each ball type.

The Instron Series IX automated materials testing system was used for our tests with crosshead speeds varying between 10 mm/min and 50 mm/min. The loads were increased to about 10 kN where the total ball deflection was about 10 mm. The balls were all new at the start of the test and were loaded in five directions by rotating the ball about a mutually perpendicular diameter between each test.

The results are shown in figures 1, 2 and 3 for, respectively, two hard ionomer two-piece balls and the core from a two-piece ball. Variations in both shape and gradient are clearly visible implying that there is some permanent deformation of the balls in each test. For all subsequent results we have used the average of five such tests.

Figures 4 and 5 show the effect of crosshead speed on the load/compression curve for a two-piece hard ionomer ball and a three-piece balata ball. As expected the three piece ball is highly non-linear but there is little variation with head speed suggesting perhaps that the ball's construction allows relaxation of the stresses. The gradient of the curve for the two-piece ball is more constant but there is a significant difference in load amplitude. The differential behaviour of the two types of ball would have an effect on the mechanism of creating spin on the ball.

Figure 6 compares the averaged load/deflection curves for most of the balls tested and at the higher crosshead speed of 50 mm/min. The variation in load at a given deflection is considerable particularly above 4 mm.

Obviously an in-depth investigation of the reason for the variations between balls and between different cross head speeds would normally follow such test results but the idea of these tests was to find usable ranges values of the elastic modulus for golf balls which are then inputted to the model of the head/ball impact.

The elastic modulus can be found by taking the gradients of the averaged curves and averaging them again for two and three piece balls. The results are shown in figure 7. The trace of the two-piece ball shows very large moduli at low deflections. The modulus gradually decreases and tends towards 90 Mpa at high deflections. The values for the three-

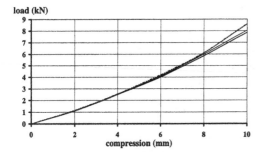

Figure 1. Compression test #1, two-piece ball.

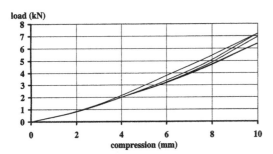

Figure 2. Compression test #2, two-piece ball.

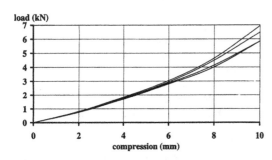

Figure 3. Compression test for two-piece ball core.

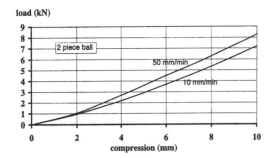

Figure 4. Compression test for a two-piece ball at different crosshead speeds.

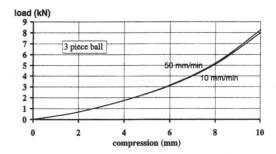

Figure 5. Compression test for a three-piece ball at different crosshead speeds.

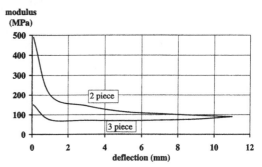

Figure 6. Compression test for two- and three-piece balls.

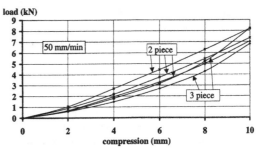

Figure 7. Elastic modulus for two- and three-piece balls.

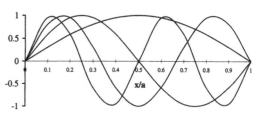

Figure 8. Harmonic shapes for a rectangular plate.

piece ball vary much less, are generally lower than those for the two piece ball and again tend towards a figure at higher deflections of slightly less than 100 MPa..

In the calculations that follow we shall assume, therefore, that the elastic modulus for a two-piece ball is 100 to 120 Mpa and that for a three-piece ball is 80 to 100 Mpa. These figures are similar to those of Chou et al (1994) and somewhat higher than Thomson et al (1990).

One further piece of information is available from these tests. The model often assumed for the collision of elastic bodies, and one which we also shall assume in the next section, is that attributable to Hertz (see for example Goldsmith (1960)). For the impact of a sphere of radius R_2 with a flat plate, his basic formula reduces to:

$$F = \frac{4}{3} \frac{E_2 \sqrt{R_2}}{1 - v^2} \cdot \alpha^{\frac{3}{2}}$$

where F is the force, E_2, R_2 and v are the elastic modulus, radius and Poisson's ratio for the ball respectively, and α is the difference between the displacements of the plate and the ball. We derived the index value for α for our results from the averages of all of the tests and found it to be closer to 1.2 than 1.5. It is this lower value that is used in our theoretical model.

4 THEORETICAL MODEL OF IMPACT

As we have stated, Goldsmith (1960) suggested a model for the impact of an elastic sphere and a flat plate. This model has the advantage for us in that it includes the derivation of the detailed movement of the plate. The forced vibration ender the action of a time dependent load w(x,y,t) is given by

$$\overline{w} = D\nabla^4 w_1 + 2\rho b \frac{\partial^2 w_1}{\partial t^2}$$

where 2b is the thickness of the plate and

$$D = \frac{2b^3 E}{3\rho(1 - u^2)}$$

The solution to this equation exists in the form

$$\alpha = w_2 - w_1(c) = v_0 t - \frac{1}{m} \int_0^t dt \int_0^t F dt - w_1(c)$$

where w_2 is the deflection of the ball and w the deflection of the plate. The three terms in the equation can be thought of as the translation of the system, the compression of the ball, of mass m, under the force F and the deflection of the plate.

The latter can be written in terms of the harmonics of vibration of the plate. These are in the form of an integral equation

$$w_1\langle x_0, y_0, t \rangle = \frac{1}{2b\rho} \sum_{i=1}^{\infty} \sum_{k=1}^{\infty} \frac{\overline{H_{ik}^2}\langle x_0, y_0 \rangle}{\omega_{ik} \int \int \overline{H_{ik}^2} \, dx \, dy}$$

$$\times \int_0^t F\langle \tau \rangle \sin \omega_{ik}(t - \tau) \, d\overline{\tau}$$

where x_0, y_0 is the central point of impact of the plate. The functions H_{ik} can be found from the eigenfunctions of the plate. For a rectangular plate which is simply supported at the edges

$$\overline{H_{ik}} = \sin \frac{i\pi x}{m} \sin \frac{k\pi y}{n}$$

where m and n are the lengths of the two sides of the plate and

$$\iint \overline{H_{ik}} \, dx \, dy = \frac{mn}{4}$$

The values of ω_{ik} can be found from the frequency equations

$$\Omega_{ik}^2 = \left(\frac{i\pi}{m} \right)^2 + \left(\frac{k\pi}{n} \right)^2$$

and

$$\omega_{ik} = b \sqrt{\left[\frac{E}{3\rho(1 - u^2)} \right] \Omega_{ik}^2}$$

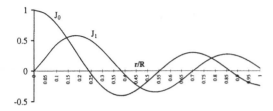

Figure 9. Harmonic shapes for a rectangular plate.

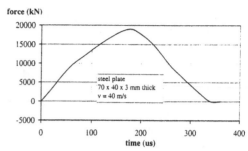

Figure 11 (a). Theoretical force versus time for a golf ball projected at a steel plate at 40 ms⁻¹.

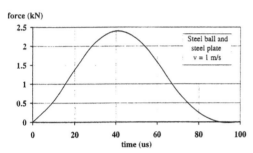

Figure 10 (a). Theoretical force versus time for a steel ball dropped on a steel plate at 1 ms⁻¹.

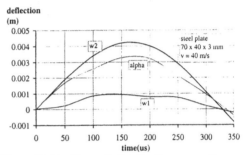

Figure 11 (b). Theoretical deflection versus time for a golf ball projected at 40 ms⁻¹ at a 3 mm thick steel plate.

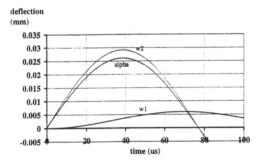

Figure 10 (b). Theoretical deflection versus time for a steel ball dropped on a steel plate at 1 ms⁻¹.

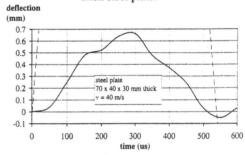

Figure 12. Theoretical deflection versus time for a golf ball projected at 40 ms⁻¹ at a 30 mm thick steel plate.

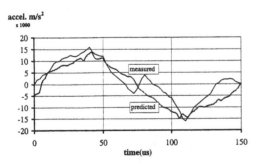

Figure 10 (c). Measured and predicted accelerations for a steel ball dropped on a steel plate at 1 ms⁻¹.

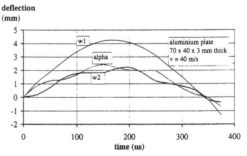

Figure 13. Theoretical deflection versus time for a golf ball projected at 40 ms⁻¹ at a 3 mm thick aluminium plate.

226

If the plate is circular with radius R and clamped at its edges then the above equations become

$$\overline{H_{ik}} = \left[J_k(\Omega_i r) - \frac{J_k(\Omega_i R)}{I_k(\Omega_i R)} I_k(\Omega_i R) \right] \sin(k\theta + \theta_0)$$

and

$$\iint \overline{H_{ik}^2} \, dx \, dy = 2\pi R^2 J_k^2(\Omega_i R)$$

where r, θ are the cylindrical co-ordinates of the point on the plate..

The frequency equation is more complicated than for the rectangular plate and becomes

$$\frac{I_{k+1}(\Omega_i R)}{I_k(\Omega_i R)} + \frac{J_{k+1}(\Omega_i R)}{J_k(\Omega_i R)}$$

Note that the relationships given in Goldsmith for the circular plate are incorrect.

The harmonics shapes for a rectangular plate are as shown in figure 8. Without loss of generality we can assume that for a central impact only the odd harmonics are excited.

For a circular clamped plate the corresponding harmonic shapes are as shown in figure 9. Here we can apply the restriction that only those radial modes associated with the zero order of the Bessel functions will be excited for a central impact. Further it may be argued that we can also let $k = 0$ for a central impact. Further work will obviously look at the non-central impact for both the rectangular plate and the circular clamped plate where higher order modes will be excited.

The equation for α can be written for a rectangular plate as follows:

$$\alpha = \left[\frac{F}{k_2} \right]^{\frac{2}{3}} = v_0 t - \frac{1}{m} \int_0^t dt \int_0^t - \frac{2}{\pi^2 b^2 mn} \sqrt{\frac{3(1 - \mu^2)}{\rho E}}$$

$$\times \sum_{i=1,3,5..}^{\infty} \sum_{i=1,3,5..}^{\infty} \frac{1}{\left(\frac{i^2}{m^2} + \frac{k^2}{n^2} \right)} \times$$

$$\int_0^t F(\overline{\tau}) \sin \pi^2 b \sqrt{\frac{E}{3\rho(1-\mu^2)} \left[\frac{i^2}{m^2} + \frac{k^2}{n^2} \right]} (t - \overline{\tau}) d\overline{\tau}$$

The inherent problem with this equation is that the time dependent force, $F(\tau)$, cannot be found directly. Therefore the solution proceeds by using a time step method assuming that $F(0)$ is known and finding the new value for $F(\Delta\tau)$ assuming that k_2 is known.

The equation can then be re-written in the time-step form given in, say, Knowles (1990). The whole is then coded into Quick Basic which enables the plotting of the results for force and deflection directly and quickly.

It remains to find how many harmonics should be included before the solution converges. Tests on the equation showed that the solution will converge for only a few harmonics in the (x) and (y) directions. For completeness 70 harmonics were used. Note that damping is not included in the solution. For the circular plate (solutions not given in this paper) some 10 harmonics will suffice to produce a solution of sufficient accuracy.

5 COMPARISON OF THEORY AND EXPERIMENT

In the first test a steel sphere was dropped onto a steel plate at a velocity of 1 m/s. The ball diameter was 25.4 mm and the plate was 120 mm x 90 mm x 15 mm thick. The results of the theoretical calculations are shown in figure 10A for the force pattern and figure 10B for the deflection of the ball and the plate. Figure 10C compares the measured and calculated accelerations for the plate. There is a reasonable correspondence between the two.

In the next test a special ball firing rig was used. This consists of two large rotating discs with high inertias which propel the ball at speeds up to 50 m/s and at angles up to 50° to the vertical. Impacts for most speeds and face angles that are found in the game of golf can therefore be modelled.

Figures 11A and 11B show the theoretical results for the impact of a golf ball at 40 m/s ontyo a steel plate of dimensions 70 x 40 x 3 mm. The peak force is over 19 kN, and the ball obviously deflects much more than the plate. The contact time is about 325 µsecs and the plate has returned to its zero position at the same time thus increasing the velocity of the ball. It is ambitious to state in practice what this increase might be but the theoretical figures suggest about 10% of the incoming ball velocity.

The results in figure 12 show what happens when the plate is thickened (to 30 mm). The plate deflection is much less (the ball deflection has been truncated to show the details of the plate movement) and it has moved forward of the null position when the ball leaves. The contact time has now increased to over 500 µsecs.

If an aluminium plate is used in place of steel, then figure 13 shows that the deflections are much larger and almost half of that of the ball. It is likely that the stress limit of this plate has been exceeded.

Table 12 shows how the four parameters, force, ball and plate deflection and contact time vary as the values of the input parameters are increased.

Table 12

	Force	Ball def.	Plate def.	Contact time
Velocity	up	up	up	up
Plate thickness	up	up	down	down
Modulus plate	up	up	down	down
Modulus ball	up	down	up	down

Note that the model predicts that the contact time varies with the four input parameters which is contrary to popular belief.

Many more combinations of plate thickness and material are being examined and rig tests continue.

6 FUTURE WORK

Work is in hand on the calculations and experiments using a circular plate and these will be available shortly. Also in development are the cases for a clamped rectangular plate and those to include energy dissipation in the ball and a distributed load.

In addition pseudo-golf club heads - that is to say ones with the characteristics of a golf club but not playable - are in manufacture one of which, according to the theoretical model should propel the ball at higher velocities and one at a lower velocity to test the validity of the model.

Material combinations are being looked into including composites, ceramics, titanium and the usual materials in use in modern golf clubs.

7 CONCLUSIONS

A theoretical model has been constructed of the impact of the golf ball and a flat plate. Values of the elastic modulus have been found from experiments on static load/deflection curves and these are in close agreement with values found by others.

The models used enable a prediction to be made of

the plate deflection and the possible effect on the ball movement. The model also predicts that the contact time varies with ball velocity, elastic modulus of ball and plate and plate thickness.

8 REFERENCES

Riccio, L. (1990) *Statistical analysis of the average golfer.* Science and Golf. Ed Cochran. E & F.N. Spon. p 153

Mather, J.S.B. (1996) *The role of club response in the design of current golf clubs.* Proc.14th IMAC, Dearborn. Volume 1. P 397

Knowles, S. Mather, J.S.B., Brooks, R. (1996). *Novel design of sports equipment through impact vibration analysis.* Proc. 14th IMAC. Dearborn. Volume 1. p390.

Cochran and Stobbs (1968) *The search for the perfect swing.* Heineman

Gobush, W. (1990) *Impact force measurements on golf balls.* Science and Golf. Ed Cochran. E & F.N. Spon. p219

Scheie, C. (1990) *The golf ball collision - 50000g's.* Science and Golf. Ed Cochran. E & F.N. Spon. p237

Ujihashi, S. (1994) *Measurement of dynamic characteristics of golf balls and identification of their mechanical models.* Science and Golf II. Ed Cochran and Farrally. E.& F.N. Spon p302

Chou, P.C., Gobush, W., Liang, D., Yang, J. (1994) *Contact forces, coefficient of restitution and spin rate of golf ball impact.* Science and Golf II. Ed Cochran and Farrally. E.& F.N. Spon. p296

Iwata, M., Okuto, N., Satoh F., (1990) *Designing golf clubs by Finite Element Method analysis.* Science and Golf. E & F.N. Spon. p 274

Thomson, R.D., Whittaker, A.R., Wong, K., Adam, A. (1990) *Impact of a golf ball with a rigid clubface.* Manual of applications of Abacus.

Goldsmith, W. (1960) *Impact, theory and behaviour of colliding solids.* Arnold Ltd.

The Engineering of Sport, Haake (ed.) © 1996 Balkema, Rotterdam. ISBN 90 5410 822 3

The development of protection systems for rock climbing

R.A.Smith

Department of Mechanical and Process Engineering, The University of Sheffield, UK

ABSTRACT: The historical development of protection systems for rock climbing is summarised. A simplified theory for the dynamic loads on ropes subjected to falling masses is presented. Application of these ideas to further kinds of protection systems are discussed and the paper ends with an outline of ideas that can be used to stimulate and inform the teaching of engineering students.

Climb if you will, but remember that courage and strength are nought without prudence, and that a momentary negligence may destroy the happiness of a lifetime. Do nothing in haste; look well to each step; and from the beginning think what may be the end.

(Whymper, Scrambles Amongst The Alps, 1871)

1 REVIEW OF THE DEVELOPMENT OF ROCK CLIMBING PROTECTION SYSTEMS

1.1 *Beginnings*

Rock climbing, a branch of the wider sport of mountaineering, involves an element of risk. This is one of its attractions. In the approximately 120 years since the inception of rock climbing as a sport, the equipment used to protect participants from death or injury has developed from extremely rudimentary to scientifically sophisticated, enabling participants to push up standards of performance whilst still maintaining some element of risk. This subconscious adjustment of an individuals risk 'thermostat', known as risk compensation, has been noted in other activities, such as driving which becomes bolder when the driver is protected by, for example, air bags, anti-lock brakes and seat belts.

To the uninitiated the joining of a team of climbers together on a rope represents a source of danger, such that if one slips, the remainder are pulled off. Indeed some of the earliest mountaineering accidents on the Alps seems to substantiate this idea, and the term rope was used as a noun to describe the system of the climbers and their connecting rope. The classic accident occurred on 14 July 1865, when on the descent from the first ascent of the Matterhorn, Croz, Hudson, Lord Douglas and Hadow fell from the mountain to their deaths and would have dragged Whymper and the Zermatt guides, the Taugwalders, father and son after them, had the rope not broken, see Fig. 1. In Whymper's epic book, Scrambles Amongst The Alps (1871), he observes that there is no good reason for employing a rope on easy rocks, because its use is likely to promote carelessness, but on steep rocks it should be used by adopting the plan of moving only one at a time. He reported that a committee of the (English) Alpine Club tested ropes for mountaineering purposes in 1864 (see Kennedy et al, 1864) and approved two types, of manila and Italian hemp; both of which could sustain 168 lbs falling 10 feet, or 196 lbs falling 8 feet, and break at a dead weight of two tons. The manila rope weighed 6.4 lbs per 100 feet. (In order to avoid constant conversion in the text of original units, note that 1 lb = 0.45 kg, 1 foot = 0.030 m and 1 ton ≈ 1000 kg). It is worth noting that the above figures are equivalent to an average sized climber and the weight of his equipment falling 3m or a heavier climber falling just under 2.5m. Manila hemp ropes were made from a fibre obtained from

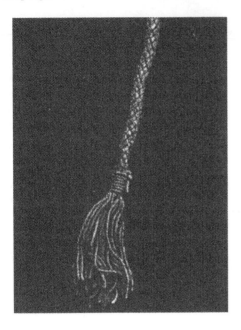

Fig. 1. Rope broken on the Matterhorn
(Whymper, 1871)

the Philippine abaca plant. Hemp fibre, generally, from the cannabis plant also came from India and Italy. Flax ropes were made from the fibre of the herbaceous plant of the same name. Until the advent of artificial fibres all ropes were made from these natural sources.

1.2 *The First Half of the 20th Century*

The later years of the nineteenth century and early years of this century, saw a rapid increase in the severity of rock climbs taken by an increasing number of climbers. Typically the climbers moved out of the security offered by gullies and chimneys, to the more open faces of steep crags. The situation at the start of the First World War, is recorded in a classic instructional book of the era, by G D Abraham (1916). "The parting of a rope to which a climbing party is tied..... is a frequent accompaniment of an accident. Yet this generally means that the leader has fallen, and but for the breakage of the rope the rest of the party must have been dragged down". Abraham then describes the system in which the leader climbs to a resting place or anchorage, whilst the second

man "carefully watches the leader's upward progress, and slowly pays out his rope, probably around some outstanding knob of rock, known as a belay or belaying pin".

Fig. 2. The "belaying pin"
(Abraham, 1916)

Abraham further describes how the leader after running out a length of rope may be able to find a stone wedged in a crack, such that, "it is often possible to untie the rope end from the waist, thrust it up behind the stone, *from below*, be it noted, and retie on again". This is an early description of what later became known as a running belay. He says that "the new English Alpine Club rope - and no other should be used - is tested to hold a twelve - stone man falling ten feet through mid-air. Since 12 stone = 168 lbs, this is exactly the same figure quoted by Whymper from 1864, that is 52 years previously.

Abraham described an accident on Eagles's Nest Ridge on Great Gable in 1909 and noted other similar ones, when the rope broke at the position of the direct belay, after a leader had fallen. In addition to the accident described by Abraham, we might note that Owen Glynne Jones, a frequent climbing companion of the Abraham brothers, was probably the first to use the threaded belay in a climb with the Abrahams in 1886. Jones was killed in 1899 when the leading guide fell on the Dent Blanche, pulling three other climbers, including Jones, to their deaths. The rope broke leaving one climber still on the mountain. The first fatal accident in the Lake District occurred in 1903 when a party of four fell to their deaths roped together from Scafell Pinacle. It is surprising that in the face of these and similar accidents, the lack of real rather than illusory protection to roped climbing parties continued for so many years. Abraham suggested that a double rope might offer a better safeguard, but states "the leader must never slip......If a leader has ever been known to fall, the writer would emphatically advise all climbers not to accompany such a one unless he takes on an inferior position on the rope". It is worth noting that if such advice were to be followed today, there would be a distinct shortage of leaders!

No review of the development of climbing techniques can ignore, Geoffrey Winthrop Young's "Mountain Craft", published in 1920. Young distinguished between an 'anchor' , that is a loop of inactive rope with which a stationary climber secures himself to a rock point, in order to protect himself and the rest of the party while somebody is climbing and a 'belay' which is the rock-and-rope attachment by which the active rope of a moving man is protected while it is running out or being pulled in. Further, distinction was made between a 'direct belay' where the rope in action connects directly onto or round rock and an 'indirect belay' where some from of human spring is interposed between the active rope and the solid rock. Young recognised that the *direct belay* was unsound to protect the *leader*, because of the danger of the rope breaking. He then stated that "a long rope may take up much jerk in its elastic spring, but a short rope cannot. This should be more widely known". In a chapter of Young's book on Equipment, written by Farrar, the properties of rope are discussed. Because of fatalities due to rope breakages, Farrar had some tests performed and reported that flax rope in terms of strength and extension, surpassed weight for weight any other rope. For 1.4 inches circumference rope breaking strengths of 1904 and 1992 lbs were reported for flax and manila respectively and corresponding extensions of 16.3 and 12.3% on a 5 foot length of rope. Further the work required to break a test length of 5 feet was 451 and 332 foot pounds. As far as the author is aware, these are the first references in the literature to the importance and quantification of rope's energy absorbing properties.

These figures represent very low energy absorbing capabilities: as an approximation, noting that the work required to break the flax rope is 90 foot pounds per foot, then an extremely short drop of a 12 stone man (168 lbs) on a dead belay will be sufficient to break the rope. We note that these figures seem small compared with the results for the Alpine Club drop tests mentioned earlier.

1927 saw the publication of a paper, in the Rucksac Club Journal by Bower remarkable for its penetration combined with humour; it will be referred to later in the section on pedagogical applications. The tests reported by Young are recalled, together with later tests which gave a value of 152 foot lbs/foot for the resilience of manila rope. However, the conclusion remained the same, that "the rope will break no matter what its length may be, when it is fixed at one end to a belay just above a ledge from which the ecstatic experimentalist escapes to Erebus". In the light of subsequent developments, there follows an interesting suggestion "if a leader contemplates making a speciality of 'first descents' he had better invest in an oil-filled shock-absorber, in which the kinetic energy of the fall is absorbed by the oil being forced past the clearance between the piston and the cylinder. The latter is attached to a special waist belt, and the climbing rope to the piston rod. The more dashing juvenile spirits will then be readily identified from a distance by a sporty smell of Castrol pervading their neighbourhood". Despite the humour, the conclusion is serious and, by now familiar: "The Moral of Morals then is: DO NOT FALL".

Bird in the Climbers Club Journal of 1931, published an article called "The Strength of Ropes" which summarised the current knowledge and added a little more, including the important consideration that the maximum load generated on the climber by the rope should not, 'from anatomical consideration' exceed 1000 lbs. It should be remembered that the method of attaching the rope to the climber was very simple - a bowline knot round the waist. Thus shock loads were transmitted through a very small area and could themselves cause considerable damage to the falling climber - yet another reason for the conclusion of this paper, (the results) "confirm the oft- repeated dictum that the leader must not fall".

Complete lack of belaying by the leader coupled with an inadequate follower's belay caused a fatality on Dow Crag in April 1932 and an injury to one of the climbers, H W (Bill) Tilman, which hampered him throughout his famous career, (Chorley, 1932).

A further book by G D Abraham (1933), disappoints by adding nothing new to his earlier cited work as regards ropes and belaying methods, but a paper by a climber then at the height of his powers, A T Hargreaves in the Fell and Rock Club Journal (1935), illustrates, see Fig. 3, a classic indirect belay. He offers advice about a better attachment to the rope for the leader, by making a rudimentary harness under the armpits and over the shoulder and illustrates a free running thread belay, in which a spare loop of rope is threaded round a clockstone and the rope is run through this loop rather than behind the clockstone. In end notes added by the Journal editor (G R Speaker), a

STANCE & BELAY

Fig. 3. Attachment to an anchor
(Hargreaves, 1935)

discussion is held on the ageing on ropes (advised to retire after 100 climbing hours) and the need to scrap a rope after a fall, together with some estimates of the reduction in rope strength caused by various kinds of knots (as much as 40%). The Munich method of using an independent waist rope is claimed to overcome several weaknesses - this is an early (the first?) example of acknowledgement of superior practice from overseas. Belaying by a loop and karabiner (an oval metal ring with a sprung side opening) was mentioned by Peacock in his textbook 'Mountaineering' first published in 1941.

1.3 *After the 2nd World War*

The 2nd World War saw the introduction of a new 'wonder'; material, nylon, and commando soldiers were trained to climb using it and any other aid that made scaling cliffs possible: the ethics of sport had no place in the serious business of war. The popularity of rock climbing after the war, merited the publication of a Pelican hand book, 'Climbing in Britain' by J E Q Barford (1946). Details of hemp and manila ropes were given and it was added that "experiences in the services has shown the virtue of nylon and it is not improbable that soon this may become the standard rope". In austerity Britain, it was hard to come by as was money for any kind of climbing equipment: the author suggested that "As a measure of economy it is permissible to cut out an injured section of the rope and splice the join" !

Another handbook was published in 1955, The Technique of Mountaineering by J E B Wright. Details of hemp rope were still being quoted. Nylon was quoted as having a tensile strength of 4000 lbs for full weight (5.50 lbs/100 foot) rope, but "although nylon remains more flexible than hemp when wet, one of its great disadvantages is that it melts quickly under friction heat". For this reason a thin hemp waist line, wrapped four or five times round the body before being knotted was used in conjunction with a karabiner to attach the climber to a rope. (The present author was introduced to climbing in 1963 when this method was still in common use). Wright recognised that "good mountain walking, climbing and the use of mechanical devices is, in the main, applied dynamics" and stated that "two dynamic theories are widely accepted; the dynamic theory of Kant which claims that energy is dependent upon mechanical activity and the doctrine of Leibnitz that all substance involves forces". One supposed he knew more about climbing than dynamics! He did however, introduce to a wide audience the work of Tarbuck (1949) who introduced a sliding friction knot to provide elasticity in the belay chain (*cf*. the dash-pot of Bower) and Wexler (1950), who published theoretical calculations on the dynamic theory of belaying, which will be introduced in the next section of this paper.

Developments since the mid-1950's have been rapid and effective. The following summarises what was already a brief summary, originally published to accompany a television programme 'Fear of Falling' in 1993.

The first type of nylon rope was hawser laid

formed from three strands. In the 1950's a German company invented the *kernmantel* rope, which used narrow nylon strands running the length of the rope, the 'kern', protected from abrasion and dirt by a braided sheath, the 'mantel'. This type of rope is less prone to linking than the hawser laid nylon rope, has great strength; typically 11 mm diameter has a breaking strength of 2,300 kilograms and is now the only type of rope used for climbing.

In the 1960's the problem of tying on via a waist loop was addressed. Using this method the force of a fall was concentrated around the waist, when the soft internal organs, the ribs and the spine could be damaged. Further, after ten minutes or so hanging in such a device, the climber would lose consciousness, thus despite the improvements in the rope the traditional maxim still held. The first solution was to use wide waist-belts. The first ones were made in leather from machine-belting from old woollen mills in the author's home district, Saddleworth. The leather was replaced when flat nylon webbing became available, but in 1970 Don Whillans invented the sit-harness, a belt with integral leg-loops which transferred some of the load to the legs. Various improvements and modifications have subsequently been made and modern harnesses are now lightweight but comfortable and efficient in distributing load between thigh muscles and the pelvic girdle.

If no clockstone existed for running or main belays, climbers inserted their own small stones into cracks, this, in the main, avoiding the use of pitons, metal spikes hammered into cracks, which had been developed by the Munich school before the war but were generally thought unsporting by British climbers. Artificial clockstones made of steel, then aluminium became available in the late 60's, in various sizes, hexagonal and tapered wedge designs to suit most cracks. During this period, the strength of karabiners improved dramatically with a combination of better lighter materials (aluminium and titanium alloys) and better design to eliminate bending loads and strengthen the gate. Nylon webbing tape became available to attach the belay, natural or artificial, via a karabiner, to the main climbing rope or climber. These improvements in belaying opportunities, coupled with the improved rope and harness, meant that leaders might be able to fall and get away without serious injury. Although not the topic of this present review, it is worth noting that boots had remained unchanged for more than a century:

heavy nailed boots being replaced by rubber soled "Vibrams", then thin but rigid Klettershuhe and finally light but stiff smooth rubber 'PAs' or 'EBs' by around 1970, see Brigham (1976).

1.4 *Recent Developments*

One the most important developments in the last fifteen years has been the introduction of friction belay devices. The device clips via a karabiner to the harness and a rope is passed through the device in such a way as to generate a large frictional force due to a large angle of wrap. Holding a fall is thus made much more straightforward for the second man and, further, should the leader be injured, it is much easier for the second to transfer the load directly to the belay, release himself and go to the aid of the leader.

A further ingenious development has been the moving cam belay device which can be inserted into parallel or flared cracks and can provide protection from high shock loadings. Introduced from America, 'Friends' have made running belay placements both easier and more reliable.

Rising prosperity has enable climbers to buy and use large quantities of protection equipment. Climbs which were previously difficult to protect can now be 'stitched' together with running belays. Systems of double rope operation are in wide use on more difficult climbs and the gear used is generally much more reliable. Ropes have adequate strength and resilience, belay devices are tenacious and strong. If the protection gear is correctly placed, the leader can be reasonably confident of surviving a fall. Indeed, many would say that if a leader does not fall, he is not trying a hard enough climb! In view of what was said earlier on risk compensation, it would be interesting to see if accident statistics have decreased with the availability of superior equipment. This study needs access to considerable historic and current data and, as far as the author is aware, has not yet been satisfactorily completed. A separate branch of the sport, performed on pre-bolted routes or even on indoor artificial climbing walls, has developed which requires outstanding agility and gymnastic strength, but can be performed at nearly no risk should a fall occur.

2 SIMPLE THEORY OF DYNAMIC LOADING OF CLIMBING ROPES

As previously noted an appropriate theory of the

dynamic loads generated in ropes by falling climbers was produced by Wexler (1950). Although this theory contains several simplifications, it produces some sound general conclusions and is worth reproducing in part. Consider the situation shown in Fig. 4(i): A climber has moved above an anchor, A, past a running belay placed, at B, by a distance H/2 to C. At this time the total length of rope run out is L. The climber then falls freely and vertically, Fig. 4(ii). At a distance H/2 below the runner, D, the rope becomes taut and begins to extend. At position E, the rope has stretched by its maximum amount, δ and the climber is momentarily at rest. The sequence of events can be described in terms of energy exchange: the total energy (TE) of the system remaining constant during the process. At C the energy of the system is the potential energy (PE) of the climber due to his elevated position. As he falls PE is exchanged for kinetic energy of motion (KE), until the point D is reached. The rope's stretching then begins to store energy in the form of strain energy (SE). At E, the PE has been reduced to a minimum, the KE to zero and the strain energy is a maximum. If we assume that the rope is *elastic*, Fig. 4(iii), then the load is induced in the rope, P, is proportional to the extension, x. It is convenient to express the load as $P = kx/L$ where k is a measure of the rope's elasticity. The strain energy stored is the (shaded) area under the load/extension line and is given by $SE = kx^2/2L$. These energy changes are shown on the sketch of energy/position in Fig. 4(iv), on which the PE datum corresponds to the height at D. Since $PE = mgh$, where h is the distance from the datum m is the mass of the climber and g gravitational acceleration, the PE decreases linearly from C to E, whilst strain energy begins to accumulate with the square of the stretch from D to E. Notice that the sum of the PE and SE follows the solid line drawn between ED and that the distance between the constant overall energy, TE, and this complete solid line is the kinetic energy, KE. The maximum speed occurs at extension d when the force in the rope just balances the downwards force on the climber due to gravity in $mg = kd/L$, the minimum of the PE + SE sum, position O on Fig. 4(iv).

By equating the total energy at C to the loss in PE and the gain in SE at E, we can write:

$$mgH = -mg\delta + k\delta^2/2L$$

The solution to this quadratic equation for the maximum rope stretch δ can be written

$$\delta = \frac{mgH}{k} \left[1 + \sqrt{1 + \frac{2k}{mg} \cdot \frac{H}{L}} \right] \qquad (1)$$

Alternatively in terms of the maximum force P_{max} corresponding to the stretch δ,

$$P_{max} = mg \left[1 + \sqrt{1 + \frac{2k}{mg} \cdot \frac{H}{L}} \right] \qquad (2)$$

Notice that the static force required to support the climber's weight is mg, so that the square root term in Equn(2) represents the magnification factor due to dynamic loading, and for a give rope stiffness, k, and a given climber, m, depends on the ratio H/L, that is *the height of the fall divided by the length of rope run out*. This ratio has been termed the *Fall Factor*. Note that the Fall Factor can vary from 0 to 2. A value of 1 corresponds to a fall past a running belay halfway between an anchor and the maximum height reached by the climber and a value of 2 corresponds to a fall past a fixed belay or anchor. *The absolute values of the height fallen are unimportant*: the forces generated are governed by the Fall Factor (FF) ratio.

Equation (2) has been evaluated for the case of an 80 kg climber falling on three different kinds of rope. The required stiffnesses, k, for manila and hemp were taken from data in a report in the Alpine Journal (Anon. 1931), that for nylon from Wexler (1950). As the fall factor increases, the dynamic loads increase. In each case the force needed to break the rope has been added to the graph. Notice that the relatively high stiffness for manila and hemp, means that high dynamic forces are generated and Fall Factors of approximately 0.5 and 0.75 are sufficient to break these ropes. The superiority of modern kernmantle is clearly shown: the low stiffness generates lower dynamic loads, such that even at FF = 2, the dynamic load is about 2½ times less than the strength of the rope.

When these results were originally published, they caused considerable concern. Wexler (1950) showed how the dynamic loads could be reduced by resilient belays or by letting the rope slide on impact. Tarbuck (1949) used the same arguments and suggested the use of an eponymous knot which could slide and absorb extra energy. Although resilient belays are still used, modern belay devices use sliding friction to attenuate dynamic loads. The simple analysis resulting in equations (1) and (2), can be used to estimate the time taken for the load to rise during dynamic loading. If the equation of motion for the mass is written as

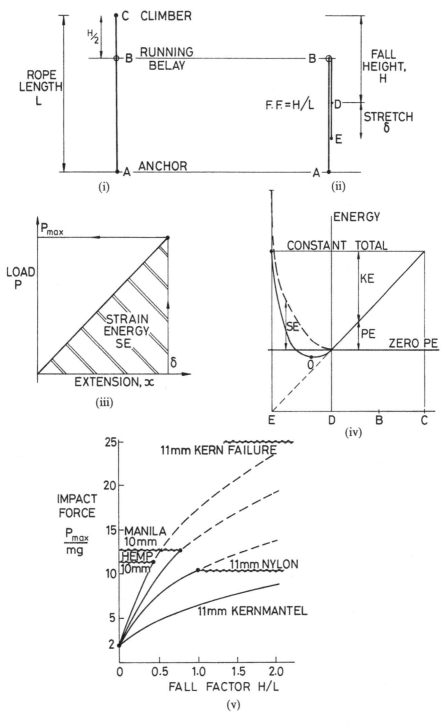

Fig. 4. Theoretical forces generated in a fall compared with rope strength
(see text for details)

$$m \frac{d^2 x}{d t^2} = mg - \frac{kx}{L}$$

and reorganised to the standard form, it is easily recognised that it represents a single degree of freedom system subjected to a restoring force. If the rope could accept both tension and compression, the system would execute oscillations with frequency $\omega^2 = k/mL$. As an approximation, then, the rise time T_R to maximum load is given by the period of vibration/4. Thus

$$T_R \approx \frac{\pi}{2} \sqrt{\frac{mL}{k}}$$

Now the time taken for a climber to fall freely under gravity from 3m above an anchor, to 3m below is simply $\sqrt{2H/g}$ with $H = 6m$ and $g = 9.8 m/s^2$ ie. 1.1 seconds. Assuming an 80 kg climber falling on a kernmantle rope, $k = 1,100$ kg, then the load rises to a maximum in 0.23 seconds. This is an extremely short time: it is essential that the second man concentrates! The awful suddenness of the events subsequent to a fall has to be experienced to be believed.

It should now be clear that the high impact forces can injure a fallen climber - the stretch of the rope serves to limit these forces to an acceptable level. Military research on the opening impact of a parachute on the human body have suggested that 12 kN is the maximum force a body can withstand without injury. This figure has been used as the basis of a standard test laid down by the Union Internationale des Associations d'Alpinisme (UIAA). In this test a drop weight is used to simulate a leader's fall. The weight is dropped free for 5m with 2.8m of rope in use, ie. a Fall Factor of 1.78. The weight used for a single rope is 80 kg and during the first fall the impact force must not exceed 12 kN. If this test is repeated after a short time interval, it is found that the stiffness of the rope increases, thus increasing the impact force. The effect tends to saturate after some 7 or 8 falls - a good rope will still then have an impact force of less than the maximum allowed value.

3 APPLICATIONS TO OTHER TYPES OF PROTECTION SYSTEMS

Developments in rock climbing gear have enabled many apparently dangerous industrial jobs to be performed in comparative safety. The maintenance of towers and cables of suspension bridges, the care of power lines and towers, the inspection of narrow flues in chemical plants, the cleaning of the exterior of high buildings and the investigation of birds nesting sites on steep cliffs are all examples of situations where 'personal protection gear' against falls is vital.

As this paper is being written two groups of climbers are pitched against each other at the site of the proposed Newbury by-pass. One group have used their climbing skills to climb up trees which are to be uprooted: the other group have been hired at £250/day to evict them!

In general workers in these exposed situations wear a nylon webbing harness and are either protected by a climbing rope in the normal manner or are attached via a short webbing strap and karabiner to an anchor. The difficulty arises in that a fall onto this short attachment generates a Fall Factor of 1 and therefore high dynamic loadings which can injure the user. If the karabiner can slide downwards before it comes up against a stop then Fall Factors higher than 2 can be generated: a situation which can also occur on fixed cables such as those in the Dolomites known as "Via Ferrata". To guard against these situations, special energy absorbing tape has been designed in which a loop of tape is stitched together and folded in a zig-zag manner to occupy a short length. On impact loading the tape 'unfolds' absorbing energy whilst its length increases, thus providing a kind of damper to absorb the shock. Such industrial safety lanyards have been designed to comply with detailed European Standards which require, inter alia, that the length of the lanyard including the energy absorber shall not exceed 2m, that it should withstand a dynamically applied force of 100 kg with a Fall Factor 2, such that the breaking force shall not exceed 6 kN and the arrest distance shall not exceed 5.75m.

Many readers will recall seeing film of the spectacular land-diving, or *naghol* ceremony, held in the village of Vanuatu on the Pentecost Island in the South Pacific. This is part of an age old ceremony, the Festival of the Yams, held every April and May to celebrate and bless the crop. Village men and youths leap from an 80 feet high wooden tower with only two springy liana vines tied to their ankles. If they judge the distance right, their foreheads will brush the soft soil, symbolically refertilising it. Clearly, this is a very dangerous activity - even during a very well planned demonstration held to honour the visit of the Queen on 1974, a diver was killed. Strange then that people all over the world have copied this practice and re-named it *bungee jumping*. Naturally, devotees of this strange sport do not rely on the uncertain strength of liana vines, but use a specially developed bungee rope.

These ropes (or cords) are typically much thicker than climbing ropes, often in the order of 23 mm diameter. The internal structure consists of a large number (~400) of elastomeric filaments which run the length of the rope. These filaments are held together by two outer layers of a woven synthetic material. The strength of such ropes is considerably lower than climbing ropes, at about 600 kg, but the extension to failure is about an order of magnitude higher at about 170%. This very high elasticity gives the characteristic yo-yo motion at the end of a bungee jump and, of course, acts to reduce the dynamic loads in the rope. Typically, recalling that a bungee jump has a fall factor of 1, the jump of an average sized person will generate a peak dynamic load of about four times body weight and an extension of about 90%. Strong internal damping within the rope acts to reduce the amplitude of the oscillation at the end of the jump.

4 PEDAGOGICAL APPLICATIONS

Application of the principles of mechanics and materials to climbing offer the opportunity to inject relevance and excitement, as well as practical experience, into the teaching of young students of engineering. At my own University of Sheffield, we note that many students are attracted to our course because they (rightly) regard Sheffield as the leading centre of the UK climbing scene and they avail themselves of the many opportunities to climb on gritstone outcrops in the surrounding Peak District and, latterly, on several indoor artificial climbing walls.

Any serious student should be directed to the paper by Bower (1927). This paper deals in a classical theatrical manner with the theory of climbing equilibrium using geometric and trigonometric analyses, the mechanics of slab and wall climbing, together with land traversing are discussed, together with the previously mentioned notes on belaying and rope strength. It is sobering to reflect that the quality of writing and the subtle humour, combined with sound practical conclusions, is not now generally seen in climbing club journals. Students may care to ponder on the complexity of the analysis and the brilliance of the use of English which was attained some 70 years ago and use this paper as a model for their own efforts!

More recently Hudson and Johnson (1976) produced an excellent article in a teaching journal in which they demonstrated the use of the principles of friction and equilibrium to climbing positions and discussed the mechanics of arresting a fall, including rope friction effect. This paper was the basis of an article designed for a much wider audience (Walker, 1989) which subsequently appeared in Scientific American.

Some standard text books have included problems involving the dynamic loadings of climbing ropes: Sandor (1987) is a splendid example and also includes many examples of the application of mechanics principles to skiing, another area of considerable interest to many students. Jones (1993) devoted a chapter to this topic in a text book of case studies. The analysis of dynamic loading of ropes is, of course, not new. It appeared in standard engineering text books long ago, eg. Goodman (1899). He examined the case of a weight falling onto a collar at the end of a vertical bar. The now well known 2 times magnification of load due to a suddenly applied load dropping through an infinitely small distance was derived. In a curious way over the years this idea has become transmuted into the Goodman law of fatigue, which relates permissible levels of mean and cyclic stresses. Further discussion of this interesting point awaits a future publication. The dynamics theory of rope loading can, of course, be applied to problems such as the winding of heavy cages in mine shafts. When the author, early in his career, installed new gears in the winding mechanism of such systems and was 'invited' to be the first man down, he had more than an academic interest in the strength of wire ropes!

Further applications of climbing gear technology to teaching programmes include discussions of materials developments (polymers, alloys, heat treatments) stress/strain relationships, including non-linear behaviour and strain rate dependence, the effects of stress concentration, and, particularly in the design of karabiners, shape optimisation and the minimisation of bending effects.

CONCLUDING REMARKS

For near 100 years since the inception of the sport of rock climbing, very little progress was made to improve the chances of surviving a fall. Since the

2nd World War rapid improvements in equipment and technique have been made and the level of risk involved has been substantially reduced. With correctly placed protection equipment, leaders now fall and live to tell the tale!

REFERENCES

Abraham, G. D. 1916. *On Alpine Heights and British Crags*, London: Methuen

Abraham, G. D. 1933. *Modern Mountaineering*, London: Methuen

Anon. 1931. Report on Rope. *Alpine J.* 43, No 243: 325-329

Barford, J. E. Q. 1946. *Climbing In Britain*, Harmondsworth, Penguin Books
(Barford was killed by a stonefall in the Dauphiné on 23 July, 1947, in a party with W. H. Murray and M. Ward, see Alpine J. 56, No 275, p 190, 1947)

Brigham, B. 1976. Underfoot information. The Story of the Climbers Boot. *Alpine J.* 81, No 325: 133-142

Bird, A. L. 1931. The Strength of Ropes. *Climbers Club J.* 17, New Series 4, No 3: 192-196

Bower, G. S. 1927. Climbing Mechanics. *Rucksac Club J.* 6, No 1: 65-78

Chorley, K. C. 1932. Editor's Notes, Accidents. *J. Fell & Rock Climbing Club* 9, No 2: 205-206

Goodman, J. 1899. *Mechanics Applied to Engineering*. London, Longmans

Hargreaves, A. T. 1935. Rope Management. *J. Fell & Rock Climbing Club* 10, No 2: 232-242
Hudson, R. R. & Johnson, W. 1976. Elementary Rock Climbing Mechanics. *Int. J. Mech. Eng. Education* 4, No 4: 357-367

Jones, D. R. H. 1993. *Engineering Materials 3: Materials Failure Analysis*. Oxford, Pergamon.

Kennedy, E. S. 1864. Report of the Special Committee on Ropes, Axes, Alpenstocks. *Alpine J.* 1, No 7: 321-331.

Sandor, B. I. 1987. *Engineering Mechanics:*

Dynamics. Englewood Cliffs, NJ: Prentice-Hall

Stevenson, S. 1993. *Fear of Falling*. London, Channel 4 Television

Tarbuck, K. 1949. Safety-Methods with Nylon Rope. Series in *Mountain Craft* No 3 to No 14 (1952). (Also published in collected form as a Special Supplement)

Walker, J. 1989. The Amateur Scientist: The Mechanics of Rock Climbing, *Scientific American*, 267, No 6: 92-95

Wexler, A. 1950. The Theory of Belaying. *American Alpine Club J.* 7, No 4: 379-405. (Published in the UK as a separate Special Supplement of *Mountain Craft*)

Whymper, E. 1871. *Scrambles Amongst the Alps in the Years 1860-69*. London: Murray

Wright, J. E. B. 1958. *The Technique of Mountaineering*. 2nd Ed. London: Kaye

Young, G. W. (Ed). 1920. *Mountain Craft*. London: Methuen

The Engineering of Sport, Haake (ed.) © 1996 Balkema, Rotterdam. ISBN 90 5410 822 3

Experimental mechanics and artificial turf

C.A.Walker
University of Strathclyde, Glasgow, UK

ABSTRACT: The viscoelastic nature of artificial turf pitches has been evaluated, and the decay constant (0.25 s) was found to be too long for time - dependent pitch properties to affect footfall reaction forces. It has been suggested that pitch design parameters and pitch testing, should be derived from quasi-static tests. In the light of this, recommendations have been made for the design and maintenance of pitches, and for the choice of footwear to avoid the risk of long-term pounding injuries.

INTRODUCTION

The test procedures for artificial turf pitches may be divided into two groups:- firstly there are the functional elements, such as the height of bounce, the rolling resistance, the bias (i.e. the deviation off a straight line which a rolling ball experiences), and the damage potential, which simulates a player falling and striking his head; secondly there is a further set of tests which relate to the "feel" of the pitch - how hard/easy it is to run and turn on, the extent to which a player's limbs are jarred or cushioned, and the long-term effect of playing on artificial pitches.

Such long-term studies as have been carried out relate mostly to American Football (Powell et al, 1992, Keene et al, 1980). It was found that the injuries recorded over 10 years, on grass, and on artificial turf, really failed to show a case for or against either type of pitch. On the other hand, it might be imagined that in the UK, with its lush growth of grass, and the generally plentiful rain, that on average grass pitches would in general be softer and more cushioning. There have been a number of studies carried out over the years, to assess the compliance of pitches, both natural and artificial turf, and these h;have mostly used a variation of the impact test. (e.g. Martin, 1990; Rogers & Waddington ,1990) The DIN Standard test method, the "Stuttgart Athlete" utilises a 50 kg drop test weight, and is typical of the impact testing systems that have been used. The acceleration of the weight striking the pitch on a 100 mm diameter "foot" is used as a measure of performance: the time scale of the impact is approximately matched to the impact of a foot on the pitch. Over a period of time, this has been used widely, but the DIN standard is currently under review, at least in part due to the limitations of this approach. These limitations will now be discussed at length.

A CRITIQUE OF IMPACT TESTING SYSTEMS FOR PITCH EVALUATION.

The basic idea behind the design of impact testers is that they should simulate the interaction of the foot with a pitch. Under normal running conditions, the foot will experience peak accelerations of 2.5g, on a well-sprung surface, over a time-scale of 50 milliseconds. All impact testers simulate the footfall by dropping a weight on to a foot-sized plate; a spring is used to match the impact force profile to the 50 millisecond time scale. The reaction forces on the drop mass are recorded by an accelerometer, while in some systems, a transducer on the "foot" will record the pitch deformation.

The data are usually analysed in terms of a force-time curve, with the peak deceleration being used as the quality index - the lower the better. Alternatively, a Severity Index (S.I.) may be calculated, viz.

$$S.I. = \int a^{2.5} dt$$

Valid as these measurements may be for their own purposes, a dynamic analysis of the pitch-limb system (McMahon & Greene, 1979) indicated that the relevant pitch stiffness parameter should be the tangent modulus at the point on the load-deflection curve (Fig 1).

The point of this analysis is that the peak acceleration may be predicted directly if the tangent modulus is known, and since the tangent modulus may be readily measured from quasi-static tests, this parameter may be used for the purposes of pitch design. From the standpoint of designing an experimental system to assess pitches, it can be seen that, owing to the non-linear nature of the load-deformation curve, it is important to match the load per unit area on the pitch during measurement to that actually experienced by a foot. If the "foot" in the measurement is too small or too large, then the wrong value- of the tangent modulus will be calculated. The simplest approach is to use a heel-sized, heel shaped impact plate; for the measurements in this study, a spherical cap, 50 mm in diameter, and 50 mm in width was chosen as the "standard heel". Associated with this point is the need to scale the applied load to the size of the chosen impact plate. For the chosen dimensions, close to life-size, loadings of typical athlete size will be required, with the usual dynamic multiplier of 2.5 for well-sprung surfaces, and up to 4 for hard surfaces.

Such loads are achievable with impact masses at the upper end of the range of those used - say 50 kg. The correct load levels cannot be attained with impact weights in the range 1.5 kg, such as have been used

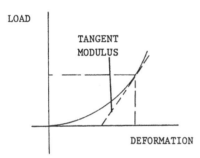

Fig 1. Definition of tangent modulus

in some studies. The point is that such light impacts will give comparative readings from one surface to another, which may be misleading, unless they are combined with other data which allow an extrapolation to higher loads. Plainly, it is possible to scale down the size of the "standard foot" which is to be used for measurement, but a point is quickly reached at dimensions just below a life-size heel, where the indentation into the turf becomes excessive.

In summary, then, impact experiments should aim, as far as is possible, to duplicate the conditions of an actual footfall - there are no easy shortcuts; however, this is not the last word, as will become plain on further consideration.

PROBLEM STATEMENT

The real problem which exists is one of particular relevance to hockey, for while soccer has turned its back on artificial pitches, all first class hockey is played on carpet, even at school level in many instances. While most modern pitches are well-designed, many older pitches were laid down to minimise maintenance and maximise usage, by

compromising as dual football/hockey pitches. These tend to become badly compacted and unyielding to play on. Many footballers simply play in training shoes, with inbuilt cushioning, while hockey players often play in "Astro" shoes, with no cushioning.

The question arises, therefore, as to what test method should be used; what variables are relevant which pitch design will protect players' legs in the long term and what maintenance is required. This investigation set out to look at the elements in an artificial pitch, and to evaluate the elements singly and in combination. In particular, the aim was to look at the problem, recognising that the elements formed a viscoelastic/plastic system, and to attempt to construct guidelines for testing, for maintenance and for advice to players as to the appropriate footwear.

THEORETICAL CONSIDERATIONS

When the design of systems for testing artificial pitches have been considered, it has been presumed that a dynamic measurement was required to represent the dynamic nature of a heel impact, where the typical contact time will be of the order of 50 milliseconds. The often unspoken assumption is that the surface is visco-elastic, and that this property modifies the nature of the impact of the foot. The situation may be illustrated by reference to a load-deflection curve on an artificial turf model (Fig 2), where the model surface was loaded in a servo-hydraulic test frame. It can be seen that the load-deflection curve of a typical artificial turf is (a) non linear and (b) time dependent, in that a 20 second pause at the peak load results in a 20% drop in the applied load.

These results would seem to justify the use of a testing routine which closely mimics the actual foot strike. Further reflection, however, would indicate that only a limited range of visco-elasticity actually is of relevance (Fig 3). If the decay constant is very long, then the actual foot strike dynamics will be unaffected, since the time scale of the foot strike is such that a significant change must occur within the 50 milliseconds time scale. In other words, decay constants longer than, say, 50 milliseconds are of academic interest only. This is a crucial inference, since if it can be shown that the artificial turf has a time constant longer than this figure, then one may use quasi-static measurements alone there is no need to mimic the time scale of the footfall, even if, as has been discussed earlier, there is a need to match the load, and the load per unit area.

MATERIALS AND METHODS

Measurements of model surfaces were made on a standard hockey carpet, filled with no-fines sand as per specification. The carpet was tested on its own, and in combination with two thicknesses of rubber crumb underlay - 12 mm and 20 mm. These were also assessed on their own.

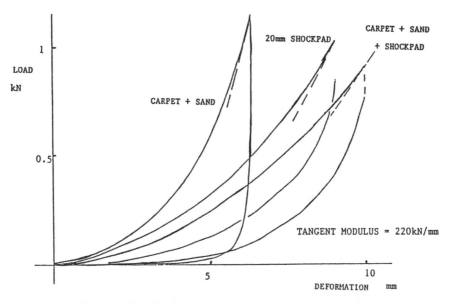

Fig 2. Load - deflection curves for artificial turf and components

The testing procedure used a 50 mm diameter spherical indenter, mounted in a 100 kN capacity servo-hydraulic load frame. The model pitches were mounted in a steel box to contain the sand. The load and ram displacement were recorded on an x-y plotter; each loading phase took around 60 seconds, so that the loading could be considered quasi-static; after reaching the peak load used - 1.2 kN, the ram was kept stationary for 20 seconds.

In all tests the load fell; the ram was withdrawn and the load returned to zero load and zero deformation - in other words the pitch components showed no permanent set - the deformation was totally recoverable.

Typical load-deflection curves are shown in Fig 3. It will be seen that the carpet alone quickly compacts, and at the 1.2 kN level is really quite harsh. This is for an uncompacted pitch. After prolonged use, the sand compacts and similar levels of resistance will be reached at lower loads. The main shock absorbance is provided by the rubber underlay; the combination of carpet and underlay in fact perform better together than their individual elements would lead one to suppose.

The time dependent nature of the mechanical properties artificial turf was assessed by impressing a step function of deformation on to the hemispherical indenter (50 mm diameter), intended to mimic a heel impact. The step function chosen was a 3 mm ram movement input by the hydraulic test frame. The load was monitored as a function of time. While the load does decay as a function of time, in line with expectation, the decay constant is of the order of 0.5 seconds. The peak value of the load will not vary, due to viscoelastic decay, by more than a few percent from its value for a purely elastic medium. For most purposes, therefore, one may adopt the quasi-static

elastic values of elastic behaviour as the most useful for design consideration.

DESIGN CONSIDERATION

(1) Testing

As far as the testing of individual pitches is concerned, one can see that the complexities of the impact testers may be avoided by using a static loading device to record an in situ load-deformation curve, bearing in mind that loads equivalent to 2.5-3 times the typical athlete need to be applied over a foot-shaped area (see below).

(2) Pitch Specification

From the work of McMahon & Green an artificial surface should have a stiffness of the order of $250 \, kNm^{-1}$, measured at the tangent to the load-deformation curve at the peak load encountered. Since the load-deflection curves for artificial pitches are non-linear, and peak loads of 2.5 to 3 times body weight will be encountered, a figure of 2 kN over a foot sized area (approximately 75 cm^2) should be used to calculate the stiffness. By this means, different pitch specifications may be compared. For the previous measurements, with the heel-shaped indenter, a load level of 1200N was calculated as the appropriate level for a player of body mass 60 kg, on a well-sprung surface.

EFFECT OF SURFACE COMPACTION

After a pitch has been in use for some time, the sand beds down (or in the case of a water-based pitch, the carpet pile will be flattened). The performance as far

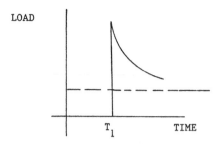

Fig 3. Viscoelastic response to a step input at time T_1.

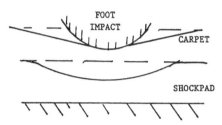

Fig 4. Impact on a compacted carpet. Note the lack of indentation by the foot.

as cushioning is concerned, according to anecdotal evidence, is greatly diminished, although no qualitative assessment has as yet been published. That the overall stiffness will change may be understood from the capability of a compacted pitch to transmit shear forces around the point of contact, and so for an element of the carpet to act like a beam (Fig. 4). The implication for maintenance routines is that the sand layer should not be allowed to become compacted to the point at which it can act in a rigid manner.

CHOICE OF FOOTWEAR

In general, hockey players use special-purpose "Astro-shoes", with a close array of small studs. The feature of these is their grip and their lack of cushioning, although some models do now incorporate a cushion midsole. The choice of a suitable shoe is really crucially important, bearing in mind that distance runners now run constantly, and mostly on roads, distances which may exceed 100 miles per week. Most avoid major injury, so that it is possible to run large distances, and yet avoid excessive wear and tear. Hockey players should view their choice of footwear with the same degree of attention as do runners, since a small amount of prevention is worth any amount of cure. Hockey is now a more mobile game, and more intense. Inevitably impact injuries will arise, but chronic pounding injuries are avoidable. It has been shown that training shoes with quite rudimentary cushioning do reduce to acceptable levels the peak level of forces

applied upwards through the heel on running (Stacoff et al., 1988). There is anecdotal evidence for players using training shoes rather than "Astro-boots" even at the highest level of competition, free of chronic injury over many seasons. For players who find the extra grip of the multi-stud sole to be essential, a model with inbuilt cushioning may well be the best long-term answer.

CONCLUSION

Modern sand-based hockey pitches fulfil the criteria for intensive long-term usage, with the prospect of causing minimal damage to players legs due to repeated jarring impacts. Nevertheless, in view of the mobility of the modern hockey game, players should try to maintain their legs injury-free by an appropriate choice of footwear - either trainers or "Astro" shoes with inbuilt cushioning. As far as the maintenance of individual pitches is concerned, efforts should be made to maintain the sand in a mobile condition i.e. it should not be allowed to compact into a hard mass.

The pitches themselves show a rational trend in design, towards a 15-20 mm thick shockpad under the carpet. Static measurements of stiffness will serve in most cases to define the pitch characteristics with adequate precision. There remains a need for an accurate on-site testing method for artificial turf pitches.

REFERENCES

DIN Standard

Keene, J. S., et al, 1980, Tartan Turf on trial. American *Journal of Sports Medicine*, **8**, 1, 43-47.

McMaholl T. A. and Greene, P. R. ,1979. The influence of track compliance on running. *J. Biomechanics*, **12**: 893-904

Martin. R. B. 1990. Problems associated with testing the impact absorption properties of artificial playing surfaces. *ASTM STP 1073*, R. C. Schmidt et al., (eds), 77-84

Powell J. W. & Schootman, M., 1992. A multivariate risk analysis of selected playing surfaces in the NFL: 1980-1990. *American Journal of Sports Medicine*, **20**, 6, 686-694.

Rogers III, J. N. & Waddington, D. V. 1990. Portable apparatus for assessing impact characteristics of athletic field surfaces. *ASTM STP 1073*, RC Schmidt et al, (eds)

Stacoff, A., Denoth, J., Kaelin, X. & Stuessi, E. 1988. *International Journal of Sports Biomechanics*, **4**: 342-357.

7 Modelling of sport

The Engineering of Sport, Haake (ed.) © 1996 Balkema, Rotterdam. ISBN 90 5410 822 3

Parametric modelling of the dynamic performance of a cricket bat

C. Grant & S. A. Nixon
Department of Mechanical, Materials and Manufacturing Engineering, University of Newcastle upon Tyne, UK

ABSTRACT: A simple rigid-body model of a cricket bat predicts an idealised performance not achieved in practice due to vibrations set up in the bat during its impact with the ball. There are three significant flexure modes of vibration in a typical excitation spectrum. A Finite Element model is used to compare the modal performance of a variety of designs that incorporate popular design features. Most of these features are ineffective in raising the frequency of the highest mode out of the excitation spectrum. Those features that do enhance performance essentially increase the stiffness of either the handle or the blade. Whilst it is feasible to design out the influence of the highest significant flexure mode, the remaining two modes could remain a problem with a rigid handle.

1 INTRODUCTION

Sport and the manufacture of sporting equipment is now seen as an important economic activity. As a result, sophisticated engineering methods are increasingly employed to make marginal improvements for the user.

Unlike equipment used in other sports, the cricket bat has seen little improvement in design during the present century(Bailey,1979). Previous works by Grant and Theti(1994) and Grant and Baird(1995) have shown that a simple rigid model predicts an idealised performance not achieved in practice. Real bats vibrate after impact with the ball, and this can cause discomfort to the batsman. The vibrations also affect the performance of the bat by absorbing energy that might otherwise be available to propel the ball efficiently.

Modern bats invariably have sprung handles that consist of a number of alternate cane and rubber strips. The purpose of the rubber is to make the handle more compliant in an attempt to isolate the batsman's hands from painful vibrations. The effect is to lower the frequency of natural vibrations. Grant and Theti(1994) suggested that improved performance may be possible if the frequency of one or more of the significant modes of natural vibration could be raised above the excitation spectrum. Comfort and performance therefore appear to be conflicting design requirements.

Innovative features of commercial designs of bats are restricted to variations in the geometry of the back of the blade. The rules of the game (see Lewis, 1994) insist that the blade be constructed in wood. English Willow is invariably used as it is relatively light and tough. Extravagant claims are often made by manufacturers when advertising competing designs, but few of these claims are ever quantified. Supporting scientific evidence is difficult to find.

Variations in the weight and performance of similar bats can be considerable. For a given design, all bats would be manufactured to the same dimensions. The whole range of available weights is left to natural variations in the wood. Furthermore, a typical response to a manufacturer who supplies bats to professional cricketers is that only one in three identical bats, selected

Table 1: Physical properties of
English Willow*

Young's Modulus E (MPa)	6600
Density ρ (Kg/m^3)	417

* after Lavers(1983)

for the same weight, might be regarded as 'good' in terms of performance. No information is available regarding the physical differences between 'good' and 'bad' bats. Indeed there are no accepted scientific measures of performance available to form a basis for comparison. Assessment of performance is always subjective.

Manufacturing quality control must therefore rely solely on the superficial appearance of the timber used in a bat's construction. Bats have to be marketed on aesthetic appeal rather than on measurable performance criteria. Only limited data such as weight and pick-up are available prior to purchase.

The aim of this work is to determine the effectiveness of typical geometrical features widely employed in commercial designs. The criterion used for comparison is based on modal performance, and in particular on the frequencies of the first three flexure modes of vibration. This provides an objective metric that is independent of the point of impact. A good design is defined as one that would at least raise the third mode of flexural vibrations above the excitation spectrum. A poor design might lower additional modes into the excitation spectrum.

Rather than attempt to compare the modal performance of actual bats, which would be prohibitively expensive and limited in scope, a Finite Element modal model is developed. The effectiveness of each design feature may thus be tested independently. Since Finite Element modelling is itself time consuming and therefore expensive, the necessary mesh is defined in terms of a handful of design parameters. In this way, meshes may be generated automatically to incorporate a variety of individual design features. Finally a composite design is formulated and tested which incorporates the most effective features.

2 THE BASIC MODEL

The Finite Element model is based on the geometry of a traditional design of bat for which experimental modal data are available. Physical data for Young's Modulus and density of English Willow, as used in the model, are shown in Table 1. Although wood is a visco-elastic material, its time dependent properties are negligible during a typical impact period of about 1ms (Daish, 1972).

Before the effect of the various design features were investigated, the basic model was correlated against available experimental data. Using the standard properties of Table 1, the first three flexure modes of vibration are illustrated in Figure 1, Figure 2 and Figure 3 respectively. The corresponding modal frequencies given in Table 2 are about 20% greater than the observed data. Also, the calculated mass of the model is only two thirds that of the actual bat.

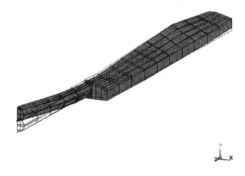

Figure 1: Flexure mode 1

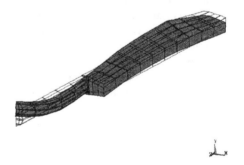

Figure 2: Flexure mode 2

Figure 3: Flexure mode 3

Blades are heavily rolled during manufacture and then 'knocked in' with repeated blows from a hard ball or mallet. The purpose is to toughen the face of the bat so that is better able to withstand severe impacts during service without suffering serious cracking or indentation. The effect is to increase the density of the timber especially in the region near to the face.

After raising the density of wood in the model by 50% to make the overall mass of the bat correct, the predicted modal frequencies then align very closely with the observed frequencies as shown in Table 2. Modes 1 and 2 are virtually identical, whilst mode 3 is about 4% lower than that observed. With this modification, the model is used as a standard to compare the effect of incorporating various design features.

A simple factor on density to correct the model is far from satisfactory. If the density were increased uniformly by rolling, then Young's Modulus of Elasticity E would also increase in proportion (see Gibson and Ashby, 1988). The reason is that the microstructure consists essentially of long hollow fibres or cells. Cell walls are principally cellulose that

would be squashed into a smaller space. The axial load carrying capacity and stiffness of the fibres would be largely unaffected.

In an unsupported prismatic beam, the square of the natural frequency of any particular flexure mode of vibration is proportional to the flexural stiffness EI of the beam divided by its density ρ. Uniformly squashing the timber would have little influence on the natural frequencies *for beams of similar section* since density *and* stiffness would increase in proportion.

Rolling is likely to compress a thin layer close to the surface rather than produce a uniform compression. The density of the cell wall is, according to Gibson and Ashby(1988) about 1500 kg/m^3, giving scope for a localised compression factor of up to 3.6 in English Willow. This denser, stiffer surface layer will have a disproportionate effect on the flexural stiffness of the section.

3 THE PARAMETRIC MODEL

The standard bat used in the model is of a traditional design with no scoops in the blade, and a single spring handle.

Provision is made in the model for

a) a central ridge of variable length and height, with the latter held at the same height as the apex,

b) up to three springs in the handle, of variable thickness. Either a single central spring or no spring was used in the trials.

c) a variable handle radius,

d) twin symmetrical scoops of variable length and thickness. The position of the scoops was held constant,

e) variable edge and toe thickness.

During a typical impact period of 1ms, forces of up to 50kN are likely to be generated. The batsman cannot hope to react with more than an insignificant fraction of such forces via his grip on the handle. In keeping with the experimental

Table 2: Modal model frequencies

	Mass (kg)	Mode1 (Hz)	Mode2 (Hz)	Mode3 (Hz)
Experiment*	1.191	111.5	365.0	611.0
FE Model (417.0 kg/m^3)	0.794	133.5	437.8	698.5
FE Model (621.1 kg/m^3)	1.191	112.2	365.9	585.4

*after Grant and Theti (1994)

Table 3: Fixed geometric parameters

Parameter	Symbol	Value (mm)
Blade length	BL	550
Handle length	HL	165

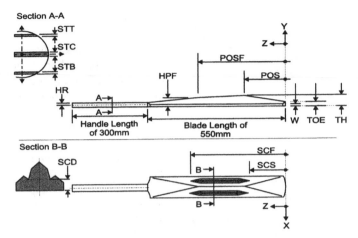

Figure 4: Geometric parameters

data of Grant and Theti(1994), the Finite Element model is assumed to be unrestrained. Figure 4 illustrates the geometric parameters used in the finite element model. The overall length of blade and handle were held constant as shown in Table 3.

Variable parameters are given in Table 4 along with the range of values employed in the trials, the standard bat value and the maximum effect of varying each parameter individually. The latter represents the maximum improvement in mode 3 frequency compared to the standard bat over the given range. The range of each parameter was based on providing a reasonable strength at the low end, and a maximum weight

of 1.36kg at the high end.

Perhaps the most striking aspect of the results in Table 4 is that most of the popular design features have very little effect on the modal model. In particular ridges, scoops and perimeter weighting are virtually ineffective at increasing the natural frequencies. Inclusion of scoops actually reduces the mode 3 frequency since the decrease in flexural stiffness has a greater influence than the corresponding weight reduction. Only ridge height and handle radius, which both imply a heavier bat, have a significant effect.

A reduction in spring thickness also had very little effect. Only the complete elimination of the spring caused the significant effect shown in

Table 4: Variable geometric parameters

Parameter	Symbol	Range (mm)	Standard value (mm)	Effect %
Apex position	POS	55-385	165	0.3
Ridge length	POSF	0-275	0	2.8
Ridge height	TH	20-80	50	17.5
Scoop length*	SCF	110-385	0	-1.6
Scoop height+	SCD	10-25	-	-2.2
Toe thickness	TOE	20-60	30	0.6
Edge thickness	W	5-35	20	2.0
Handle radius	HR	10-20	16.1	19.7
Centre spring	STC	0-5	3	46.4

* scoop height SCD= 15mm
+ scoop length SCF=275mm

Table 5: Optimised modal frequencies

	Mass (kg)	Mode 1 (Hz)	Mode 2 (Hz)	Mode 3 (Hz)
Experiment	1.191	111.5	365.0	611.0
Standard model	1.191	112.2	365.9	585.4
Optimised model	1.247	178.0	616.6	1039.0

Table 4. It might be argued that the removal of springs from the handle is a retrograde step since these were introduced to protect the batsman's hands from painful vibrations. Conversely, if the bat can be made stiff enough so that vibrations are sufficiently reduced, the batsman could be protected without compromising the bat's performance.

4 OPTIMISED DESIGN

In the light of the results given in Table 4, an optimum design was formulated to incorporate the most effective features. This design is inevitably a compromise between maximising the mode 3 frequency and restricting the overall mass. The optimised design has the following features:

a) no spring in the handle
b) increased thickness of 75mm at the apex
c) no scoops in the back of the blade

The handle radius of the optimised bat had to be kept the same as the standard bat in order to avoid an unacceptable increase in mass. Also, the increase in mass due to increased apex height had to be offset by a new scalloped profile of the toe. Whilst this had a marginal effect on the natural frequencies, it did result in a useful weight reduction. Performance of the optimised design is compared with the standard bat in Table 5. The goal of increasing the mode 3 frequency above 1kHz has just been met with a small but acceptable increase in mass.

5 CONCLUSIONS

The frequency of the third flexural mode of vibrations represents a useful criterion for comparing different designs of bat.

Correlation of the natural frequencies of flexural vibration between experiment and the Finite Element model required that the density of English Willow used in the model be increased by 50% above its normal uncompressed value. The assumption of a uniform increase in density without a commensurate increase in stiffness is unrealistic.

Most of the popular geometric features found in modern designs are ineffective in increasing the frequencies of flexural vibrations. Scoops, ridges and perimeter weighting all fall into this category. Among the factors that do make a significant improvement are:

a) increasing the thickness at the apex
b) increasing the diameter of the handle
c) removing all springs from the handle

Taking these and possibly other measures such as the introduction of new materials to stiffen the handle, it is feasible to design out the troublesome third flexure mode. Whilst this would be likely to improve the dynamic performance of a bat, it remains to be determined if the lower modes of vibration would remain troublesome, especially with a rigid handle.

6 REFERENCES

Bailey T., *A history of cricket*, Allen & Unwin, London, 1979.

Daish C.B., *The physics of ball games*, English Universities Press, London, 1972.

Gibson L.J. and Ashby M.F., *Cellular solids: Structure and properties*, Pergamon Press, 1988.

Grant C. and Baird A.D., *Modelling for an improved cricket bat*, Institute of Physics Annual Congress, Physics of Sport, Telford, 1995.

Grant C. and Thethi P., *Recent Advances in Experimental Mechanics*, Vol. 1, Silva Gomes J.F. et al (eds), Balkema, Rotterdam, 1994, 669-674.

Lavers G.M., *The strength properties of timber*, 3rd Edn., HMSO, London, 1983.

Lewis T., *MCC coaching manual*, Weidenfield & Nicholson, London, 1994.

The Engineering of Sport, Haake (ed.)© 1996 Balkema, Rotterdam. ISBN 90 5410 822 3

Normal impact models for golf balls

Stanley H. Johnson
Mechanical Engineering and Mechanics, Lehigh University, Bethlehem, Pa., USA

Burton B. Lieberman
Polytechnic University, Brooklyn, N.Y., USA

ABSTRACT: An existing theoretical treatment of normal impact between a homogeneous, imperfectly elastic sphere and a planar barrier is unable to represent the behavior observed experimentally when golf balls are fired at a steel block. A generalization of Simon's model improves its fidelity, but substantial modification is required to achieve accurate simulation of approximately homogeneous golf balls and major modifications would be required to extend the model to inhomogeneous golf-ball normal impact.

A combination of static and dynamic testing allows estimation of the parameters of a new model. Rebound velocities calculated from the new model agree with measurements within 1.5% over a range of approach velocities for two-piece, or approximately homogeneous, balls and 3% for wound, or inhomogeneous, golf balls. The new model correctly portrays the variation of time of contact, ball deformation, and coefficient of restitution with approach velocity for both ball constructions.

The testing program was carried out using an air cannon at the Research and Test Center of the United States Golf Association in Far Hills, NJ. As the international governing bodies for the game of golf, the United States Golf Association and the Royal and Ancient Golf Club of St Andrews conduct such research and disseminate information for the betterment of the game and the protection of its traditions.

1. THE SIMON MODEL

The Simon (1967) model of normal impact is

$$\ddot{y} = -\frac{k}{m} y^{3/2}(1 + \alpha\dot{y}) \qquad (1)$$

and the two parameters of this mode, k and α, are determined by fitting m$\ddot{y}$ to measured force vs time data for one dozen golf balls at two nominal approach velocities. The experimental setup is similar to that of Gobush (1990). The resulting two pairs of parameters obtained by averaging over a dozen two-piece balls and a dozen wound balls are given in Table 1.

For two-piece balls, the Simon model with the appropriate parameters from Table 1 predicts rebound velocities that are about 5% high at 120 and 140 ft/sec. These are larger discrepancies in rebound velocities than other models, despite having average errors that are comparable to other models with more degrees of freedom. The Simon model results are superior to other models in variability of rebound velocity predictions, the standard deviations being less than 1%. The agreement between 120 ft/sec and 140 ft/sec parameters is better than other models.

However there is a problem with the Simon model on theoretical rounds since it requires that the displacement and the acceleration be zero simultaneously. Therefore, at the end of impact, the displacement should be zero. Photographs taken

Table 1. Parameters for the Simon model. The errors are the average errors over fifty points of comparison between experimental data and force on the barrier computed from the model using the k and α listed.

Nominal velocity	k (2pc) lbs/ft$^{3/2}$	α (2pc) sec/ft	error lbs	k (wound) lbs/ft$^{3/2}$	α (wound) sec/ft	error lbs
120 ft/sec	1224200	0.00206	60.3	897900	0.00219	165 6
140 ft/sec	1206500	0 00219	62.6	967300	0.00198	198.1

during impact suggest that only about 95% to 98% of the diameter of the sphere has been recovered at the end of the contact time. In addition, a single integration of (1) leads to the equation

$$\dot{y}(t) - \alpha^{-1}\ln(1 + \alpha\dot{y}(t)) + .4\alpha k y(t)^{5/2} = \dot{y}(0) - \alpha^{-1}\ln(1 + \alpha\dot{y}(0))$$

(2)

which, at loss of contact, implies that the coefficient of restitution of the sphere is independent of the Hertzian spring constant.

As an aside, Simon wrote the coefficient in (1) as

$$k = \left(\frac{4}{3}\right)\frac{E_b a^{1/2}}{\left(1 - \mu_b^2\right)}$$

(3)

the parameters E_b and μ_b are effective values of Young's modulus and Poisson's ratio and a is the radius of the ball. Using 0.5 for Poisson's ratio and the average value of k from Table 1 yields an effective Young's modulus of 18,300 psi. This is almost 40% higher than Simon's value obtained from a static test. Ball creep makes static testing problematical. Simon's two values for α in (1) are 0.00321 sec/ft or 0.00464 sec/ft depending on the ball compression. Simon relied on experimental results of Briggs (1945) and NTIS (old NBS).

2. THE MODIFIED SIMON MODEL

Lieberman has proposed a modification to the Simon model that can exhibit the approach-velocity-dependent effects observed experimentally

$$\ddot{y} = -\frac{k}{m}y^{3/2}\left(1 + \alpha\dot{y}^{\beta}\right)$$

(4)

and contains three parameters to be estimated from force vs time data. The results from tests of one dozen two-piece balls are given in Table 2.

Table 2. Parameters for the Lieberman modification to the Simon model obtained by fitting solutions of (4) to measured force vs time data for two-piece balls

ft/sec	k lbs/ft$^{3/2}$	α (sec/ft)$^{\beta}$	β
120	1252200	0.0002806	1.5003
140	1221500	0.0007399	1.2571

The fit is slightly improved by the use of three parameters, but the simulated rebound velocities are still high by 3.5%. The standard deviations are greater than 1% and the differences between the 120 and 140 ft/sec results indicate that this modification to the Simon model does not well represent the dynamic behavior of the two-piece ball. This three-parameter modification of the Simon model also indicates zero deformation when the acceleration is zero. A more versatile model structure is required to portray the observed phenomena reported by Lieberman (1990).

3. THE FIVE-PARAMETER MODEL

Lieberman and Johnson (1994) have proposed a five-parameter heuristic model. Two nonrate-dependent parameters, k_1 and α, come from low speed compression tests. At rapid rates of loading the results become rate dependent. At extremely low rates of loading the balls creep and unload themselves. In between these two phenomena, experience with two-piece balls suggests that rates in the range of 0.010 in/min to 0.030 in/min produce quasistatic results. Even so, the balls are found to be out of round after one compression when checked by a ring-gage.

Table 3. Pairs of k_1 and α for two-piece and wound balls. These loading rates were selected on the basis of experience with two-piece balls and are not as suitable for wound balls. The k_1 and α move in tandem, going up or down together. Quite different pairs can produce similar force-deformation profiles.

Ball	in/min	k_1 lbs/ft$^{\alpha}$	α
2-pc	0.010	600020	1.503
2-pc	0.030	710430	1.539
wound	0.010	43754000	2.680
wound	0.030	44391000	2.697

The remaining three parameters, k_2, β and c, are derived from high-speed recordings of impact forces.

In Lieberman and Johnson (1994), these three parameters were estimated by fitting the model's predicted value of the coefficient of restitution to the actual value at three different impact velocities. (The length unit used there was inches.) In this work a three-parameter gradient search method is used to minimize the sum of the squared errors at fifty different times during contact between the model predictions and the measured normal force exerted

Table 4. Three dynamic parameters for the five-parameter model are obtained from high-speed force vs time recordings. The 0.010 in/min quasistatic values from Table 3 were used in the parameter estimation process. The 0.030 in/min values would produce slightly different values.

ft/sec	k_2 lbs/ft$^\beta$	β (2-pc)	c(2-pc)	k_2 (wound)	β (wound)	c (wound)
120	1521800	1.6436	43.580	253600	1.3343	14.353
140	1795700	1.6687	35.269	383470	1.4043	14.105

Table 5. Experimentally determined values for the parameters of the three-parameter three-havles model for a dozen two-piece balls and a dozen wound balls

	k_1	k_2	c	error (lbs)
2-pc @ 120	828540	728020	20.4231	46.8
2-pc @ 140	804130	788640	20.8564	68.6
wound @ 120	1038900	33230000	3.1389	94.7
wound @ 140	1095180	30214000	3.3616	137.6

by a ball on a Kistler three-component force transducer mounted on a 500-lb pedestal. An air cannon is used to fire the balls at this barrier. The incoming and outgoing velocities are measured by ballistic light screens. Stroboscopic photographs verify that the airgun does not impart significant spin to the balls.

4. A NEW THREE-PARAMETER MODEL

The five-parameter model of the two-piece balls has exponents that are close to the theoretical value of 3/2 for elastic impact between a sphere and a flat barrier. The two values for the wound ball are also similar but quite different from 3/2. Another model that forces these two exponents to be 3/2 can be fitted to the force-time data. The results are shown in Table 5. The values for the two-piece balls suggest a further simplification and that is to force the spring constants to be equal and reduce the model to one having only two parameters. In two-parameter form or three-parameter form, these models are better able to reproduce experimentally observed ball performance than are the two- or three-parameter Simon models above.

5. DIFFERENTIAL EQUATIONS OF MOTION

The differential equations for the various models can be written in state-vector form for computer integration with deformation and rate of deformation as dependent variables. In this form Simon's model becomes

$$\dot{y}_1 = y_2 \qquad \dot{y}_2 = -\frac{k}{m} y_1^{3/2}(1+\alpha y_2)$$

and the five-parameter model is

$$\dot{y}_1 = y_2$$

$$\dot{y}_2 = -\frac{k_1}{m}|y_1|^\alpha \operatorname{sgn}(y_1) - \frac{k_2}{m}|y_3|^\beta \operatorname{sgn}(y_3)$$

$$\dot{y}_3 = -\frac{k_2}{c}|y_3|^\beta \operatorname{sgn}(y_3) + y_2$$

where sgn($\bullet$) has the value of +1 or -1 depending on the sign of the argument.

The three-halves model is written

$$\dot{y}_1 = y_2$$

$$\dot{y}_2 = -\frac{k_1}{m}|y_1|^{3/2} \operatorname{sgn}(y_1) - \frac{k_2}{m}|y_3|^{3/2} \operatorname{sgn}(y_3)$$

$$\dot{y}_3 = -\frac{k_2}{c}|y_3|^{3/2} \operatorname{sgn}(y_3) + y_2$$

The two-parameter version of the three-halves model is

$$\dot{y}_1 = y_2$$

253

$$\dot{y}_2 = -\frac{k}{m}|y_1|^{3/2}\text{sgn}(y_1) - \frac{k}{m}|y_3|^{3/2}\text{sgn}(y_3)$$

$$\dot{y}_3 = -\frac{k}{c}|y_3|^{3/2}\text{sgn}(y_3) + y_2$$

6. COMPARISON WITH EXPERIMENT

One dozen two-piece balls were fired at a Kistler quartz crystal strain detector, Model 9067, mounted on a stationary barrier and the resulting signal amplified, sampled and filtered. The lowpass filter had a cutoff frequency of 5.2 kHz to suppress the transducer transient response. Figure 1 shows twelve force time histories obtained at an average approach velocity of 120.23±0.65 ft/sec. The average rebound velocity was 97.32±0.58 ft/sec and the average normal coefficient of restitution was 0.8094. The nominal 140 ft/sec profiles for two-piece balls peak at 3300 lbs but otherwise look very much the same as the 120 ft/sec profiles. The quality of fit is similar as well.

The wound-ball force-time histories for one dozen wound balls fired at a nominal approach velocity of 120 ft/sec are shown in Figure 3. The actual velocity was 119.92 ft/sec. The average rebound velocity was 94.37±1.00 and the coefficient of restitution was 0.7869.

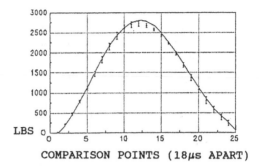

COMPARISON POINTS (18μs APART)

Fig. 2 Comparison of the five-parameter model profile with the averaged time histories for the 120 ft/sec, two-piece ball force data. Only 25 of the 50 points of comparison are shown. The particular values are k_1=600020, α=1.503, k_2=1521800, β=1.644 and c=43.58. The error bars indicate the standard deviations of the dozen profiles shown in Fig. 1

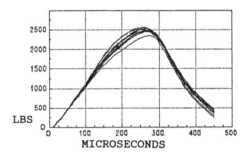

MICROSECONDS

Fig. 3 Twelve force-time histories for wound balls at a nominal direct normal impact velocity of 120 ft/sec

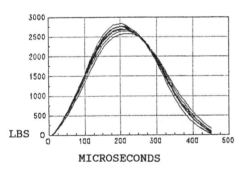

MICROSECONDS

Fig 1. Twelve force-time histories for two-piece balls at a nominal direct normal impact velocity of 120 ft/sec

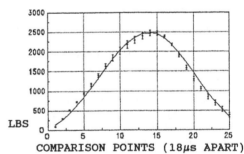

COMPARISON POINTS (18μs APART)

Fig. 4 Comparison of the five-parameter model profile with the averaged time histories for the 120 ft/sec, wound-ball force data. Only 25 of the 50 points of comparison are shown. The particular values are k_1=43754000, α=2.680, k_2=253600, β=1.334 and c=14.353. The error bars indicate the standard deviations of the dozen profiles shown in Fig. 3

254

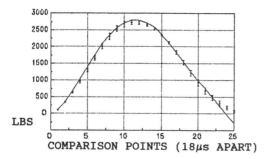

Fig. 5 Comparison of the three-halves model with averaged force data for two-piece balls at 120 ft/sec

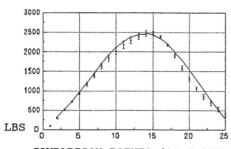

Fig. 6 Comparison of thre three-halves model with averaged force data for wound balls 120 ft/sec

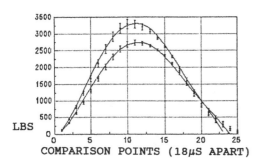

Fig. 7 Comparison of the two-parameter, three-halves model with averaged force data for two-piece balls at 120 and 140 ft/sec

7. DISCUSSION

The original, two-parameter Simon model does a remarkably good job of representing two-piece ball impact, at least at 120 and 140 ft/sec. The average errors are as low as for the five-parameter model and the standard deviations of the predicted launch velocities are the lowest of any model above, although the predicted velocities are 5% high. The coefficient of restitution is independent of the spring constant and end of contact occurs at zero deformation, in contradiction to experiments. The five-parameter model overcomes the difficulties with the Simon model but it requires two experiments, one quasistatic and one high-speed, to estimate the parameters. For homogeneous balls, the three-halves model with the exponents set to the Hertzian small-deformation value of 3/2 works well. For the particular two-piece balls tested, the two spring constants are approximately equal, within the standard deviations, and a two-parameter model does a good job of representing both 120 and 140 ft/sec force-time profiles. The Simon model produces poor fits to wound ball force-time profiles. The three-halves model is better, reducing the average error by about 35%. For the wound balls, k_2 is three times larger than k_1 so there is no opportunity to reduce the three-parameter three-halves model to a two-parameter model. The three-halves model displays most of the performance characteristics of real golf balls summarized by Lieberman and Johnson (1994); (1) energy is lost during impact; (2) the coefficient of restitution decreases with increasing approach velocity; (3) the contact time decreases with increasing approach velocity; (4) the compression phase should take less time than the rebound phase but model compression takes about 5% longer than rebound; (5) the diameter at loss of contact is deformed; and (6) the maximum deformation of the model increases linearly with velocity when experiments have shown that it should increase a little faster.

8. CONCLUSION

The various models above are able to represent the normal direct impact of homogeneous golf balls. They are less successful with inhomogeneous balls as represented here by liquid-center, wound golf balls. The Simon two-parameter model is unable to portray the complex behavior observed in tests of golf-ball impact, but a five-parameter model is quite successful for two-piece and wound balls. For two-piece balls only, the five parameter model can be reduced to a new three-parameter model that eliminates the shortcomings of the original Simon model. A three-parameter model of wound balls is less successful. In the particular case of the two-piece balls tested, the three-parameter model can be reduced to a two-parameter model that predicts velocities that are high by 3%, compared to 1% for the five-parameter model, and exhibits much the same fidelity as the five-parameter model.

ACKNOWLEDGEMENT

The authors wish to express their gratitude to Mr Frank Thomas, technical director of the United States Golf Association, for his generous support, co-operation and encouragement.

REFERENCES

Briggs, L J 1945. Methods for Measuring the Coefficient of Restitution and the Spin of a Ball. In *Journal of Research, National Bureau of Standards*, vol. 34, January 1945, pg 1. (This reference was not consulted by the authors but is reproduced here from Simon (1967) as an aid to interested readers.)

Gobush, W 1990. Impact Force Measurements on Golf Balls. In A J Cochran (ed), *Science and Golf, the Proceedings of the First World Scientific Congress on Golf*, St Andrews, Scotland, July 1990, 219-224, London, E & FN Spon.

Lieberman, B B and S H Johnson 1994. An Analytical Model for Ball-Barrier Impact, Part I: Models for Normal Impact. In A J Cochran & M R Farrally (eds), *Science and Golf II, the Proceedings of the World Scientific Congress of Golf*, St Andrews, Scotland, July 1994, 309-314. London, E & FN Spon.

Simon, R 1967. The Development of a Mathematical Tool for Evaluating Golf Club Performance, presented at the ASME Design Engineering Conf, New York, May 1967.

The Engineering of Sport, Haake (ed.) © 1996 Balkema, Rotterdam. ISBN 90 5410 822 3

A proposed mechanical model for measuring propulsive forces in front crawl swimming

M. A. Lauder & P. Dabnichki
Biomechanics Research Group, The Manchester Metropolitan University, Crewe & Alsager Faculty, Stoke-on-Trent, UK

ABSTRACT: The search for an objective measure of technique in front crawl swimming has led to much interest in the estimation of hand forces. The most popular procedure (hydrodynamic analysis) has been questioned and the limitations of this approach discussed. A proposed mechanical and computer modelling approach is presented with the aim that the combination of these models will allow a realistic estimation of the dynamic propulsive forces in the front crawl stroke.

1 INTRODUCTION

"More than in any other branch of Biomechanics, swimming has been isolated to the scientific means of investigation by a barrier due to water"

Lewillie (1971)

The above statement holds true today as it did 25 years ago. The medium of water still presents the sports biomechanist with problems in the design and implementation of analytical and experimental methods which will enable an adequate evaluation of a performer's (swimmer's) technique. For the "dry-land" or "single media" researcher, methods have been developed that allow detailed qualitative and quantitative analyses of technique to be undertaken. Such methods have been invaluable to both the sport scientist and performer in understanding key components of technique. However, as a direct consequence of the nature of the dual-media environment, many methods of biomechanical analysis have not been directly transferable. Similarly, research in the field of swimming biomechanics has not taken advantage of the opportunities that developments in technologies have presented (Hay, 1988). This has resulted in a paucity of good quality objective quantitative analyses of the swimming strokes.

In swimming, particularly at the elite level, the major determinant of performance is technique (Cappaert et al., 1995). Improvements in performance therefore are likely to be the result of recommendations made from a detailed understanding of technique. Early assessments of technique were through observation and intuition. Up until Counsilman's 1971 paper it was generally assumed (and coached) that swimmers should pull their hands directly backwards in a straight line utilising drag forces for propulsion. Counsilman (1971) identified, with the use of underwater film, that skilled swimmers executed a "S" shaped pull pattern in the front crawl. He inferred from this that lift forces could be generated by the hand in skilled swimming and that these forces could contribute to propulsion. As such a curved or "S" shaped pull pattern was coached.

Over the past few years the evaluation of technique through methods or procedures that provide objective quantitative data has been at the forefront of much scientific research. Methods that have been used to evaluate swimming technique quantitatively include measures of velocity (Kent and Atha, 1975), acceleration (Counsilman, 1981), stroke parameters (Grimston and Hay, 1986), power (Sharp *et al.*, 1982) and the estimation of hand forces (Schleihauf *et al.*, 1983; Thayer, 1990; Berger *et al.*, 1995).

2 THE ESTIMATION OF PROPULSIVE FORCES

The search for an objective measure of technique has led to much interest in the estimation of hand forces. Schleihauf (1974) reported that the hand was the single most important contributor to propulsion in the arm stroke. Two major methods have been established for hand force estimation. The first uses pressure sensors located on the hand to estimate hand forces directly however the practicality of this method has been questioned due to the large number (11) of pressure transducers that were required to provide a good estimation of the propulsive force (Thayer, 1990). The second and most popular procedure uses the combination of kinematic data derived from underwater video analysis and hydrodynamic lift and drag force coefficients for the hand/forearm obtained from laboratory experiments (Schleihauf, 1979; Berger et al., 1995).

3 LIMITATIONS OF THE HYDRODYNAMIC APPROACH

This second method appears the most ecologically valid as propulsive force production is dependant on the hand's movement and orientation relative to the water. However, the method is not without limitations as the accuracy of the procedure is dependant on both the accuracy of the three-dimensional kinematic data from the underwater video analysis and the accuracy of the drag and lift coefficients used to calculate the propulsive forces.

3.1 *The accuracy of kinematic data collection*

The protocols employed for the collection of underwater kinematic data include the use of underwater windows (Payton and Lauder, 1995), underwater housings (Vertommen et al., 1983) and periscope systems (Yanai et al., 1996).

Invariably with windows and housings the camera has to be set behind a transparent barrier of either perspex or ideally glass. Theoretically therefore the recording of images would be influenced by the effect of refraction at the air-glass-water interface. This effect would result in a violation of the colinearity assumption of the Direct Linear Transformation algorithm that is used for three-dimensional reconstruction and therefore should influence the accuracy of the underwater data collection.

The same would also be true of periscope systems where the image is reflected off at least two mirrors and often through a perspex barrier at the air water interface. Yanai et al. (1996) reported that the effect of refraction on the image when using a panning periscope system was to enlarge the pincushion distortion of the lens at the interface between the two media. However it would seem that a full analysis of the effect of this distortion on the level of error was not undertaken and as such the effect of refraction on reconstruction accuracy remains unaddressed for this particular method of data collection.

Lauder et al. (1996; *this volume*) used a scaled down underwater environment to investigate the effect of changes to the experimental set-up on the accuracy of two-dimensional and three-dimensional reconstructions from video analysis. It was found that the position of the camera and the calibration object relative to the glass interface influenced the level of error in the reconstruction of known lengths. The results provided evidence that there was a need for an augmented calibration procedure in order to minimise the influence of unaccounted random errors in underwater video analysis that uses underwater windows/housings. It was noted also by the authors that the position of the object within a calibrated volume and the influence of turbulence may also influence the accuracy of underwater data collection.

The generation of turbulence in the underwater stroke can be related to the frequency of the movement may be an important factor that effects the accuracy of underwater video analysis. The internal friction of a fluid causes a resistance to the motion of an object moving through it which is proportional to the gradient of the velocity. Increasing velocities influence the laminar boundary layer around the front of a moving object causing a transition of the laminar flow to a turbulent one (Prandtl and Tietjens, 1934). Turbulence in underwater analysis severely distorts the image of the hand. The frequency of the movement therefore becomes an important consideration when assessing error due to the relationship between the frequency of the movement and the creation of turbulence. An assessment of this relationship would be essential if establishing the accuracy of underwater kinematic data.

3.2 The accuracy of the hydrodynamic coefficients

The procedures of Schleihauf (1979) and Berger et al. (1995) determined lift and drag coefficients using a series o tests that were performed with a hand/forearm model immersed in various orientations in n open water channel under steady flow conditions. The force exerted by the hand/forearm on the water was determined and using the measured force and standard hydrodynamic equations (1 and 2) to calculate the coefficients of lift (C_l) and drag (C_d) for known orientations of the hand relative to the flow of the water:

$$C_l = \frac{2 F_l}{\rho . v^2 . A_l} \qquad (1)$$

$$C_d = \frac{2 F_d}{\rho . v^2 . A_d} \qquad (2)$$

where F_l is the lift force; F_d is the drag force; ρ is the density of water; v is the relative velocity between the hand and water; A_l is the hand surface area; and A_d is the frontal area of the hand.

This approach (*quasi-static*) to determining lift and drag coefficients has recently been questioned (Pai and Hay, 1988). Hydrodynamic forces in swimming are dependant on two important effects associated with an immersed accelerating segment, namely vortex shedding and added-mass effects (Childress, 1981). The testing protocols that have been previously used to obtain lift and drag coefficients ignore these effects, adopting the *quasi-static* approach. This has been shown to give different lift and drag profiles when compared to fluid conditions of unsteady flow, which are similar to those experienced in swimming (Pai and Hay, 1988).

Both Schleihauf (1979) and Berger et al. (1995) used a *quasi-static* approach to obtain hydrodynamic data for an immersed hand and immersed hand-forearm respectively for modelling hydrodynamic forces in front crawl swimming. Wood (1979) applied a similar approach however lift and drag data were obtained from wind tunnel experiments. These are the only studies that have published hydrodynamic force data for the hand-forearm. The relatively steady flow conditions created in these studies do not readily transfer to the flow conditions that are experienced in swimming. The hand in skilled front crawl swimming constantly changes its angle of attack and sweepback angle with respect to the water. It also accelerates, thus experiencing unsteady flow conditions.

Similarly the movement of the forearm and upper arm could be expected to influence flow conditions. It would be expected therefore that the lift and drag profiles in swimming would be different from those obtained in the models of Schleihauf (1979) and Berger et al. (1995).

Before techniques that maximise propulsive force production can be identified, there is a need validate the estimation of propulsive forces from protocols that use underwater kinematic data and hydrodynamic data derived from a *quasi-static* approach.

In order to overcome the problems that have been identified it was deemed that the solution was to adopt a direct approach. It is the purpose of this paper to propose that a mechanical model of the whole arm should be developed which would simulate the dynamic action of the arm in the front crawl stroke and measure the force profile throughout the stroke. The data from such a model would then be used in a computer model of the arm to establish the dynamic profile of lift and drag coefficients.

4 ARM MODEL DESIGN

The proposed mechanical arm model is deemed to resemble the anthropometric data of an elite swimmer and is to be covered by a prosthetic shell. The purpose of the prosthetic is to simulate as closely as possible the aerodynamic shape of the human arm. The structure is calculated to have the same mass as a relevant arm. The arm movement is to be computer controlled via double action pneumatic cylinders with optically encoded shaft pick-up providing position and acceleration feedback. The pneumatic system allows constant measurement of the required power and the estimation of the forces involved can be satisfactory matched with the measured values. The elbow joint (Fig. 1) will be driven by a two-way piston to allow flexion and extension through a typical range of motion during the underwater front crawl stroke. These movements will be controlled by a computer using kinematic data from previous recordings of the underwater stroke (Payton and Lauder, 1995) such that a realistic stroke simulation is achieved.

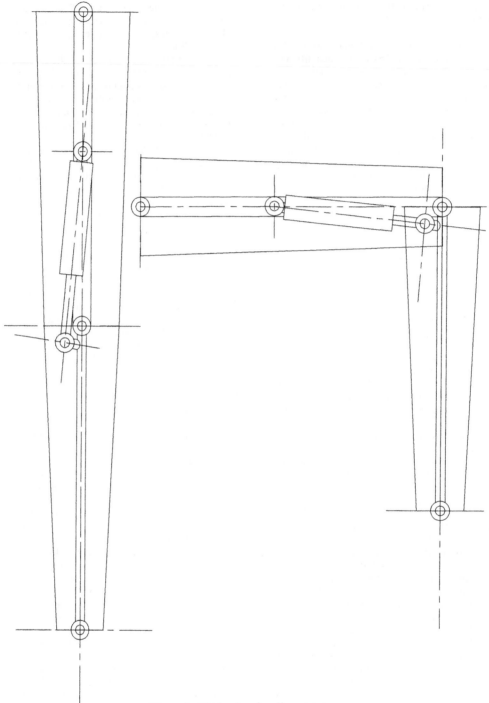

Figure 1. Mechanism for elbow joint

Theoretically the simulation of the underwater stroke using the whole arm will generate flow conditions as those experienced in swimming and therefore provide resistive force data representative of a true underwater stroke. Certainly some deviations are expected due to unaccounted degrees of freedom but as always the design is a compromise. A fully functional arm is extremely complicated even for kinematic analysis only and to some extent simplicity was intentionally sought.

5 ESTIMATION OF THE DYNAMIC LIFT AND DRAG PROFILE

In order to estimate the dynamic lift and drag profiles of the simulated underwater stroke it is proposed that a hybrid modelling approach be used. By least squares fitting of the data on the position of the arm and the torque with simple hydrodynamic equations data a representation of the dynamic profile of the stroke would be achieved. To achieve this a computer model of the arm will be created using ADAMS (Silicon Graphics Computer Systems). As both the mechanical arm and the computer model will have the capacity to simulate different strokes trials will be undertaken that will allow the output of both models to be directly compared. The approach is deemed to resemble Prof. Ewins idea of modal analysis for analysing vibrations in mechanical structures.

Through repeated experimentation it is hoped that the dynamic nature of the hydrodynamic forces experienced by the arm in the front crawl stroke will be established therefore contributing to the knowledge of effective techniques. The flexibility of the system allows to project different trajectories and force profiles and also via the feedback to monitor the peak forces involved. The last fact is very important for the estimation of the strength requirements of the upper body of the swimmer.

6 OTHER APPLICATIONS

The mechanical arm will facilitate the validation of the procedures used by Schleihauf (1979) and Berger *et al.* (1995) to estimate propulsive forces. The underwater arm movement can easily be filmed therefore validating the accuracy of the kinematic data derived from underwater video analysis. The dynamic hydrodynamic profile derived from the computed model can also be compared to the computed lift and drag profile computed using equations (1) and (2).

The computer model theoretically would allow the determination of an 'optimal' technique, based on the trajectory of the arm movement. It would also allow to quantify the arm contribution to the swimmer's movement.

7 ACKNOWLEDGEMENT

The authors would like to express their gratitude to Mr Gerald Wright for the extremely valuable suggestion to implement pneumatic system as opposed to a cam driven one and Dr Diane Taktak for the discussions and help.

8 REFERENCES

Berger, M.A.M., de Groot, G. and Hollander, A.P. (1995a) Hydrodynamic drag and lift forces on human hand/arm models. *J. Biomechanics* 28, 125-133.

Cappaert, J.M., Pease, D.L. and Troup, J.P. (1995) Three-dimensional analysis of the men's 100-m Freestyle during the 1992 Olympic Games. *Journal of Applied Biomechanics.* 11. 103-112.

Childress, S. (1981) *Mechanics of swimming and flying.* Cambridge: Cambridge University Press.

Counsilman, J.E. (1971) The application of Bernoulli's Principle to Human Propulsion in Water. In. Lewillie, L and Clarys, J.P (Eds.) *First International Symposium on Biomechanics in Swimming.* pp. 59-71. Belgium: Universite Libre de Bruxelles.

Counsilman, J.E. (1981) Hand speed and acceleration. *Swimming Technique* 18,(1). 22-26.

Grimston, S.K. and Hay, J.G. (1986) Relationships among anthropometric and stroking characteristics od college swimmers. *Medicine and Science in Sports and Exercise.* **18**, 60-68.

Hay, J.G. (1988) The status of research on the biomechanics of swimming. In: Ungerechts, B.E., Wilke, K. and Reischle, K. (Eds.) *Proceedings of the Vth International Symposium of Biomechanics and Medicine in Swimming. Swimming Science V.* pp. 3-14. Champaign: Human Kinetics.

Kent, M.R. and Atha, J. (1975) A device for the on-line measurement of instantaneous swimming velocity. In: Lewillie, L. and Clarys, J.P. (Eds.) *Swimming II. Proceedings of the second International Symposium* on *Biomechanics in Swimming.* pp.58-63. Baltimore: University Park Press.

Lauder, M.A., Dabnichki, P., Bartlett,R.M. and Aritan, S. The accuracy of the kinematic data collection from underwater three-dimensional analysis. In: Haake, S. (ed) *Proceedings of the International Conference on The Engineering of Sport.* Rotterdam: A.A. Balkema.

Lewillie, L. (1971) Opening address and welcome. In: L.Lewillie and J.P.Clarys (Eds.) *Proceedings of the First International Symposium on Biomechanics in Swimming, Waterpolo and Diving.* pp. 17. Belgium, Universite Libre De Bruxelles.

Pai, Y. and Hay, J.G. (1988) A hydrodynamic study of the oscillation motion in swimming. *International Journal of Sport Biomechanics* 4, 21-37.

Payton, C.J. and Lauder, M.A. (1995) The influence of Hand paddles on the kinematics of front crawl swimming. *Journal of Human Movement Studies.* **28**, 175-192.

Prandtl, L. and Teitjens, O.G. (1934) *Fundamentals of Hydro- and Aerodynamics.* New York: Dover Publications.

Schleihauf, R.E. (1974) A Biomechanical analysis of freestyle. *Swimming Technique* 11, 89-96.

Schleihauf, R.E. (1979) A Hydrodynamic Analysis of Swimming Propulsion. In: Terauds, J. and Clarys, J.P. (Eds.) *Swimming III.* pp. 70-109. International Series on Sport Sciences, Vol.8. Baltimore: University Park Press.

Schleihauf, R., Gray, L. and DeRose, J. (1983) Three-Dimensional Analysis of Hand Propulsion in the Sprint Front Crawl Stroke. In: Hollander, A.P., Huijing, P.A. and de Groot, G. (Eds.) *Biomechanics and Medicine in Swimming.* pp. 173-183. International Series on Sport Sciences,Vol.14. Champaign: Human Kinetics.

Sharp, R.L., Troup, J.P. and Costill, D.L. (1982) Relationship between power and sprint freestyle swimming. *Medicine and Science in Sports and Exercise.* **14**, 53-56.

Thayer, A.M. (1990) *Hand pressures as predictors of resultant and propulsive forces in swimming.* Unpublished PhD. Thesis. University of Iowa.

Vertommen, L. Fauvart, H., and Clarys, J.P. (1983) A Simple System for Underwater Video Filming. In: Hollander, A.P., Huijing, P.A. and de Groot, G. (Eds.) *Biomechanics and Medicine in Swimming.* pp. 120-122.

International Series on Sport Sciences,Vol.14. Champaign: Human Kinetics.

Yanai, T., Hay, J.G., and Gerot, J.T. (1996) Three-dimensional videography of swimming with panning periscopes. *J. Biomechanics.* 29(5): 673-678.

The Engineering of Sport, Haake (ed.) © 1996 Balkema, Rotterdam. ISBN 90 5410 822 3

Analyzing championship squash match-play: In search of a system description

Tim McGarry, Michael A. Khan & Ian M. Franks
University of British Columbia, Vancouver, B.C., Canada

ABSTRACT: We consider championship squash match-play as an open system in which order spontaneously arises through self-organization as a result of information that is directly available in the environment. Perturbing an open system from its current state temporarily destabilizes the system, from which the system either returns to its former state or settles into a new state in accord with the newly imposed conditions. If perturbations exist in championship squash match-play, then they are likely a result of a shot that places one player at a disadvantage to the other and these shot perturbations should be prone to detection by independent observers. We report data from a behavioural observation study of championship squash match-play that shows that (a) the system can be reliably identified as being in either a stable or unstable state by expert observers ($\kappa \geq .80$) and, to a lesser extent, also by non-expert observers ($\kappa < .80$), and (b) the shot perturbations responsible for the transition to instability are easily evidenced. We consider future possibilities for sport analysis research in the context of open systems in our undertaking to describe and explain athletic behaviour.

1 INTRODUCTION

A key coaching objective is to effect improvement in athletic behaviour produced in sport competition. One way that this can be achieved is to use the behavioural observations from past competition to formulate match strategies for use in the next contest. The information derived from past observation is therefore used to guide future decision making, even though it is well known that the human perceptual system is limited in its capacity for information recall. Football coaches, for example, have been shown to be in error in their ability to recall observed events, sometimes by as much as 70%. Furthermore, notable observation error still remains even when training has been undertaken beforehand in order to prime the observational structure to analyze key behavioural events (Franks & Miller 1991). The inability of observers to recall pertinent match information is therefore a limitation to improving sport performance, and it is reasonable to suggest that the more accurate the information recall, the more appropriate the decision making for future performance enhancement. A number of computer-assisted systems have thus been developed for a variety of sports, to complement the coaching process by providing an objective and quantitative audit of the behaviours that comprise sport competition.

The derivation of information from one contest and its application to the next (usually against a different opponent) necessarily assumes that the information is transferable. In other words, the behaviours observed in one setting are likely to repeat in a similar future setting. This would suggest that athletes (or teams) have behavioural characteristics that typify performance and comprise a playing signature, or profile. These profiles would then provide useful a priori information to the coaching process in preparing for a future contest. We analyzed the sequence of shots observed in championship squash match-play (quarter-finals, semi-finals and final) as a Markov chain with this objective in mind (McGarry & Franks 1994, 1995). This allowed the comparison of a player's playing profile, defined as the probability of responding with any shot (by the player) to any preceding shot (from the opponent), when matched against different opponents. A two-way χ^2 analysis revealed that a player's profile varied when observed against different opponents (p<.25), although invariant shot responses increased when the conditions associated with the precedingshot were further detailed (McGarry & Franks in review-a). (The type I error is increased from traditional values since we wish to infer no difference between shot responses from a test of difference.) That a player's shot response to a preceding shot varies according to the particular opponent has two implications. First, the behavioural information derived from one contest is not necessarily pertinent to a future contest if, as is usually the case, the latter contest takes place against a different opponent. The exercise of caution in the traditional sport practice of scouting would thus seem a logical conclusion if sport performance is analyzed in this way. Second, a reduction of sport performance observed in championship squash match-play to a succession of individual shots, and their subsequent re-synthesis through putative relationships between these behaviours, might be misleading. We now consider an alternative view of squash match-play as a non-linear (open) system that is subject to dynamic principles of self organization and briefly borrow from the control of motor coordination literature in support of this proposition.

One of Bernstein's foremost contributions to the study of motor control was the recognition that a univocal (one-to-one) correspondence between the neural excitation and the evoked muscle contraction cannot exist, rather that the relationship is equivocal and influenced by both internal and external conditions. This classical problem, often referred to as the degrees of freedom problem because of the many degrees of freedom that must be managed in order to control movement if intentional action is to be successfully realised, has led to different considerations as to how this might be achieved. If control is achieved via a "motor program", then the program must somehow not only be cognisant of, and account for, the initial system states, but constantly (or periodically) update these internal states with reference to both external conditions and the final target. This requirement does not preclude such a control mechanism but it comes at a high computational cost. An alternative account of motor control considers the many degrees of freedom to be reduced by the temporal deployment of task specific muscle synergies (Bernstein 1967), or coordinative structures (Kelso, Southard & Goodman 1979). In particular, control is thought to be achieved via these structures through cooperation between information seeking (perception) and environment changing (action) systems, in which the animal acts to purposefully change its relation with the environment. This intent is well illustrated by Reed (1984, p165) in an example regarding the behaviour of jumping predatory spiders in which he states that "the spiders do not jump in order to predate, they predate by jumping" . The import of this distinction to this paper is that action is considered a consequence, or emergent property, of the animal's functional intent rather than a purposeful act per se. Thus, behaviour is not ascribed to an a priori organization, such as a central representation for the control of movement, rather it is assigned to an a posteriori consequence. We now consider squash match-play behaviour as an a posteriori consequence of the system's natural tendency to settle to coherent orderly behaviour, as opposed to an a priori prescription that seeks to liken the empirical behavioural shot responses to an already established playing signature.

If the observed shot exchanges between two competing players in squash match-play is considered to be the product of an open system that tends to homeostasis (regularity), then it follows that the stability and instability of the system should be identifiable at any time. We used perceptual judgement of both expert and non-expert observers to detect shot transitions between states of stability and instability. We refer to those shots that transit the system from stability to instability as shot perturbations, and consider them to be similar to Newtson, Hairfield, Bloomingdale and Cutino's (1987) breakpoints, which typically delineate action within the stream of behaviour as observed by rapid changes in body position. We reason that shot perturbations will likewise be readily subject to perceptual detection and consequently use human raters to measure the degree of behavioural interaction, a technique that has already been successfully employed in the study of perceived interactional synchrony between mother-infant dyads (Bernieri, Reznick & Rosenthal 1988).

It has been shown that a difference in perceptual ability between expert and non-expert (novice) observers of squash match-play exists in their ability to predict shot behaviour from advance information. The difference in perceptual ability is not a result of different visual search strategies but of information pick-up, that is how the information is used (Abernethy 1990). Thus, while the experts are not better able to acquire the available information, they are better able to use that information to form a perceptual judgement. We therefore hypothesize that expert observers will have increased perceptual sensitivity afforded by prior experience and, therefore, be better able to detect those shot behaviours that transit championship squash match play between stable and unstable states.

2 METHOD

2.1 Procedure

We recorded the court location and shot type in historical sequence for each rally within a match from video recordings of the quarter-final matches, semi-final matches and final match of the men's Canadian 1988 Open Championship (McGarry & Franks 1995). Sixty rallies were then randomly selected without replacement from the entire data set to comprise the experimental video excerpts. The first two authors then coded each shot as being stable and unstable in order to form a master transcript. We then recruited six expert observers (professional squash coaches) and asked them to identify those shots which, in their judgement, perturbed the system from stability to instability (perturbation onset) and also those shots that returned the system from instability to stability (perturbation offset). The experts were asked to do this twice, first without a video playback facility in order to determine whether the perturbations were immediately evident and second, with the unlimited use of the video playback facility. We then recruited six non-experts who were familiar with squash, having played it on occasion at a recreational level, and repeated the study. Testing occurred in a single session lasting approximately two and a half hours. Each group was tested in a one week period. The master record is coded with M, the experts randomly coded A through F and the non-experts randomly coded U through Z.

2.2 Analysis

The identification of the perturbation onsets and offsets allow easy identification of the system state at any instant. The shots which follow a perturbation onset are necessarily considered unstable and similarly, those that follow a perturbation offset are considered stable. The best statistic for use as a comparison of agreement between observers with regard to the identified system state is the kappa coefficient (κ), which is the ratio of observed to possible agreement after partialing out chance. It is represented in the following formula $\kappa = p_0 - p_c/(1-p_c)$, where p_0 is the proportion of observed agreement and p_c is the

Table 1. Example of the computation of the kappa coefficient (κ) for a single rally[a]

Shot	First Observer State	Second Observer State
Serve	Stable	Stable
Volley	Stable	Stable
Drive	Stable	Stable
XC-Boast	Unstable	Stable
Drop	Unstable	Stable
XC-Drop	Unstable	Unstable
Lob	Unstable	Unstable
Volley	Unstable	Unstable
XC-Drive[b]	Unstable	Stable
Error		

	Second Observer		
First Observer	Stable	Unstable	Total
Stable	3(1.875)[c]	0(1.125)	3
Unstable	2(3.125)	3(1.875)	5
Total	5	3	8

$p_0 = (3+3)/8 = .75;$
$p_c = (1.875+1.875)/8 = .469;$
$\kappa = p_0 - p_c/(1-p_c) = .75 - .469/(1-.469) = .529$

[a] The kappa coefficients reported in this example provide an index of agreement between two observers for a single rally only. The kappa coefficients reported in the text, however, provide an index of agreement between observers for 60 rallies. [b] The last shot of each rally was considered unstable by default and was accordingly omitted from the kappa computation. [c] Expected number of observations in parentheses.

proportion of chance agreement. Table 1 provides an example of how the kappa is computed from the available data. Note that a value κ=.53 is reported in the above example which is well below the agreement of .75 (6 from 8 observations agree) otherwise reported if chance is not accounted for.

3 RESULTS AND DISCUSSION

We determined the level of agreement between expert and non-experts with regard to the master record by reference to a criterion value $\kappa \geq .80$, which is taken as evidence of excellent agreement between two observers in behavioural observation studies. Figure 1 shows that four of the six experts passed criterion ($\kappa \geq .80$) while all of the non-experts did not ($\kappa < .80$). (The data are presented from highest agreement to lowest agreement for ease of interpretation only.) An independent t-test confirms that the difference between expert and non-expert agreement is significant ($t_{1,10} = 3.27$; $p < .01$), and consequently supports the hypothesis that experts are better able than non-experts to identify the system state at any instant. That a difference in agreement in the expected direction between experts and non-experts is established supports the interpretation that the reported measures provide a genuine index of behavioural synchrony. In addition, the results do not seem to be an artifact of pseudo synchrony (i.e., synchrony observed as a result of coincident behaviour; see Bernieri et al. 1988, for further detail), nor do they seem to be a product of adherence to some relatively simple observation rule, since in neither case would a difference between experts and non-experts be expected.

The system state can be validly identified by the experts which would suggest that the shot perturbations that transit the system from stability to instability can also be identified. Figure 2 provides examples from three selected rallies in support of this proposition. Each shot perturbation onset identified by each observer is represented by a point and the last shot of each rally by a diamond.

If good agreement between observers with regard to an identified shot perturbation exists, then the point sequence will approximate a horizontal line parallel to and below the rally end line, while poor agreement between observers will be reflected by a sporadic point sequence below the rally end line. Figure 2a shows good agreement between observers in that no shot perturbation was identified, except

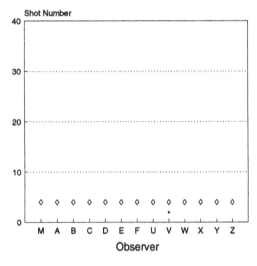

Figure 2a. Evidence for no shot perturbation in the rally.

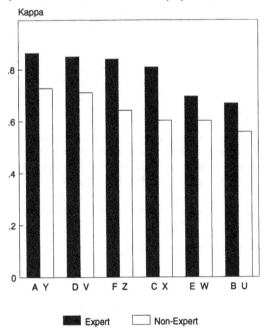

Figure 1. Inter-rater kappa coefficients for expert and non-expert observers when compared to the master record.

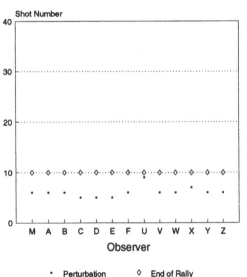

Figure 2b. Evidence for a single shot perturbation in the rally.

for V. Figure 2b also shows good agreement for the existence of a shot perturbation, as does Figure 2c which illustrates multiple shot perturbations within a rally, evidenced by the multiple point sequences below the rally end line. Note that the rally in this particular instance necessarily transits between stable and unstable phases throughout its time course since the rally must be perturbed from a stable state at each observation. Importantly, the shot perturbations were evident to the majority of observers in the majority of instances, which supports the consideration of championship squash match-play as an open system that periodically transits between stable and unstable phases.

Table 2 details the shot perturbations agreed upon by three or more experts for each of the 60 rallies. (We excluded the remaining data for both the experts and non-experts for clarity.) These data again show good general agreement in identifying the shot perturbation, especially if one allows a leeway of a shot either side. For example, rally 2 contains 9 shots in which a shot perturbation is identified as shot 6, 5, 6, 6, 7, 6, by each expert A through F respectively, from which we interpret evidence for a shot perturbation at shot 6. We consider a leeway of plus or minus one shot to be reasonable since the identification of a shot perturbation will likely be open to interpretation since adjacent shots are necessarily inter-dependent. The identification of a weak shot, say shot 6, or a strong predecessor (shot 5) or successor (shot 7) is not a disagreement that a perturbation has occurred,

but a difference as to the reason credited for why that perturbation has arisen. Note that in the above example the shot perturbation occurred on shot 6 and while shot 5 and shot 7 each agree since they bridge one shot either side, shot 5 and shot 7 do not agree with each other. Thus, five of the six experts agree that a perturbation exists although there is no obvious reason in this example to favour either shot 5 or shot 7 as part of the agreed upon perturbation.

4 GENERAL DISCUSSION

The system state is readily identified by human perception as evidenced by good agreement between the observers and is, moreover, sensitive to the degree of expertise of the observer ($p<.01$). This result is in keeping with reasoned expectation (cf. Abernethy 1990) and gives testimony to the genuine existence of shot perturbations in championship squash match-play. Thus, squash competition can be considered a series of behavioural shot exchanges that intermittently transits between phases of stability and instability. This is in line with Newtson et al.'s (1987) view of action and interaction (the action of a dyadic pair) as a repetitive wave with variation rather than as a succession of discrete actions. This is in contrast to current sport analysis that has until now exclusively treated sport performance as a composite of discrete successive actions (or behaviours).

Previous researchers have traditionally collected the behavioural data available from a sport competition and subsequently sought to elicit playing patterns from the data in its entirety, a technique that credits each behavioural action (e.g.,shot) as equally informative. This study has shown that an alternative analytic framework yields dividend in that sport (squash) competition can profitably be analyzed as a non-linear system that fluctuates between stable and unstable behaviours. The implication is that the behavioural actions are not equally informative, since those shots that perturb the system would presumably be considered key athletic behaviours that markedly influence the pattern of play and, likely, the game outcome. These shots would thus be considered critical game events and accordingly offer valuable a priori information for the development of optimal strategies, providing a posteriori consequences can be subsequently associated with them.

Usual dependent measures of dynamic systems are frequency, amplitude and period of oscillation as the system settles into a rhythmic behavioural function. A good example is that of four legged

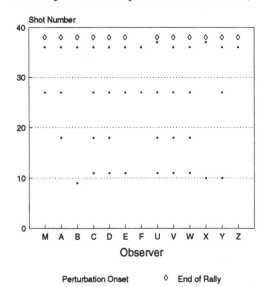

Figure 2c. Evidence for multiple shot perturbations in the rally.

Table 2. Shot perturbations identified in each rally as agreed upon by three or more experts.

R#	S#	A	B	C	D	E	F
1	6	4	4	4	4	4	5
2	9	6	5	6	6	7	6
3	9	1	2	1	2	2	3
4	44	3	2	3	3	3	3
		43			43		43
5	41	4	3	3	4	3	4
		23			23	23	23
		38	37	38	38	38	38
6	20		7		7		7
		16		17	16	16	17
7	8	6		6	7		
8	4	1	2	2	2		2
9	4						
10	10	4	6	4	4	4	5
11	21	1		2	1	1	2
		17	18	17	16	16	17
12	6	3	4	4	4	4	5
13	4	2	3	2	3	3	
14	53	3	3	3	3	3	4
		16	15	15	16		
		24	26	24	25		
				28		28	28
15	15	3	2	3	3		
			8	9		9	
16	61	5	5		5		5
		9		9	9		9
		23	23	23	23		23
		40		40	40	41	40
17	4	3	3	3	3	3	3
18	5	2	2	2	2	2	
19	15	5			5	5	5
			11	11		12	
20	10	7	8	7	7	7	7
21	4	3	3	3	3	3	
22	12	10	10	10	10	10	10
23	3	2	2	2	2	2	2
24	11	3			3	2	4
25	20	4	6	4	4	4	4
26	15	1	2	2	2	1	2
		14		14	14		14
27	7	2	2	2	2		2
28	6	3	3	3	3	3	3
29	3	2	2	2	2	2	2
30	29	2			2	2	2
		12	12	12	12	12	12
		19	21	20	18	19	19
31	6	4	4	4	4	4	
32	19	8	10		8	10	10
		18		18	18		18
33	11	2		2	2	2	2
		7	7	9	8	9	9
34	42	11	10		10		
			16	16	16	16	
			24	22	22	24	22
			32	30	32	31	32
			40		40	40	40
35	9	7	8	7	8		7
36	6						
37	19	4	4	4	4	4	4
			12	13	12	13	13
38	19		13		13		12
				18	18	18	
39	5	2	2	2	2	2	2
40	3						
41	10	9	9	9	9		9
42	13	6	6	6	6	6	6
43	38		9	11	11	11	
			18		18	18	
			27		27	27	27
		36	36	36	36	36	36
44	4	4				3	
45	18	12	12	12	12	12	12
46	10	6	6	5	5	5	6
47	33	1	2	2	1	1	1
			18		18	18	
			23		23	23	23
			28	28		28	28
48	9	8	6	7	7	7	7
49	4	2	2	2	2	3	2
50	8	5	6	5	5	5	5
51	4	2	3	2	2	2	2
52	3						
53	36	11			12	11	
			16	16	16		
			31	31	30	30	30
54	9			5	5		5
55	24		23	23	23	23	23
56	4		3	2	3		2
57	7	5	6	5	5	6	5
58	3						
59	31		14		12	12	12
					21	21	21
			30	29	29	30	29
60	4						

R# = Rally number. S# = Number of shots in rally.

gait (walk, trot, canter, gallop) whose rhythmic behavioural measures change as the velocity is scaled. We suspect that such dependent measures might be less meaningful in a competitive sport context since the system behaviour results from many external constraints such as individual playing habits, the technical and physiological capacities of the competing players, the scoring system, the stage of the game and match strategies. We prefer to consider the system to be in a permanent state of flux, where the behavioural perturbations and their related behaviours are written onto the system's tendency to stabilize as a result of what we have called a "physiological attractor" (McGarry, Khan & Franks in review-b), which is simply a tendency of each player to seek (or revert to) a state of physiological comfort.

This study has demonstrated the utility of an alternative analytic framework for describing sport competition. These intriguing results lead to the possibility of ultimately explaining the organizing principles that underlie sport competition from a unifying theory of open systems. African termites, for example, build nests by depositing nest material, initially at random, along with a scent that increases the likelihood of attracting other termites and, consequently, further nest deposits. The cyclical process of increasing scent concentration, increasing likelihood of deposit and increasing deposit results in a nest that is uniquely built in accord to principles of self organization (see Kugler & Turvey 1987, for further detail). It might be that sport competition also has self-organizing principles that enable orderly self regulating behaviours to produce unique sports contests. This study has shown that championship squash match-play behaves as an open system at a descriptive level, our task is to now investigate the system further in an attempt to explain the organizing principles to which the observed behaviours adhere.

REFERENCES

Abernethy, B. (1990) Expertise, visual search, and information pick-up in squash. *Perception.* 19:63-77.

Bernieri, F.J., J.S.Reznick & R.Rosenthal 1988. Synchrony, pseudosynchrony, and dissynchrony: Measuring the entrainment process in mother-infant dyads. *Journal of Personality and Social Psychology.* 54:243-253.

Bernstein, N. 1967. The problem of the interrelation of co-ordination and localization. In H.T.A. Whiting (ed), *Human motor actions.*

Bernstein revisited: 77-119. Amsterdam: North-Holland.

Franks, I.M. & G.Miller 1991. Training coaches to observe and remember. *Journal of Sports Sciences.* 9:285-297.

Kelso, J.A.S., D.L.Southard & D.Goodman 1979. On the nature of human limb coordination. *Science.* 203:1029-1031.

Kugler, P.N. & M.T.Turvey 1987. *Information, natural law, and the self-assembly of rhythmic movement.* Hillsdale, NJ: Lawrence Erlbaum Associates, Inc.

McGarry, T. & I.M.Franks 1994. A stochastic approach to predicting competition squash match-play. *Journal of Sports Sciences.* 12:573-584.

McGarry, T. & I.M.Franks 1995. Modeling competitive squash performance from quantitative analysis. *Human Performance.* 8:113-129.

McGarry, T. & I.M.Franks in review-a. In search of invariant athletic behavior in sport: An example from championship squash match-play.

McGarry, T., M.A.Khan & I.M.Franks in review-b. Sport competition as an open system: Evidence for behavioral perturbations in championship squash match-play.

Newtson, D., Hairfield, J., Bloomingdale, J., & Cutino, S. 1987. The structure of action and interaction. *Social Cognition.* 5:191-237.

Reed, E.S. 1984. From action gestalts to direct action. In H.T.A. Whiting (ed), *Human motor actions. Bernstein revisited:* 157-168. Amsterdam: North-Holland.

The Engineering of Sport, Haake (ed.)© 1996 Balkema, Rotterdam. ISBN 90 5410 822 3

Derivation of a rope behaviour model for the analysis of forces developed during a rock climbing leader fall

Martyn Pavier
Department of Mechanical Engineering, University of Bristol, UK

ABSTRACT: Rock climbers have to trust their climbing ropes since falls while climbing are common and the consequences of rope failure are usually serious. Climbing ropes are designed to withstand numerous falls but have a finite life, therefore old ropes should be retired from use. Since no quantitative measure exists of the effect of a fall on the remaining life of a rope, work has been carried out to develop a theoretical analysis to predict the severity of a fall. This theoretical analysis relies on an understanding of the mechanical behaviour of a rope, therefore this paper describes experiments designed to measure this behaviour and how a model of the behaviour is included in the analysis of the climbing fall.

The mechanical behaviour of climbing ropes is difficult to measure by conventional methods since the rope can suffer large strains without failure and exhibits a strong strain rate dependency. The test method that has been followed here is to load the rope dynamically via a falling mass and record the load versus strain response of the rope. Tests can be carried out with different masses and fall distances to vary the strain rate. Although strain rate varies throughout each test during the initial part of the loading the strain rate is relatively constant and sufficient data can be derived to formulate a rope behaviour model.

The theoretical analysis of the climbing fall uses expressions relating increments of tension and strain for rope segments between two adjacent karabiners, allowing large strains and the dynamic non-linear behaviour of the rope to be accounted for. These expressions are combined with relationships governing slip of the rope past karabiners and through the belay device enabling an incremental solution for the tension in each rope segment during a fall. Predictions of forces developed during a fall show good agreement with experimental measurements.

1. INTRODUCTION

The modern rock climber puts considerable trust in his rope: falls while climbing are commonplace, particularly for experienced climbers attempting difficult routes, and the consequences of rope failure are usually very serious. Consequently climbing ropes are of a sophisticated construction, combining low mass with high strength and energy absorbing characteristics. The modern climbing rope has a construction known as kernmantle where a twisted nylon kern, the load bearing part of the rope is protected from abrasion by a woven nylon mantle. The kern is made up of a number of cords, the cords in turn being made of several strands twisted together. Each strand is then formed from nylon threads, each of about 25 μm diameter. A typical rope has of the order 5 million threads. The

rope is designed to decelerate a falling climber without subjecting the climber to very high forces.

Climbers should understand the effect on their rope of any climbing falls so that they are able to decide whether the strength of the rope has been reduced sufficiently for it to be sensible to retire the rope. Currently, climbers rely on rule of thumb methods to assess the effects of falls on their ropes, but the understanding of the physical processes involved during falls is often poor. Fortunately, climbing rope failures are not common, and when they do occur are usually because of other factors, for example the rope running over the sharp edge of a rock. Nevertheless, studies have shown (Microys 1983) that even experienced climbers may have little appreciation of the current state of their ropes and in some cases their ropes may be uncomfortably near the end of their useful lives.

The research described here aims to provide a more precise description of a climbing fall and therefore give a better appreciation of the tension developed in the rope. It is hoped that such an appreciation may be combined with a model for the degradation of the rope to give a life prediction for a rope.

Little previous work of this nature exists, however Schubert (1986) provides some data of tensions developed in climbing ropes and a description of what influences these tensions although the work is largely descriptive. In addition, there have been some previous experimental measurements of tensions developed in climbing ropes during falls (Perkins 1987, Reddy 1989).

First the procedures involved in climbing and the circumstances leading to a fall are described in more detail. This is followed with a simple analysis of a climbing fall to provide an introduction to the subsequent, more complete analysis. Following this a series of tests are described with the aim of generating a model for the dynamic behaviour of the climbing rope. Next the results of experimental tests measuring the tensions in ropes during laboratory simulations of climbing falls are presented. Finally the full theoretical analysis is detailed with predictions of tensions developed in the rope during the experimental tests.

2. BACKGROUND

The procedures involved in rock climbing are first described in more detail. Usually climbers climb as a pair: a leading climber (or leader) and a second climber (or second). The second belays himself to the rock face, by attaching himself via his harness to anchors fixed to the rock. These anchors take various forms but typically they may be nuts (metal wedges) jammed in cracks in the rock, pitons (metal spikes) hammered into cracks or slings draped over a rock spike. The leader climbs above the second to the top of the rock face, or until he can find a safe place to belay himself. Once the leader is belayed, the second removes his belay and climbs to rejoin the leader. As the second climbs, the leader takes in the rope so as to protect the second should he slip.

The critical part of the climbing procedure is as the leader climbs above his second. To make this stage more safe the leader will run the rope through karabiners (metal clips) attached to additional anchors fixed to the rock. These additional anchors are known as running belays (or runners). The

situation during one climb may then be as shown in Figure 1, where the leader has fixed three running belays. Should the leader fall from the rock, the distance he falls until the rope begins to decelerate him will be twice the distance from his last running belay, assuming this was vertically beneath him. As the leader climbs the second pays the rope out through a belay device attached to his harness. The belay device is designed so as to make it easy for the second to hold the rope tight in the event of a fall. The rope from second to leader is referred to as the live rope. The remaining part of the rope yet to paid out by the second is known as the dead rope.

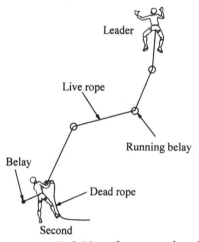

Figure 1. Definition of terms used to describe a typical climbing situation.

Experienced climbers undertaking a difficult climb may expect to take a number of falls during the climb. These falls inevitably lead to a deterioration in the rope and it is therefore useful for the climber to have an appreciation of the factors controlling the magnitude of the tensions developed in the rope. The traditional measure of the severity of a fall has been the fall factor, that is the ratio of the length of the fall to the length of rope between the leader and second. The fall factor varies from zero when the leader falls just after placing a runner, to two when the leader falls before he has been able to place any runners. This fall factor approach implicitly assumes the rope behaves in a linear elastic manner and takes no account of the slip through the belay device and friction of the rope over the runners. Before describing the theoretical procedure that has been developed for taking into account these effects an expression will first be derived for the maximum tension in the rope using a fall factor approach.

3. A SIMPLE ANALYSIS

The conventional analysis of a fall assumes no slip at the belay, no friction at the runners and a linear elastic rope. (Wexler 1950) This analysis leads to the conclusion that the maximum tension in the rope is controlled by the fall factor, the mass of the climber and the stiffness of the rope, but not the absolute length of the fall. The leader of mass M, falls from a distance d above his last runner as shown in Figure 2.

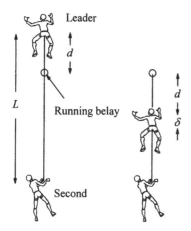

Figure 2. Geometry of a simple climbing fall.

The fall factor F is given by

$$F = \frac{2d}{L} \tag{1}$$

where L is the length of rope run out. The climber falls a distance $2d$ before the rope begins to tighten plus a further distance δ due to the stretch of the rope before coming to a halt. The loss of potential energy of the climber due to the loss of height is $Mg(2d + \delta)$ which is transferred into strain energy stored in the rope of $k\delta^2 / (2L)$ where g is the acceleration due to gravity and k the elastic modulus of the rope. From the conservation of energy we calculate the further distance as:

$$\delta = \frac{MgL}{k}\left[1 + \sqrt{1 + \frac{2kF}{Mg}}\right]$$

and hence find the maximum tension in the rope to be

$$T = Mg\left[1 + \sqrt{1 + \frac{2kF}{Mg}}\right] \tag{2}$$

For the minimum fall factor of zero the maximum tension in the rope is twice the weight of the climber whereas at the other extreme with a fall factor of two the maximum tension is

$$T = Mg\left[1 + \sqrt{1 + \frac{4k}{Mg}}\right] \tag{3}$$

4. DESCRIPTION OF APPARATUS

The experimental apparatus used to measure forces developed in the laboratory simulation of climbing falls consists of a trolley running on a vertical steel H section column as shown in Figure 3. The mass of the trolley can be varied from 40 to 80 kg by adding extra lead weights. A catch at the top of the steel column holds the trolley in the raised position until a test is carried out. Two strain gauged belay mounts were used, one to represent a running belay bolted to the column, the other to represent the second's belay, bolted to the floor. The belay mounts were manufactured so as to have a similar cross section to a standard karabiner. Strain gauges were attached either side of the belay mounts in a half-bridge configuration and connected to Fylde type FE-492-BBS bridge conditioners and FE-254-GA amplifiers. The outputs from the strain gauge amplifiers were recorded during the test on a Datalab DL1200 data logger with a sampling interval of 500 µs. The belay mounts were calibrated statically using a tensile test machine. The tensions in the two rope segments during the test were derived from the loads measured by the belay mounts.

The rope used was Edelweiss M/W V.31 of diameter 8.7mm. The rope was tied to the trolley, fed through the running belay then back to the belay device. The grip of the second's hand on the rope was simulated using a pneumatic gripper as shown in Figure 3. The gripper allowed different tensions to be applied to the rope running through the belay device and different angles to be set for the rope either side of the belay device. The gripper was designed so as to be able to apply a higher load than a human second would find possible.

In tests carried out to measure the behaviour of the rope itself, the rope was tied to the upper belay

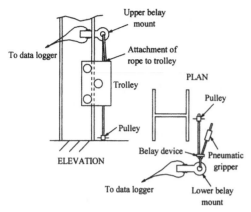

Figure 3. Details of experimental apparatus

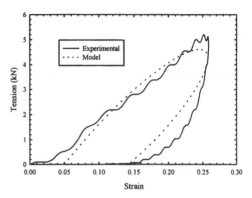

Figure 4. Tension versus strain from a dynamic test on a rope compared with the model

mount, rather than fed through the upper belay mount and then through the belay device.

5. MEASUREMENT OF ROPE BEHAVIOUR

The mechanical behaviour of climbing ropes cannot be measured using a conventional tensile test machine since the rope can suffer large strains and exhibits a strong strain rate dependency. The method that has been used here is to load the rope dynamically via the falling trolley and record the tension in the rope as a function of time. The acceleration of the trolley may therefore be calculated and hence the position of the trolley derived by a process of double integration. Recording only the tension in the rope thus allows the tension versus strain response to be derived.

Tests can be carried out with different trolley masses and fall distances to vary the conditions, in particular the maximum tension and strain experienced by the rope.

The tension versus strain response for a typical test is shown in Figure 4. This test used a 2.1 metre length of rope, a fall factor of 1.3 and a trolley mass of 40 kg. One test provides the complete response as the rope is loaded to the maximum tension and then relaxes to zero tension as the trolley returns upwards. Particular points to note are that the rope behaves in a significantly non-linear fashion and has considerable hysterisis. The non-linearity of the rope is most noticeable at low strains due to the fibres is the rope aligning themselves with the loading direction. The hysterisis is a result of the friction between adjacent fibres and the damping behaviour of the nylon fibres themselves. The experimental trace shows a superimposed oscillation which is believed to be a result of stress waves in the rope.

For the purposes of the theoretical analysis of climbing falls conducted here it was felt to be sufficient to model the rope using the linear visco-elastic model shown in Figure 5. An initial strain offset was also included in the model so that the rope only begins to take tension once a certain value of strain has been exceeded. Values for the spring and dashpot parameters in the model were chosen to give adequate simulations to the measures tension versus strain curve in Figure 4.

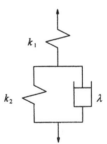

Figure 5. Visco-elastic rope model.

The values for the chosen parameters are

$k_1 = 35.0$ kN

$k_2 = 20.0$ kN

$\lambda = 3.0$ kN / s

$\varepsilon_o = 0.05$

where ε_o is the initial strain offset. These parameters will only give an adequate model for falls where the maximum tension is of the order of the maximum of Figure 4, that is up to 5 kN. Current work is being carried out to develop a model using non-linear parameters that will be valid for a wider range of fall conditions.

6. RESULTS FOR EXPERIMENTAL FALLS

Experimental tests have been conducted with varying trolley masses, drop distances and fall factors. Three different types of belay device have also been used: A Stitch plate, a Tuber and a Figure of Eight, each with varying belay angles and tensions. In addition a number of tests were carried out where the rope was tied off to the lower belay mount, rather than fed through a belay device. The tests with the rope tied off show increasing maximum tensions with increasing fall factor, also a slight increase with distance of fall for the same fall factor. For the tests using a belay device the maximum tensions are much lower except for the Figure of Eight belay device.

Figure 6 shows a typical trace for the case with a Stitch plate, a trolley mass of 40 kg, 4.9 m of rope, and a fall factor of 0.54. The tension in the rope between the upper belay mount and the trolley quickly to a peak of 1.9 kN, then drops slightly to 1.6 kN before falling to zero. There is then a small second peak after which the tension remains at a constant value equal to the weight of the trolley. The cause of the small initial peak is due to the behaviour of the pneumatic gripper, where the tension to initiate slip through the gripper is higher than that for slip to continue.

7. THEORETICAL ANALYSIS

The theoretical analysis of climbing falls will now be presented, including slip at the belay, friction at runners and a visco-elastic model for the behaviour of the rope. The analysis uses an incremental approach, necessitating the use of a computer to give a complete analysis of a climbing fall but allowing the analysis of a fall situation with an arbitrary number of runners. The results give tensions in individual rope segments, loads at runners, slip through the belay device and the position of the falling climber.

First the geometry of a pitch is described using a two-dimensional Cartesian co-ordinate system with the second at the origin (Figure 7). The leader has placed n-1 runners, with runner i located at x_i, y_i. The leader falls from a point, distance d vertically above his last runner. The original length l_i of rope segment i between runners $i-1$ and i, the angle of lap θ_i at runner i and the total length of rope run out L are calculated using the runner locations. The fall factor F is given by

$$F = \frac{2d}{L} \tag{4}$$

A maximum tension ratio η_i is calculated for each runner from

$$\eta_i = e^{\mu_i \theta_i} \tag{5}$$

where μ_i is the coefficient of friction at runner i. Experimental results not described here support the exponential form of equation (5).

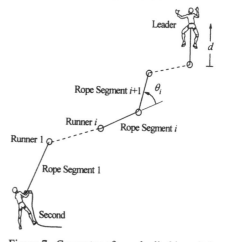

Figure 7. Geometry of a rock climbing pitch

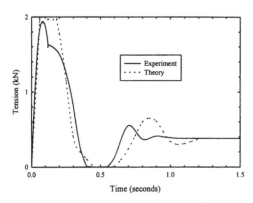

Figure 6. Graph showing tension in the rope versus time from experimental test compared to theoretical simulation.

The relationship between the strain in each rope segment and the total slip at each runner is now considered. Taking segment i (Figure 8)

$$s_i(1+\varepsilon_i) - s_{i-1}(1+\varepsilon_i) = \varepsilon_i l_i \qquad (6)$$

where s_i is the total slip at runner i, positive in the direction from the second to the leader and measured with zero strain in the rope. An incremental form of equation (6) may be derived:

$$\Delta s_i(1+\varepsilon_i) - \Delta s_{i-1}(1+\varepsilon_i) = \Delta\varepsilon_i(l_i + s_{i-1} - s_i) \qquad (7)$$

where $\Delta\varepsilon_i$ is the incremental strain in rope segment i and Δs_i the incremental slip at runner i. Note that equations (6) and (7) are also valid for the first rope segment if s_0 and Δs_0 are taken to be the slip and incremental slip through the belay device. The final rope segment, that is between the leader and his last runner, gives a modified version of equations (6) and (7):

$$s_n - s_{n-1}(1+\varepsilon_n) = \varepsilon_n d \qquad (8)$$

$$\Delta s_n - \Delta s_{n-1}(1+\varepsilon_n) = \Delta\varepsilon_n(d + s_{n-1}) \qquad (9)$$

where s_n and Δs_n are the position and incremental position of the leader measured positive downwards relative to the point at which the rope becomes tight. That is the origin for s_n is a distance d vertically below the last runner. For small increments Δs_n is given by

$$\Delta s_n = \dot{s}_n \Delta t \qquad (10)$$

where $\dot{s}_n$ is the velocity of the leader and Δt the size of the increment of time.

Figure 8. Tension in rope segment i

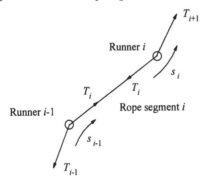

A matrix equation can now be formulated giving the vector of incremental strains in each rope segment $[\Delta\varepsilon]$ in terms of the vector of incremental slips $[\Delta s]$

$$[C][\Delta s] = [\Delta\varepsilon] \qquad (11)$$

where $[\Delta s]^T = \begin{bmatrix} \Delta s_0 & \ldots & \Delta s_i & \ldots & \Delta s_n \end{bmatrix}$
and $[\Delta\varepsilon]^T = \begin{bmatrix} \Delta\varepsilon_1 & \ldots & \Delta\varepsilon_i & \ldots & \Delta\varepsilon_n \end{bmatrix}$
Note that matrix $[C]$ has dimension n by $n+1$. A typical row of matrix $[C]$ relating the incremental strain in rope segment 4 to the incremental slips at runners 3 and 4 has the form:

$$\begin{bmatrix} & 0 & 0 & \\ & \vdots & \vdots & \\ 0 & \cdots & C_{43} & C_{44} & \cdots & 0 \\ & \vdots & \vdots & \\ & 0 & 0 & \end{bmatrix} \begin{bmatrix} \Delta s_0 \\ \vdots \\ \Delta s_3 \\ \Delta s_4 \\ \vdots \\ \Delta s_n \end{bmatrix} = \begin{bmatrix} \Delta\varepsilon_1 \\ \vdots \\ \Delta\varepsilon_4 \\ \vdots \\ \Delta\varepsilon_n \end{bmatrix}$$

where $C_{43} = \dfrac{-1-\varepsilon_4}{l_4 + s_3 - s_4}$ and $C_{44} = \dfrac{1+\varepsilon_4}{l_4 + s_3 - s_4}$

The incremental strain in any one rope segment may be related to the incremental tension by use of the visco-elastic rope model (Figure 5). In which case we have for rope segment I

$$\Delta T_i = k_1 \Delta\varepsilon_i + \frac{\Delta t}{\lambda}\left[k_1 k_2 \varepsilon_i - T_i(k_1 + k_2)\right]$$
$$= k_1 \Delta\varepsilon_i + \Delta T_i^0 \qquad (12)$$

where k_1, k_2 and λ are the constants in the visco-elastic rope model. A matrix equation is again written giving the vector of incremental tensions $[\Delta T]$ in terms of the vector of incremental strains

$$[K][\Delta\varepsilon] = [\Delta T] - [\Delta T^0] \qquad (13)$$

where $[\Delta T]^T = \begin{bmatrix} \Delta T_1 & \ldots & \Delta T_i & \ldots & \Delta T_n \end{bmatrix}$
and $[\Delta T^0]^T = \begin{bmatrix} \Delta T_1^0 & \ldots & \Delta T_i^0 & \ldots & \Delta T_n^0 \end{bmatrix}$
Matrix $[K]$ is of dimension n by n and is given by $[K] = k_1[I]$, where $[I]$ is the n by n unit matrix. It should be pointed out that any model for the rope may be used, provided it can be written in the general form of equation (12) relating increments of tension to increments of strain.

At each runner slip may occur depending on the magnitude of the tension in the rope segments either side of the runner. If slip does occur the incremental tensions vary according to the tension ratio for the runner. We express these conditions for runner i (Figure 9) as:

$\Delta T_i = \eta_i \Delta T_{i+1}$; $\Delta s_i < 0$; $\sigma_i = -1$ if $T_i = \eta_i T_{i+1}$

$\Delta s_i = 0$; $\sigma_i = 0$ if $\eta_i T_{i+1} < T_i < \dfrac{T_{i+1}}{T_i}$ (14)

$\Delta T_i = \dfrac{\Delta T_{i+1}}{\eta_i}$; $\Delta s_i > 0$; $\sigma_i = +1$ if $T_i = \dfrac{T_{i+1}}{\eta_i}$

where σ_i is used to specify the slip condition.

Figure 9. Tension in rope segments across runner i

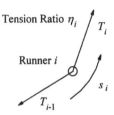

Tension Ratio η_i

T_i

Runner i

T_{i-1}

s_i

Slip is assumed to occur at the belay if the tension in the first rope segment is equal to a critical value. Thus the condition for slip at the belay is expressed as:

$\Delta s_0 = 0$; $\sigma_0 = 0$ if $T_1 < S$

$\Delta T_1 = 0$; $\Delta s_0 > 0$; $\sigma_0 = +1$ if $T_1 = S$ (15)

where S is the critical tension in the first rope segment.

Equations (14) and (15) can be used to derive a matrix equation relating the incremental tensions

$[L][\Delta T] = 0$ (16)

Each row of matrix $[L]$ corresponds to a slip condition where σ_i is non-zero. For example if the slip condition at runner 3, σ_3, is equal to +1, the corresponding row of matrix $[L]$ has the form:

$$\begin{bmatrix} & 0 & 0 & \\ & \vdots & \vdots & \\ 0 & \cdots & 1 & -1/\eta_3 & \cdots & 0 \\ & \vdots & \vdots & \\ & 0 & 0 & \end{bmatrix} \begin{bmatrix} \Delta T_1 \\ \vdots \\ \Delta T_3 \\ \Delta T_4 \\ \vdots \\ \Delta T_n \end{bmatrix} = \begin{bmatrix} 0 \\ \vdots \\ 0 \\ 0 \\ \vdots \\ 0 \end{bmatrix}$$

In the computer program a modified form of equation (16) is used to avoid instabilities when the size of increment is other than very small.

$[L][\Delta T] = -[L][T]$ (17)

where the vector of rope tensions $[T]^T = \begin{bmatrix} T_1 & \cdots & T_i & \cdots & T_n \end{bmatrix}$. Equations (16) and (17) are identical provided equation (16) has been satisfied exactly for all previous increments.

Equations (11), (13) and (17) may now be combined to give a set of linear equations that can be solved to find $[\Delta s]$.

$[L][K][C][\Delta s] = -[L]([T] + [\Delta T^0])$ (18)

At the end of each increment the incremental strains are calculated from the incremental slips by equation (11). The incremental tensions are then found by equation (13) and the current rope tensions incremented, taking account that the rope tensions must be always greater than or equal to zero. Finally the new velocity of the leader can be found from $\dot{s}_n + (g - T_n / M)\Delta t$.

A computer program has been written to carry out the calculations described above. The size of the time increment Δt is set small enough to achieve good accuracy, but it is found that the numerical calculations are well-behaved and largely insensitive to the size of the increment.

8. THEORETICAL RESULTS

The rope model parameters for the theoretical analysis were derived as has been described. Meanwhile, the coefficient of friction was found by examining the ratio between the adjacent rope segments in the tests. The ratios varied form 1.8 to 2.2, so an average value was used giving a coefficient of friction μ of 0.22.

The critical load S in segment 1 for slip to initiate at the belay was measured from the relevant test results. For the analysis of cases where the rope was tied off to the belay the value of S was set sufficiently high for no slip to occur.

The theoretical technique may now be used to provide predictions of tension versus time which can be compared with the experimental results. Figure 6 shows one such comparison. The theoretical technique assumes a constant value for the critical tension in segment 1 and is therefore unable to predict the small drop in load after the initial peak that occurs in the test results. The theory is thus unable to provide a wholly accurate simulation of experiment when slip through the belay occurs although giving reasonable predictions of maximum tensions.

The theory may now be used to give predictions of tensions for more complicated pitches than it is possible to instrument. Figure 10 shows such a pitch with 4 runners, each with a coefficient of friction of 0.22, and a critical tension in segment 1 of 1.0 kN. The same rope properties were used as derived previously. Figure 11 shows the predicted maximum tensions in the final rope segment, that is the segment attached to the leader, for different fall factors where the variation in fall factor was achieved by varying the distance d (Figure 10) above the last runner. Figure 11 shows in addition the effect of varying the leader's mass. As the fall factor is increased, the maximum tension soon reaches a maximum value, defined by the critical load for slip at the belay, the number of runners and the friction at the runners. The fall factor at which the tension achieves the maximum factor is lower for increasing climber mass.

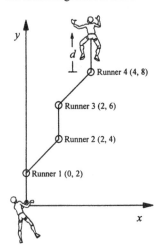

Figure 10. Geometry used for theoretical experiments

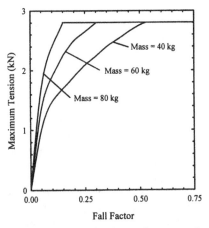

Figure 11. Predicted maximum tensions from theoretical experiments

9. DISCUSSION

The theory that has been developed includes the major phenomena influencing the tension developed in a climbing rope. Differences between experiment and theoretical prediction are largely because a simple model has been used for the behaviour of the belay device: that slip occurs once a critical tension has been reached. Certainly a more sophisticated model could be used, by relating tension to slip for example. Indeed the behaviour of a human second could be simulated, perhaps including varying reaction times.

The visco-elastic model used for the rope is characterised by 3 constant properties. However, it is known that the elastic properties of a rope are non-linear, especially at low strains where stretch of the rope occurs through alignment of fibres rather than stretching. Furthermore, an important mode of energy dissipation within the rope is by friction between fibres in addition to damping in the fibres themselves. Neither elastic non-linearity or frictional hysterisis were included in the current rope model, but provided the necessary data were available, such modifications to the model could be easily made.

The mass of the rope itself was not included in the analysis, that is the model was unable to predict the size of tension waves in the rope. For a linear elastic rope the magnitude of these tension waves can be predicted from:

$$T = vcm \qquad (19)$$

where T is the magnitude of the wave, v is the velocity of the climber just before the rope becomes tight, c is the velocity of propagation of waves in the rope given by $c = \sqrt{k/m}$ and m the mass per unit length of the rope (Timoshenko and Goodier 1982). Using typical values the magnitude of the tension waves may be found to be much less than the predicted tensions neglecting the mass of the rope. Conversely, tension waves in polymer ropes used for mooring ships are significant, but the behaviour of these ropes are very different from the climbing rope (Leeuwen 1981).

The analysis of climbing falls has concentrated on the prediction of maximum tension in the rope. Since the aim of the work is to allow the assessment of the damage suffered by a rope in any given fall situation, a method of relating the maximum tension to damage must be found. However, work on fatigue of nylon ropes (Kenney 1987) suggests that an important factor is the duration of the load on the rope in addition to the maximum value. Thus, although a large fall may not subject a rope to higher tension than a small fall, the tension in the rope will last for a longer time and therefore will cause more damage. The current theory is able to the duration of the fall but must be combined with data on failure of ropes before rope life can be assessed.

10. CONCLUSIONS

A theoretical method to analyse climbing falls has been developed, allowing good predictions to be made of tension in the rope. The method includes slip of the rope through the belay device, friction at runners and a visco-elastic model for the rope and could be extended to account for more general non-linearity in the rope and a more sophisticated simulation of the belay device.

The analysis predicts that except for very small falls the critical factors influencing the maximum tension developed in the rope are critical tension for slip in the belay device and the friction developed at the runners. The mass of the climber and the fall factor, that is the ratio of the length of fall to the length of rope run out, do not have a significant effect.

Further work must be carried out to understand the effects causing damage of the rope before the analysis described here can be used to predict the life of the rope.

REFERENCES

Kenney, M.C. 1987. Fatigue Behaviour of Synthetic Fibres, Yarns and Ropes, J Matl Sci, 20

Leeuwen, J.H.V. 1981. Dynamic Behaviour of Synthetic Ropes, Proc 13th Offshore Technology, Texas, 453-463

Microys, H.F. 1983. Letter to Summit, 31, (July-August)

Perkins, A. 1987. Development of Fall Arrest Equipment using Textiles, PhD thesis, University of Leeds

Reddy, T.Y. 1989. Private Communication, UMIST

Schubert, P. 1986. Ausrustung, Sicherung, Sicherheit, BLV Verlagsgesellschaft, Munich

Timoshenko, S.P. and Goodier, J.N. 1982. Theory of Elasticity, 3rd Edn. McGraw-Hill, New York

Wexler, A. 1950 The Theory of Belaying, American Alpine Journal

The Engineering of Sport, Haake (ed.) © 1996 Balkema, Rotterdam. ISBN 90 5410 822 3

Symbolic dynamics for motion analysis in sports

A.J.Subic & S.B.Preston
School of Engineering, University of Ballarat, Vic., Australia

ABSTRACT: Symbolic dynamic modelling of complex human motion is gaining increasing importance in engineering of sports. In particular, the development of general dynamic models which describe a variety of human motor activities represents a fundamental stage in obtaining more sophisticated forms of optimal motion. This paper gives a brief overview of the symbolic dynamics modelling of take-off in jumping and diving, based on Kane's method in association with Amirouche's algorithms for inverse dynamics. Examples of computer models with initial results are illustrated and some practical aspects of correlation with experimental results are discussed.

1 INTRODUCTION

The notion of analysing, quantifying and predicting motion of humans has intrigued researchers for many years. Accordingly, there have been several techniques employed by researchers, in an attempt to achieve this goal. Specific areas of the complete array of possible human motions have been investigated by different researchers, who have all had the same ultimate goal: to better understand the way in which humans move.

To achieve this goal, various approaches have been utilised. Exhaustive experimental methods, (Bobbert, 1988; Anderson, 1993) have been used, along with, analytical techniques, (Amirouche, 1990; Pandy, 1990), and the inverse dynamics approach, (Happee, 1994; Gershkovich, 1995).

This paper reports on research in progress which focuses on specific human motions involving take-off, such as diving and jumping. The inverse dynamics approach has been used for parameter sensitivity analysis and computer simulation of the mechanics of human take-off, for particular cases of vertical height jumping and racing dive in swimming. The modelling procedure applied in this research, focuses on the development of dynamic equations of motion of the human skeletal system to predict relevant biomechanical parameters for real world performance criteria. Experimentally obtained data is used as input in this inverse dynamics approach.

The main dynamic modelling techniques which can be used for analysis of articulated rigid multibody systems, (where the human body is a subset), are the Newton-Euler method, Lagrange's method and Kane's method. Newton-Euler and Lagrange are more traditional techniques, whereas Kane's method is a contemporary technique which is gaining popularity due to its vector based construction which leads to a more computationally efficient linearised system of equations of motion (Kane, 1985). The reason that Kane's method is computationally quicker than other methods is because unlike the Newton-Euler technique which utilises an iterative approach and Lagrange's method which results in a non-linear system of differential equations, Kane's method is based on a linearisation of the generalised velocities, which results in a linear system of differential equations.

The equations of motion determined through dynamic modelling enable prediction of biomechanical parameters through the inverse dynamics approach. This is a procedure in which experimentally determined kinematic data, (angular and linear displacements, velocities and acceleration), is input to the equations of motion to determine the dynamic behaviour, (joint torques and external forces), of the body performing the motion. Thus, from the equations of motion, which describe the complete dynamic behaviour of the system, it is then possible to predict the behaviour of the model.

After the quantification of human motion through development of the system of equations of motion, the next phase of this research is to improve or optimise the motion in order to increase specific aspects of human performance. This is initially achieved by performing parametric sensitivity studies such that parameters that have the greatest effect on the motion can be identified and optimised. The limits of optimisation are of course constrained by the limits of human performance in terms of achievable motion.

2 DYNAMIC MODELING OF TAKE-OFF

When an in depth analysis of human take-off is to be undertaken, some form of dynamic modelling is the most systematic and scientifically viable method of achieving a high level of accuracy in the analysis. Dynamic modelling is preferred over traditional empirical trial and error approaches where the coach, athlete and investigator are unable to separate, accurately measure or vary any biomechanical parameter or set of parameters during experimentation. Thus, they are also unable to determine the degree of effect on the final motion each parameter, or change of parameter exerts.

Dynamic modelling, however, through the development of equations of motion enables all relevant biomechanical parameters to be identified. Further to this, the dynamic model then allows investigation or sensitivity studies to be carried out on specific parameters or sets of parameters so that their impact on performance can be accurately gauged. Thus, a dynamic model of human motion allows for optimisation of that motion to take place, as parameters describing the motion can be accurately defined.

As mentioned previously, researchers use a variety of dynamic modelling techniques to develop the equations of motion which describe a particular human motor activity. Both Newton-Euler and the Lagrange method are applicable to the modelling of human motion, but each of these has limitations when complex multidegree of freedom systems are being modelled. The Newton-Euler technique,

which is based on Newtons Laws of motion, is computationally laborious as it uses an iterative approach to solving the resulting equations. Lagrange's equations on the other hand require the solution of second order nonlinear differential equations. The nonlinearity aspect of the resulting equations means that the solution to these equations is also computationally labour intensive. Kane's method, on the other hand, linearises the generalised velocity resulting in a computationally efficient linear system of differential equations which, comparatively, is easier to solve.

The simplified models of the human body, used in the dynamic modelling, are shown in Figures 1 and 2. The dynamic modelling process, from definition of the system, through to equations of motion, for an articulated rigid multibody system is illustrated in the flow chart, (see Figure 3). This procedure shown in the flow-chart is based on Kane's equations.

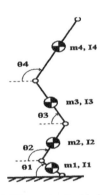

Figure 1. Human vertical height jump model.

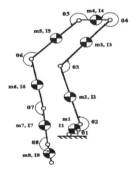

Figure 2. Human diving model.

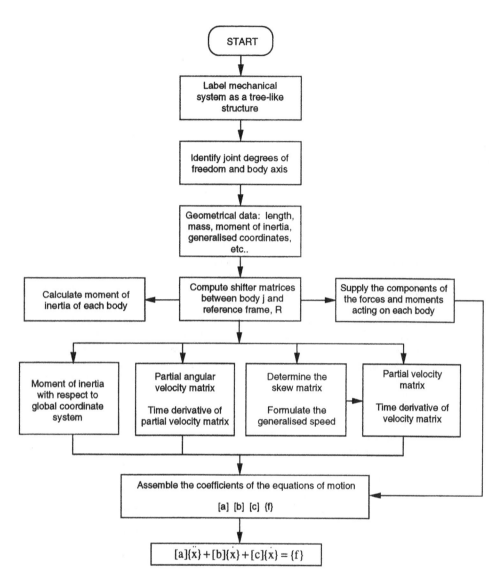

Figure 3. The Dynamic Modelling Process, Utilising Kane's Method, (Adapted from Amirouche, 1992).

3 KANE'S METHOD

The basis of Kane's equations, for an unconstrained system is,

$$f_l + f_l^* = 0 \qquad l = (1, .. , n) \tag{3.1}$$

where f_l and f_l^* denote the generalised active and inertia forces respectively; n represents the number of generalised coordinates, (Amirouche, 1992).

Also,

$$f = W(q, t)u + X(q, t) \tag{3.2}$$

$$\dot{q} = Wu + X \tag{3.3}$$

where W and X represent a linear approximation of f to the generalised speeds, u; $\dot{q}$ is the first derivative of the generalised coordinates, q, (Mingori, 1995).

The linearisation of the generalised velocity results in a system of linear differential equations which describe the behaviour of the system. Kane's method also has the added benefit of being able to be formulated in matrix form, which also simplifies the computer manipulation of the terms during formulation of the system of equations of motion.

As can be seen in Figure 2, the equations of motion for an articulated human linkage are based on the following expression, in matrix form,

$$[a]\{\ddot{x}\} + [b]\{\dot{x}\} + [c]\{\dot{x}\} = \{f\} \tag{3.4}$$

$$\text{where } [a] = \sum_k m_k \left[V_w^k\right]\left[V_w^k\right]^T + \sum_k \left[\omega^k\right]\left[I_{ko}\right]\left[\omega^k\right]^T$$

$$[b] = \sum_k m_k \left[V_w^k\right]\left[\dot{V}_w^k\right]^T + \sum_k \left[\omega^k\right]\left[I_{ko}\right]\left[\dot{\omega}^k\right]^T$$

$$[c] = \sum_k \left[\omega^k\right]\left[\Omega_x^{ok}\right]\left[I_{ko}\right]\left[\omega^k\right]^T$$

$$\{f\} = \sum_k \left[V_w^k\right]\{F_x\} + \sum_k \left[\omega^k\right]\{M_k\}$$

$$\{x\} = \{x_1, x_2, x_3, ..., x_{3N}, x_{3N+1}, x_{6N}\}$$

$$\{\dot{x}\} = \{\dot{x}_1, \dot{x}_2, \dot{x}_3, ..., \dot{x}_{3N}, \dot{x}_{3N+1}, \dot{x}_{6N}\}$$

$$\{\ddot{x}\} = \{\ddot{x}_1, \ddot{x}_2, \ddot{x}_3, ..., \ddot{x}_{3N}, \ddot{x}_{3N+1}, \ddot{x}_{6N}\}$$

where a is the generalised mass matrix; b and c are the non-linear velocity contribution, (coriolis, quadratic terms in the velocities, etc.); b and c are also known as the dynamic damping of the system; f is the generalised force; $\dot{x}$ is the first derivative of the generalised coordinates; $\ddot{x}$ is the second derivative of the generalised coordinates.

The resultant set of equations can be solved using the inverse dynamic approach so that the biomechanical parameters of the human skeletal model can be evaluated. The equations of motion for both the jump model and the dive model have been evaluated in symbolic form, using software MAPLE, (Subic, 1996).

4 INVERSE DYNAMICS

The inverse (or forward) dynamics approach to mechanism analysis, is a technique which combines the theoretical and experimental aspects of dynamic modelling. Once a mathematical model of the human skeletal system has been formulated, it is then possible to input experimentally determined data into equations of motion, solve the resulting system of equations and determine (or validate) further the behaviour of the model. That is, the kinematic experimental data can be used to solve the equations of motion and determine the dynamic experimental values, such as ground reaction forces and joint torques.

Figure 4. Jumping Motion Trace.

The motion of the jump and dive models can also be represented by trace diagrams, (see Figures 4 and 5), which show the time history of the motion. The kinematic experimental data, which when used in conjunction with the equations of motion through the inverse dynamics approach, leads to the determination of generalised active forces which are present in the model during motion.

The inverse dynamics approach is beneficial to human motion analysis because it allows prediction of either forces or trajectories after experimental data has been substituted into the equations of motion. This means that the performance of the model can be increased or optimised by optimising the parameters of the model.

Figure 5. Diving Motion Trace.

Validation of the results determined from the inverse dynamics approach is obtained by comparing the results with those recorded during experimentation. That is, when the kinematic data is used to determine the dynamic forces acting in the model, it is possible to compare the experimental hand and foot forces with those predicted analytically from the dynamic model.

5 EXPERIMENTAL INVESTIGATION

Experimental investigation is undertaken in this research for two main reasons:

1. to obtain intrinsic kinematic data which is to be integrated in the inverse dynamics procedure.

2. to correlate analytical and experimental results and refine the dynamic models.

A variety of different measurement are obtained using the following techniques and equipment (see Table 1).

Table 1. Experimental Data and Corresponding Techniques in Diving Take-Off.

	Kinematic Data	Dynamic Data
Parameters Measured	Linear & Angular • Displacement • Velocity • Acceleration	Take-Off Forces • Feet • Hands
Measuring Equipment	Video Filming and Digitising • Ariel Performance Analysis System (APAS)	Multi-function Starting Block • Kistler Force Plate • Dynamic Strain Gauging

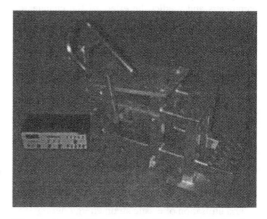

Figure 6. Multi-functional Experimental Starting Block, (Designed by Aleksandar Subic and Manufactured by William Barrett & Sons, Ballarat).

Figure 6 shows the multi-functional starting block used for measurement of take-off forces. The separation of take-off forces is achieved by using the Kistler force plate to measure the reaction force present at the feet, and strain gauge measurements on the handles of the starting block for the hand component of the reaction forces. By assembling different grip elements on the sides and at the front of the starting block it is possible to investigate the effect of different take-off styles on the overall performance during diving motion.

6 CONCLUSIONS

The benefit of using symbolic dynamic modelling and in particular Kane's method, in biomechanics of human motion is in obtaining dynamic equations of motion which describe completely the motion of the system of articulated linkages representing the human skeletal system. Combining the equations of motion with experimental data by means of either forward or inverse dynamics and utilising constraint equations to ensure the limits of human performance are enhanced but not physically exceeded, gives it is possible to obtain optimal human motion in sports.

The advantages of this technique over the traditional empirical trial and error approach are numerous. Perhaps, the most significant gain is in the ability to identify, separate and simulate the change of any biomechanical parameter or set of parameters and predict the effect of the changes on the overall performance of the human dynamic system.

REFERENCES

Amirouche, F., Ider, S. and Trimble, J. (1990). Analytical Method for the Analysis and Simulation of Human Locomotion. *J. Biomech. Eng.* **112**, pp. 379-386.

Amirouche, F. (1992). Computational methods in multibody dynamics. Prentice-Hall, New Jersey.

Anderson, F. C. and Pandy, M. G. (1993) Storage and utilisation of elastic strain energy during jumping. *J. Biomechanics.* **26**(12), pp. 1413-1427.

Bobbert, M. F. and Van Ingen Schenau, G. J. (1988). Coordination in vertical jumping. *J. Biomechanics.* **21**(3), pp. 249-262.

Gershkovich, V., Glover, B., Subic, A., Yearwood, J., and Preston, S. (1995). *Mathematical Modelling of the Deformable Bodies Problem.* Int. Conf. on Computational Techniques and Applications, Melbourne.

Guimaraes, C. S. and Hay, J. G. (1985). A Mechanical Analysis of the Grab Starting Technique in Swimming. *Int. J. Sport Biomechanics.* **1**, pp. 25-35.

Happee, R. (1994) Inverse Dynamic Optimization Including Muscular Dynamics, A New Simulation Method Applied to Goal Directed Movements. *J. Biomechanics.* **27**(7), pp. 953-960.

Kane, T. R. and Levinson, D. A. (1985). Dynamics: Theory and Applications. McGraw-Hill, New York.

Mingori, D. L. (1995). Lagrange's equations, Hamilton's equations and Kane's equations: Interrelations, Energy Integrals and a Variational principle. *Journal of Applied Mechanics.* **62**, pp. 505-510.

Pandy, M. G., Zajac, F. E., Sim, E. and Levine, W S. (1990). An Optimal Control Model For Maximum-Height Human Jumping. *J. Biomechanics.* 1185-1198.

Subic, A, Glover, B, Gershkovich, V. and Preston, S. (1996). *Applications of Deformable Body Dynamics to Take-Off Mechanics in Aquatic Sports.* First Australasian Congress on Applied Mechanics 96, Melbourne.

8 Motion analysis

The Engineering of Sport, Haake (ed.) © 1996 Balkema, Rotterdam. ISBN 90 5410 822 3

Static and dynamic accuracy determination of a three-dimensional motion analysis system

L.W. Alaways & M. Hubbard
Mechanical and Aeronautical Engineering, University of California, Davis, Calif., USA

T.M. Conlan & J.A. Miles
Biological and Agricultural Engineering, University of California, Davis, Calif., USA

ABSTRACT: Video motion analysis techniques are used in various applications to determine the three-dimensional kinematics of objects in motion. The typical procedure first determines the position of the cameras and the coefficients for the Direct Linear Transformation (DLT) method of Abdel-Aziz and Karara, (1971) using images of control points with known positions. Then the digitized images of a tracked target from multiple cameras are reconstructed with the DLT to obtain three-dimensional space coordinates. The purpose of this study was to determine the accuracy of this procedure for a four-camera video motion analysis system and to investigate the possible causes of any systematic errors in the process. Sixteen control points enclosing a three-dimensional control volume of 8m×6m×3m were surveyed to an accuracy of 2.1 mm and were used to determine the positions of the four cameras. As a measure of static accuracy, some of the same control points were digitized at 60 Hz and compared statistically to the known survey results. A dynamic accuracy study was conducted by comparing the theoretical trajectory of a spherical pendulum moving in various types of motion to experimental 60 Hz digitized data. Studies were carried out in various lighting conditions and with different threshold values for the digitization. In the static tests the known "model" was that the objects were motionless. In the dynamic tests, a spherical pendulum was chosen for its fully three-dimensional, yet highly precisely known, motion. Its dynamic model included the effects of aerodynamic drag on both the bob and suspending wire, an accurate determination of gravity, the rotational effects of the earth, and accurately measured physical properties of the bob and wire. The initial conditions and other imprecisely known quantities were determined using the parameter estimation techniques of Hubbard and Alaways (1989). The results of the study show that great care must be taken, not only in data collection, but also in determining control point locations, lighting conditions and threshold values.

1 INTRODUCTION

The purpose of this investigation was to determine the dynamic accuracy and to investigate the possible causes of any systematic errors that may arise while using a video based motion analysis system. A common practice in biomechanical analysis is to obtain three-dimensional space coordinates from two-dimensional video images using the Direct Linear Transformation (DLT) method of Abdel-Aziz and Karara, (1971). DLT parameters are obtained by recording the locations of control points that enclose the control volume in which the motion is to take place. In many cases the control volume can be rather large. For example, to study the entire trajectory of a pitched baseball the control volume would have one dimension of at least 19 meters. Researchers have shown that camera location is important for accuracy (Gosh, 1979), that the larger the number of control points the better (Fraser, 1982), that significant inaccuracies in three-dimensional reconstructions are likely to occur if the target points lie outside the control volume (Wood & Marshall, 1986) and that targets and control point diameters should be 1/50th

to 1/75th of the field-width (Motion Analysis Corporation, 1989). It is also common practice to place ring lights around the camera lens or shine light from directly behind the camera in the belief that this will eliminate shadows on the target. However, in acquiring the trajectory of a target that is moving inside a large control volume, defined with many accurately located control points, while using optimal camera placement and target size, few if any studies have investigated the effects of target illumination and thresholding. This study considers the effects that illumination and thresholding have on the accuracy of trajectory determination.

2 EXPERIMENTAL METHODS

2.1 Equipment Description

A motion analysis system, manufactured by the Motion Analysis Corporation (MAC) of Santa Rosa, California, was used for this investigation. The system includes four multispeed MAC cameras with fixed 12.5 mm lenses, a MAC VP320 video processor, and a Sun SPARCstation 330 running MAC ExpertVision (EV) version 3.2 software. Each camera was encompassed by four 30 Watt spotlights with additional fill lighting provided by four stand-alone 500 Watt flood and four 500 Watt spotlights.

2.2 Calibration

The system was calibrated by hanging sixteen retro-reflective coated spheres or control points from the laboratory ceiling on long stiff rods. To determine the position of each control point as precisely as possible and to help maintain integrity in the procedure, a weight was attached to the lower end of each rod and submersed in a fluid bath to help prevent any movement caused by air circulation and to provide damping to the rods. Each control point was then surveyed from three different locations, with a precision of twenty arc-seconds, using a surveyor's theodolite. The entire calibrated space enclosed a volume of 8m×6m×3m and the furthest control point from any survey location was 12 meters. Using triangulation techniques the location of each control point was then determined to an accuracy of 1.2 mm in each coordinate direction for a total position accuracy of 2.1 mm. These control points were then used to determine the 11 DLT parameters of each camera.

The cameras were arranged in pairs such that each camera in a pair was located at roughly the same horizontal location but separated by one to two meters in the vertical direction. Each pair was then oriented to look into the center of the calibrated space with the separation angle between the pair set to approximately 90 degrees. Two 500 Watt flood and two 500 Watt spotlights were then arranged around each pair of cameras to fill in any shadows and to help in providing even lighting throughout the entire range of motion of the tracked object. Each camera also had four 30 Watt spotlights located 10 cm radially from the center and encircling the lens. The image from each camera was then fed directly into and controlled by the VP320.

The calibration was completed by recording two seconds of 60 Hz data and using the supplied EV calibration routine to determine the location of, and the 11 DLT parameters for, each camera. The EV software gives an indication of the calibration "accuracy" by calculating for each camera a residual norm, in pixels, at the completion of its operation. This number is the norm of the position errors calculated from differencing the average centroidal location of a control point, determined over the entire number of frames used for calibration, and the frame-by-frame centroid positions. A general rule of thumb is that this value should be less than 1.0. For this study the maximum norm was approximately 0.8.

2.3 Static accuracy determination

The static accuracy was determined by tracking the trajectories of only those control points that were seen from all four camera locations and comparing the average locations to the surveyed values. The standard deviation of the random noise was also determined and this was used as a goal to shoot for in the dynamic accuracy investigation.

To investigate how lighting conditions influence the static and thus dynamic accuracy of the system, a simple experiment was conducted to see how lighting affects the apparent position of a target. Since the three-dimensional position is determined by the reconstruction of two-dimensional images, the accuracy of the target's position is inherently related to the apparent position in the two-dimensional view. To examine how light influences this position, a single target (5 cm diameter sphere) was digitized using one camera located 2.87 meters away and at the same vertical height as the target. However, no line-of-sight lighting was used, instead a single 500 Watt spotlight was positioned at 14 locations along an arc of radius 2.87 m as shown in Fig. 1. The light was also placed at the same vertical height as the camera and target. The threshold on the video processor and f-stop on the camera were adjusted to give the best digitized image when the light was at position A. These settings were not adjusted during the data collection. Data was collected for five seconds at 60 Hz and the two-dimensional position was determined using the EV software. The horizontal position of the apparent centroid was then plotted.

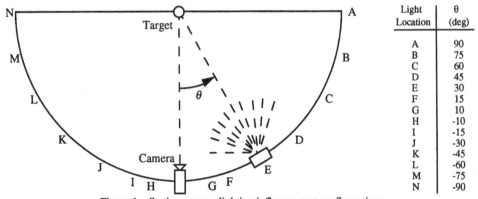

Figure 1 – Static accuracy lighting influence test configuration.

2.4 Dynamic accuracy determination

The dynamic accuracy investigation was carried out by tracking the trajectory of a spherical pendulum. The pendulum, consisting of a lead bob with a mass of 439.45 grams and coated with retro-reflective tape, was suspended from a fixed point on the ceiling of the laboratory using a 1.52 mm diameter wire approximately 5.9 meters in length. The bob was released with the bob swinging in several motions: in a plane bisecting the intersection angle of the cameras, in the plane perpendicular to that plane, and in a three-dimensional, roughly spherical trajectory. Data was collected at 60 Hz for 15 seconds using different threshold values on the video processor, under different lighting conditions and with various amplitudes of motion. The trajectory of the bob was determined using EV supplied routines with data from all four cameras and with data from selected pairs of cameras. The trajectory of the bob was then compared to a theoretical trajectory determined by integrating the three-dimensional mathematical model.

3 THEORY

Because it is somewhat unsatisfying to use static accuracy calculations as estimates of the tracking accuracy for targets in motion, we need a test target which moves. In addition, its motion must be known very precisely. The target can then be tracked experimentally and the trajectory compared with the theoretical motion to determine directly the dynamic tracking accuracy. The known trajectory should be fully three-dimensional motion and have frequency content comparable to the targets for which the standard is being developed. A spherical pendulum fits these requirements.

3.1 Spherical pendulum dynamic model

It is necessary to make a model for pendular motion which takes into account all factors which can cause substantial position changes. The spherical lead sphere (bob) was suspended from a fixed point on the laboratory ceiling by a very thin steel wire. Its diameter of 1.52 mm was chosen as a tradeoff which minimized aerodynamic wire drag yet simultaneously minimized wire compliance and the resulting sag under the varying dynamic loads. The top of the wire was clamped to a steel girder and thus can be presumed to be fixed relative to the earth. Further we assume the bob itself is exactly spherical (though the name "spherical pendulum" comes from the spherical three-dimensional motion rather than the bob surface shape) with mass m, radius r, consequent centroidal moments of inertia $I = 2mr^2/5$, and wire length L and diameter b.

Such a pendulum has several vibratory motions near its equilibrium position hanging directly below the support point. First, angular motions of the bob about its vertical axis generate a torsional restoring torque in the wire proportional to the angular deflection of the bob and thus result in sinusoidal twisting oscillations. These torsional vibrations have a natural oscillatory frequency $\omega_\tau = (\pi b^4 G / 64IL)^{1/2}$, which for the particular pendulum chosen is $\omega_\tau = 0.664$ rad/sec $= 0.107$ Hz. Here $G = 8.08$ MPa is the shear modulus of steel.

Thus this motion is not only very slow because of the small torsional stiffness of the thin wire, but it also results in zero net motion of the bob center of mass and is therefore imperceptible to the motion analysis system if the bob surface is uniformly reflective.

Two other attitude oscillation modes involve angular motion of the bob around its two horizontal axes. These contain small motion of the wire-bob contact point but even smaller (indeed negligible) motion of the bob center of mass. The frequency of these modes is $\omega_\rho = (5g/2r)^{1/2}$ (107.4 rad/sec = 17.09 Hz) where g is the gravitational acceleration. Of course, the most familiar pendular oscillation is that of the planar simple pendulum, in which the attitude changes of the bob are slightly larger than, and in phase with, the angular oscillations of the supporting wire. These occur at an intermediate frequency $\omega_\sigma = (g/L)^{1/2}$ (1.28 rad/sec = 0.204 Hz). Finally, the frequency of a purely vertical translational vibration is given by $\omega_v = (k/m)^{1/2}$ (38.0 rad/sec = 6.05 Hz), where the wire stretch stiffness $k = \pi b^2 E / 4L$ and $E = 2.068$ MPa is the modulus of elasticity of steel.

The four frequencies $\omega_\tau < \omega_\sigma < \omega_v < \omega_\rho$ of the vibratory motions discussed above are determined from linearizations of the full equations of motion of the bob when it is restricted to small motions in a plane or on a line rather than three-dimensions, and they can be useful to understand the frequency content of the most general pendular motion. Nevertheless it is necessary to obtain the full nonlinear equations of motion as a model with which to compare the experimental data. Because the ratio $r/L \cong 0.0036$ is so small the rotational motions of the bob, which are imperceptible in any case, are negligible and an adequate model is that the bob behaves as a particle. Yet two other assumptions commonly made with the point mass assumption are not valid in this case. Because of the small wire diameter its stretch compliance is substantial and it is necessary to include wire stretch in the model. Also, it is generally assumed that the earth is an inertial reference frame. Kane and Levinson (1985) show, however, that over a 60 second period, this assumption can lead to errors of the order of millimeters in the horizontal projections of a spherical pendulum. Thus it is necessary to account for earth rotation in the model.

Figure 2 shows a schematic diagram of the system and is adapted from Kane and Levinson (1985), where the equations for a similar problem without wire stretch are developed. For ease in reference to the details of their formulation, we use identical terminology insofar as possible. The earth, E with center E^* rotates at angular velocity $^F\boldsymbol{\omega}^E = \omega\mathbf{k}$ in a non-rotating inertial frame F about its inertially fixed polar axis $\mathbf{k}$, where $\omega = 0.0000729$ rad/sec. The pendulum particle P of mass m hangs by the thin wire from a point Q fixed in E at altitude L from the earth's surface. To define the three generalized coordinates q_1, q_2, q_3 which uniquely specify the configuration of P in F we refer to two additional frames; A and B. The earth fixed frame E contains $\mathbf{e}_1$, $\mathbf{e}_2$, and $\mathbf{e}_3$ a right handed set of vectors in the south, east and up directions respectively, with origin at a point O on the earth's surface below Q with co-latitude ϕ. With P initially at equilibrium distance L directly below Q on line QE^*, rotate line QP about $\mathbf{e}_1 = \mathbf{a}_1$ through angle q_1 to arrive at the A frame. Then rotate QP about the $\mathbf{a}_2$ axis through angle q_2 to arrive at the B frame. Finally allow the wire to stretch a distance q_3 along the wire in the direction $\mathbf{b}_3$.

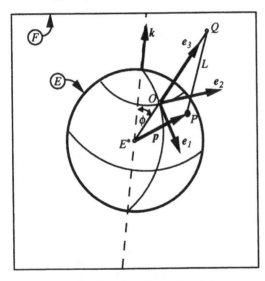

Figure 2 – Coordinate frame description
(after Kane & Levinson (1985))

Now introduce three generalized speeds u_1, u_2, and u_3 defined as the measure numbers in the B frame of the velocity of P in E

$$u_r = {}^E\mathbf{v}^P \cdot \mathbf{b}_r \qquad (r=1,2,3) \qquad (3.1)$$

Following Kane and Levinson (1985) it can be shown that the velocity and acceleration of P in frame F are given by

$$^F\mathbf{v}^P = \omega\mathbf{k} \times \mathbf{p} + u_1\mathbf{b}_1 + u_2\mathbf{b}_2 + u_3\mathbf{b}_3 \qquad (3.2)$$

$$^F\mathbf{a}^P = 2\omega\mathbf{k} \times \left(u_1\mathbf{b}_1 + u_2\mathbf{b}_2 + u_3\mathbf{b}_3\right)$$

$$+\omega^2\mathbf{k} \times (\mathbf{k} \times \mathbf{p}) + \left(\dot{u}_1 - \frac{u_2^2 s_2}{Lc_2} - \frac{u_1 u_3}{L}\right)\mathbf{b}_1 \qquad (3.3)$$

$$+\left(\dot{u}_2 + \frac{u_1 u_2 s_2}{Lc_2} - \frac{u_2 u_3}{L}\right)\mathbf{b}_2 + \left(\dot{u}_3 + \frac{u_1^2 + u_2^2}{L}\right)\mathbf{b}_3$$

where $\mathbf{p}$ is the vector from E^* to P and where s_i and c_i denote $\sin q_i$ and $\cos q_i$, respectively.

Thus the three partial velocities of P in F are given by

$$^F\mathbf{v}_r^P = \mathbf{b}_r \qquad (r=1,2,3) \qquad (3.4)$$

The pendulum is acted on by gravity, the wire tension, and aerodynamic drag. In fact the drag acts not only on the spherical bob but also on the cylindrical wire. We compute an equivalent drag force for the bob which causes the same moment about the support as the sum of the wire and bob drag.

The magnitude of the bob aerodynamic drag force is given by

$$f_{d_s} = C_{d_s} \frac{\rho v^2}{2} \frac{\pi b^2}{4} \qquad (3.5)$$

where v is the magnitude of $^E\mathbf{v}^P$ and the drag coefficient C_{d_s} for a sphere is a function (Daugherty & Franzini, 1977) of Reynolds number which is given by

$$\mathrm{Re} = \frac{\rho v b}{\mu} \qquad (3.6)$$

with ρ = air density Kg/m³; and μ = air viscosity Kg/m-sec.

In addition, a differential length dx of wire at distance x from the support experiences a differential drag force

$$df_{d_c} = C_{d_c} \frac{\rho v_w^2}{2} b dx \qquad (3.7)$$

with a similar dependence (Daugherty & Franzini, 1977) of the cylindrical drag coefficient C_{d_c} on $\mathrm{Re}(x)$ in this case because the wire velocity v_w varies linearly along its length. The drag force on the bob with moment about Q equivalent to that of the wire drag is then given by

$$f_{d_e} = \frac{1}{L}\int_0^L x df_{d_c} = \int_0^L C_{d_c}(x) \frac{\rho v(x)^2}{2} b x dx \qquad (3.8)$$

292

Finally the total aerodynamic drag force is given by

$$\mathbf{f}_d = -f \frac{{}^E \mathbf{v}^P}{v} \qquad (3.9)$$

$$f_d = f_{d_s} + f_{d_e} \qquad (3.10)$$

After using the partial velocities to calculate the generalized active and inertia forces and neglecting second order terms in ω, the three nonlinear equations of motion become

$$\dot{u}_1 = 2\omega(au_2 - bu_3) + gc_1s_2 + \frac{u_2^2 s_2}{Lc_2} + \frac{u_1 u_3}{L} - \frac{f_d u_1}{mv} \,(3.11)$$

$$\dot{u}_2 = 2\omega(cu_3 - au_1) - gs_1 - \frac{u_1 u_2 s_2}{Lc_2} + \frac{u_2 u_3}{L} - \frac{f_d u_2}{mv} \,(3.12)$$

$$\dot{u}_3 = 2\omega(bu_1 - cu_2) + g(1 - c_1 c_2)$$
$$- \frac{(u_1^2 + u_2^2)}{L} + \frac{kq_3}{m} - \frac{f_d u_3}{mv} \qquad (3.13)$$

where $\quad a = c_1 c_2 c_\phi - s_1 s_\phi$
$$b = c_\phi s_1$$
and $\quad c = -(s_\phi c_2 + c_\phi c_1 s_2)$

Together with the kinematical differential equations

$$\dot{q}_1 = \frac{u_2}{Lc_2}, \qquad \dot{q}_2 = \frac{-u_1}{L}, \qquad \dot{q}_3 = -u_3 \quad (3.14)$$

the six differential equations completely describe the dynamics of the pendulum. Given initial conditions for q_i, u_i, $(i = 1,2,3)$ the equations can be integrated numerically to predict the motion.

3.2 Parameter estimation

Although we now have a complete dynamical model for the system, it is difficult to use this model for the purpose for which it was intended because of our uncertainties in both the model parameters and the initial conditions for the state variables. Even if we are certain about the form of the model, the effects which are important and unimportant, the factors should be included and which should be neglected, still it is necessary to know exact values for model parameters before the equations which embody the model can be used to extract information. And because the function of the model in this case is to serve as a pattern against which to compare the experimental data to ascertain its accuracy, it is even more essential that the values of the parameters be exact. In addition, it is necessary to know the exact

values of the initial conditions of the state variables, which may therefore also be considered parameters.

There is a way both to determine a set of parameters which can be as exact as possible in the sense that they predict the experimental data better than any other set of values of those same parameters, and also to use the remaining residual difference between the predictions and the data as a measure of the measurement accuracy. We generate an iterative set of parameter estimates which converges to an optimal set in the sense that they minimize the residuals. Such a technique has been used by Hubbard and Alaways (1989) in the estimation of javelin release conditions. The approach taken here is similar.

Define a residual vector of size $3N \times 1$

$$\mathbf{r} = ((\mathbf{x}_m - \mathbf{x}_p)^T, (\mathbf{y}_m - \mathbf{y}_p)^T, (\mathbf{z}_m - \mathbf{z}_p)^T)^T \qquad (3.15)$$

as the concatenation, of the x, y, and z components, respectively, of the error between the measured and predicted positions of the bob over some sequence of N times.

Further suppose that the predicted positions are functions of M "parameters", arranged in a vector $\mathbf{p}$ called the parameter vector, which is a combination of true parameters and other constants, initial conditions for the states. The parameter estimation problem can be formulated (Gelb, 1974) through the question "What are the values of the parameters which make the mean squared error between the predicted and measured positions a minimum? "

As shown in Hubbard and Alaways (1988) a solution to this problem is the limit of the converging iterative sequence

$$\mathbf{p}_{k+1} = \mathbf{p}_k + \boldsymbol{\delta}\mathbf{p}_k \qquad (3.16)$$

where the initial estimate $\mathbf{p}_0$ is a simply a good guess, each correction $\boldsymbol{\delta}\mathbf{p}_k$ is given by the solution to the linear equations

$$L_k^T L_k \boldsymbol{\delta}\mathbf{p}_k = -L_k^T \mathbf{r}_k \qquad (3.17)$$

and the matrix L^T is the $M \times 3N$ Jacobian matrix (usually computed numerically) of partial derivatives of $\mathbf{r}^T$ with respect to $\mathbf{p}$. One of the advantages of this method, called the Gauss-Newton method, is that although it is second order, it requires the evaluation of only the first derivatives of $\mathbf{p}$.

We define the specific parameter vector $\mathbf{p}$ as

$$\mathbf{p} = (X,Y,Z,\rho,L,k,u_{1o},u_{2o},u_{3o},q_{1o},q_{2o},q_{3o})^T \,(3.18)$$

where X, Y and Z are the coordinates of point Q. Thus $\mathbf{p}$ contains six true parameters and six initial conditions.

We note finally that the three degree-of-freedom model detailed above can be converted to a two degree-of-freedom one with no wire stretch by

293

simply setting $u_3 = q_3 = 0$ and taking only the remaining four equations.

4 RESULTS

4.1 Static accuracy

Table 1 shows the precision and accuracy results for six control points that were simultaneously visible in all four cameras' fields of view. Precision can be defined as the inverse of the frame-by frame variance of the reconstructed target location, whereas accuracy is a measure of how well the location was determined (Koff, 1990). The values of precision listed in the table are the standard deviations in position . The values of accuracy are the differences in mean reconstructed position from the actual surveyed location. It should be noted that, for individual cameras, the distance to the target in question varied from 3 to 9 meters and widths of fields of views were as large as 4 meters at the extreme distances. The results in Table 1 were not as precise as hoped for. However, all the control points used in this run were on the outside of the control volume and away from the center of the volume where the dynamic experiments took place.

Table 1 – Static accuracy results for 6 control points.

Precision – mm	Accuracy – mm
4.45	5.4
3.71	4.1
4.34	3.0
4.27	4.7
0.79	1.7
3.46	1.6

4.2 Lighting effects

Figure 3 shows the horizontal location of the centroid of a stationary target that was digitized with variable lighting conditions. The figure shows a movement of nearly 3.5 cm or 70% of the diameter of the target. This shows clearly that uneven side lighting has an effect on the reconstructed three-dimensional position of a target. In the "optimal" camera configuration the cameras are placed at an intersection angle of 90 degrees and each camera has line-of-sight lighting, thus increasing the lighting illumination on one side of the target. The situation shown in Fig. 3 is a worst case scenario in that no line-of-sight lighting was used in the experiment. With additional line-of-sight lighting, the effect of side lighting or shadows will be less. However, when a target is moving in a plane perpendicular to the line-of-sight of the camera the effects of the uneven lighting from the other cameras will always

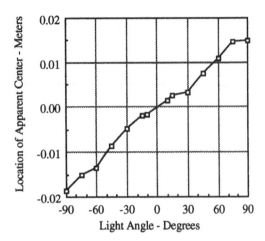

Figure 3 - Horizontal location of apparent center.

affect the accuracy of the target tracking to some degree.

4.3 Dynamic accuracy

To determine dynamic accuracy, residuals (defined to be differences between the measured and predicted trajectories) were plotted as functions of time. The root mean square (rms) residual for all three coordinates was also determined. Figure 4 shows a typical plot of the z residual for a small (i.e. total excursion < 0.5 m) horizontal amplitude planar test,

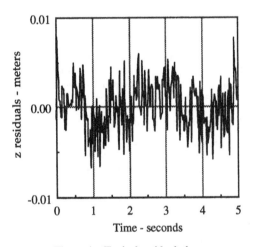

Figure 4 – Typical residual plot

in which the plane of motion bisected the intersection angle of the cameras. The z residual plotted in Fig. 4 corresponds to the vertical position of the bob only, but the residuals in the other two directions had approximately the same shape. The rms value of the residual for this run was 2.6 mm, which was also the average value for most of the small amplitude runs.

Figure 4 shows a deterministic error in the data. If the model were known exactly and there were no deterministic or systematic tracking errors, we would expect the residuals to be zero mean white noise. If the data were low-pass filtered, ideally the residual would be a straight line with zero magnitude. That this is not the case is clearly shown in Fig. 5 which is a plot of the same residual as in Fig. 4 but using data that had been filtered with a 2 Hz low-pass filter before being processed with the estimation algorithm.

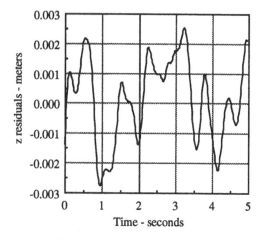

Figure 5 – Filtered residual plot

The sources of this deterministic error are not known. There are at least three different factors to which the error can be attributed. The first is an error in the system calibration, which would also be a contributing factor to the inaccuracies in the static results. The second is that uneven lighting may have existed during the experiment and the distance from the camera to the bob varied. Although the vibratory nature of the wire was not used in the model of the pendulum in this test, because the velocities are so small cable sag would not be a large factor.

Figure 6 shows the vertical position of the bob from a large amplitude test (total excursion $\cong 2$ m). It is of interest to note that the reconstructed elevations at opposite ends of the swing were not equal. A variation of as much as 5 mm is present in the figure. Again, the source for this error is not completely known. It should be noted that the rms residual for this run was approximately 5 mm. The figure does point to lighting as a possible source for the deterministic error.

To investigate lighting further, bob height for a small amplitude test is shown in Fig. 7. Because of the scale, changes in the relative peak elevations should be more noticeable than in Fig 6. However, none appear to be present. The rms residual for this run was approximately 2.5 mm, half of that in Fig. 6, though the residual still shows signs of a deterministic component.

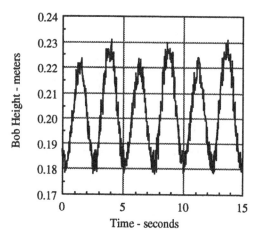

Figure 6 – Bob height during large amplitude (1 m oscillation) test

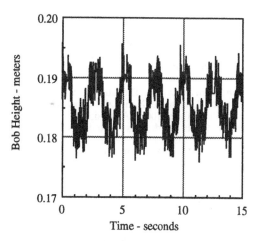

Figure 7 – Bob height during small amplitude test.

Yet another possible source of the deterministic error is a systematic user error. The threshold setting of the video processor is user adjustable and no clear guidelines or research results are available as to what the optimum should be. The typical threshold is a "feel good, look right" setting that the operator hopes will be adequate at the time of data collection. This setting is dependent on lighting conditions and camera f-stop setting. Table 2 shows the effect that two different threshold settings have on the residual and on the predicted location of point Q, the support of the pendulum. Clearly the video processor threshold setting has an effect on the reconstructed target location.

Table 2 – Threshold variation results.

Run	rms residual mm	X - m	Y - m	Z - m
1	4.1	8.617	-0.437	6.102
2	2.9	8.612	-0.441	6.087

5 CONCLUSIONS

Static accuracy of kinematic tracking systems can be determined by tracking static targets, but to determine dynamic tracking accuracy directly requires a known moving target. Because the predicted motion must be known very accurately, a spherical pendulum serves as a good dynamic test target, especially when large motions are desired. Imprecisely known parameters and initial conditions may be determined from the data using parameter estimation.

If the dynamic model is known precisely, the residuals are due entirely to measurement errors. Deterministic components in the residuals are indications of deterministic or systematic errors in the tracking system or data analysis procedure. Examples of these are 1) variation of surface lighting intensity on the target, 2) systematic calibration errors such as imprecisely known control points, and 3) variations in threshold settings.

Acknowledgments – The authors wish to thank Lawrence Livermore National Laboratory (LLNL) for the loan of the Motion Analysis equipment and to note that this work was partially supported by the National Institute for Occupational Safety and Health (NIOSH), Cooperative Agreement #T42/CCT910427–02.

REFERENCES

Abdel-Aziz, Y.I. & H.M. Karara, 1971. *Direct linear transformation from comparator coordinates into object space coordinates in close-range photogrammetry.* In ASP Symposium on Close Range Photogrammetry. Falls Church, VA: American Society of Photogrammetry.

Daugherty, R.L. & J.B. Franzini 1977. *Fluid mechanics with engineering applications.* New York: McGraw-Hill.

Fraser, C.S. 1982. *On the use of nonmetric cameras in analytical close range photogrammetry.* Can. Surveyor, **36**, 259–279.

Gelb, A. 1974. *Applied optimal estimation* Cambridge, MA: MIT Press.

Gosh, S.K. 1979. *Analytical photogrammetry.* New York: Pergamon Press.

Hubbard, M. & L.W. Alaways, 1989. *Rapid and accurate estimation of release conditions in the javelin throw.* J. Biomechanics, **22**(6,7), 583–595.

Kane, T.R. & D.A. Levinson, 1985. *Dynamics: theory and applications.* New York: McGraw–Hill Publishing Company.

Koff, D.G. 1990. *Accuracy and precision of the ExpertVision 3-D motion analysis system.* Unpublished Report, Santa Rosa, CA: Motion Analysis Corporation.

Motion Analysis Corporation, 1989. *ExpertVision user's manual.* Santa Rosa, CA: Motion Analysis Corporation.

Wood, G.A. & R.N. Marshall, 1986. *The accuracy of DLT extrapolation in three-dimensional film analysis.* J. Biomechanics, **19**(9), 781–785.

The Engineering of Sport, Haake (ed.) © 1996 Balkema, Rotterdam. ISBN 90 5410 822 3

3-D kinematic analysis of the forward stroke of white-water paddlers using a paddle-ergometer

A. Kranzl, J. Kollmitzer & E. B. Zwick

Department of Physical Medicine and Rehabilitation, University of Vienna, Austria

ABSTRACT: The forward stoke during paddling is not investigted in three dimension (3-D) up to now. Investigators studied the path, the velocity, the acceleration of the boat and the join center of the shoulder, elbow and wrist in 2 dimensions. The aim of the study was to measure 3-D kinematic data of the forward stroke during paddling. We applied a videometriy system (Motion Analysis Corporation) with 6 cameras and 21 passive markers. Subjects of this study were 10 wildwater paddlers, members of the Austrian slalom wildwater team and members of the junior slalom team. The video data were captured at 200 frames per second for 3 trials over 7 seconds each. For the marker setting and the fixation of the markers we adapted a modified application of the Cleveland Markerset for clinical gait analysis of the lower extremities. The joint angles of the shoulder, elbow and the wrist were calculated in the sagittal, frontal and transversal plane. The position of the shoulder girdle was measured relative to the lab coordinate system. To trigger the kinematic data we used light switches, on both sides of the boatto simulate the water surface. The paddle cycle was defined from the initial water contact to the next initial water contact at the same side. This cycle was subdivided in four phases. The waterphase on the ipsilateral side followed by airphase and waterphase on the contralateral side followed by air phase. The cycles were normalized in time to 100% cycle time. Mean and standard deviation of each subject was calculated. With this method it is possible to determine the path of motion for the angle in the shoulder, elbow and wrist in all three planes of movement. The combined time normalization and averaging resulted in a feasible representation of an average paddling. The timing and the range of motion are presented in a quantitative manner.

1. INTRODUCTION

The aim of this study was to evaluate the application of a lower extremity measurement method for the upper extremity to analyse the forward stroke of slalom wildwater paddlers. The method will be used for studying different shaft types. In the past studies haye been undertaken (LOGAN SM., et al. (1985); PLAGENHOFF S. (1979); MANN RV., et al. (1980)) which used two dimensional technics to describe the forward stroke of flatwater paddlers. The paths of the wrist, elbow and shoulder were presented in absolute and relative motion. This descriptive dimensional analysis were only done by analysing video observation (LOGAN SM., et al. (1985)). It is important to know that there is a difference between the forward stroke of a slalom wildwater paddler and a flatwater paddler. They use different styles, paddle length and blade forms. Thus the analysis is valide for wildwater paddle style. To provide the same conditions for all volunteers conditions we decided to test under laboratory conditions. Forward paddling is a cyclic movement. On cyclice is determined by entering the water with the blade to the next entering at the same side.

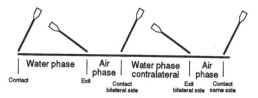

Fig 1. Phases for one paddle cycle subdivided into four phases: the waterphase and the contralateral waterphase, beginning with initial water contact to the exit of the water surface and two air phases between the water phases.

2. BACKGROUND

Biomechanic investigations in wildwater paddling are complicated by instable water conditions. In flatwater paddling investigators searched force-time characteristics (ZSIDEGH M. (1981); LENZ J. (1994)). Different blade orientations in canoeing were studied by NOLAN GN., et al. (1982) in two dimensions. One study compared flatwater paddling with paddling on a dry-land ergometer (CAMPAGNA P.D., et al. (1982)). They used one high speed camera to compare the sagittal wrist, elbow and shoulder path and found similar stoke pattern. MANN RV., et al. (1980), NOVAKOVA H., et al. (1979) and PLAGENHOFF S. (1979) used 2-dimensional analysis to describe the path, velocity and acceleration of the wrist, elbow and shoulder in absolute and relative motion. PLAGENHOFF S. (1979) used different positions to observe the movement (sagittal, transverse) with one camera. PLAGENHOFF S. (1979) and MANN RV., et al. (1980) evaluated parameters like cycle time, stroke cadence, paddle angles during initial water contact, exit of thr water surface, phases with paddle water contact (water phase) and phases without water contact (air phase). All these investigations used 2-dimensional method to measure the flatwater stroke.

3. METHOD

Austrian slalom wildwater paddlers were tested on a paddle ergometer in the laboratory. For data collection we used a videometric six camera system from Motion Analysis Corporation. This system is used in clinical gait analysis. The system inherent error and variability of the system are descriped by KOLLMITZER J. (1994).

Each camera is equipped with a red light source and a red colour filter. Thus only the markers were seen by the system. The duration of data collection is limited to seven seconds by the system. This allowed to acquire data from three complete cycles. To simulate the resistance for paddling a kayak ergometer was used which works with a friction break. The ergometer is tied with ropes, which are mounted at the outer side of the paddle shaft. To determine the paddle cycle it was necessary to generate an imaginary watersurface by light switches for each side. If the light beam of a light switch was interrupted a diode flashes which could be seen by the cameras. With this mechanism it was possible to define the phases in video frames.

For the markersettings we used a modified application of the Cleveland Markerset for clinical gait analysis with three markers for each segment. We transferred these settings from the lower to the upper extremities. On the forearm and upper arm there were marker arrays fixed. In a first data acquisition procedure we defined (calibrated) joint centres for the left and right wrist and elbow with medial and lateral markers in neutral static position.to estimate the center of rotation. The passive marker were fixed with a double sided tape on anatomical landmarks. After this the joint markers were removed and two greater markers were set on the acromion and one at the same level at the spine in a sitting position. The handmarkers were set on a specially designed marker system which facilitated marker identification and allowed greater angle calculation

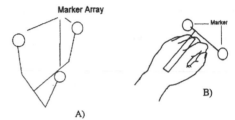

Fig.2 (a) Marker array for the arm segment
(b) Marker array for the hand segment

precision. The marker placment was done by the same person during the hole study.

The video sampling frequenze for the picture was 200 Hz for each synchronized camera. After a warm up phase 3 trials (from 7 sec long) were accessed. The data were stored in a Workstation and processed with the software EXPERT VISION and ORTHO-TRAK (Motion Analysis Corporation). After setting the watering events to define individual paddling cycles we processed the results in a databank for gait analysis (BERGHOF R., et al. (1995)). The cycles were normalized in time. As outcome parameters we calculated mean and standard deviation plots for each joint angle and the coefficient of variation (CV) over the cycle (WINTER D.A. (1984), MS EXCEL, Microsoft Corp.).

$$CV = \left(\frac{\sqrt{\frac{1}{N} * \sum_{i=1}^{N} \sigma_i^2}}{\frac{1}{N} * \sum_{i=1}^{N} |M_i|} \right) *100 \quad [\%]$$

Winter D.A. (1984)

M_i is the mean value of the angle at the i th interval, N is the number of intervals over one cycle, $i = 1$ to N, $\sigma_i =$ standard deviation of M_i at the i th interval.

We calculated the intra trial variability (one trial whith 3 cycles), the session veriability (3 trial) and the day to day variability with three sessions on different days (two days between).

Shoulder girdle: flexion - extension
                   cranial - caudal
                   internal - external rotation
This angle is measured relative to the laboratory coordination system.

Shoulder: y   flexion - extension
            x   abduction adduction
            z   internal - external rotation

Elbow	y	flexion - extension x
		abduction adduction
	z	supination - pronation
Wrist	y	flexion - extension
	x	radial - ulna deviation

These angles were measured relative to the proximal segment before.

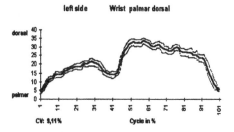

Fig. 3 Averaged sagittal angle for the wrist.

Mean and standard deviation was calculated for each of the 101 points for one trial (3 cycles). The CV of the trial to trial variability during the seven second data acquisition were calculated. The interval between the trials were 3 min to exclude fatigue factors. To test for trial to trial variability we calculated the CV for on session. This includes 3 trials with 3 cycles. The day to day variability (three days) was calculated for one person with two day between each data collection. This includes 3 data acquisitions with 3 trials.

4. RESULTS

The trial variability , session variability and the day to day variability is presented for the shoulder joint flexion - extension in Fig. 5.

The reproducibility of data collection were evaluated by using the CV. The trial to trial variability shows a low value for all CV in Table 1.

The mean value and standard deviation for one trial shows a good reproducibility. Selected kinematic results (mean and standard deviation, range of motion) are presented in Fig. 3, 4, 6, and Fig. 7.

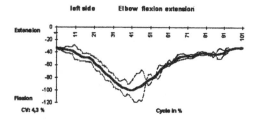

Fig. 4 Elbow flex - ext.trialvariability (4 cycles)

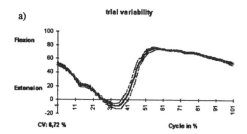

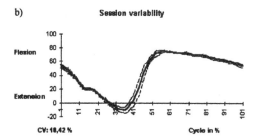

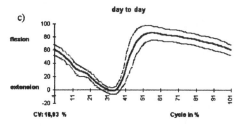

Fig 5. Shoulder angle in flexion-extension with mean and standard deviation (a) trial variability, (b) session variability and (c) day to day variability.

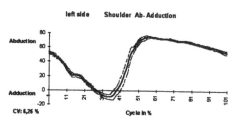

Fig 6. Shoulder Ab - Adduction session variability.

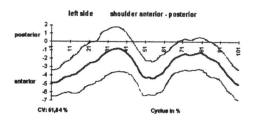

Fig 7. Averaged sagittal shoulder angle with day to day variability expand as standard deviation range.

299

Table 1 CV intra trial variability

left side		CV in %
Wrist	x	21,12
Wrist	y	12,49
Wrist	z	15,70
Elbow	x	22,30
Elbow	y	2,78
Elbow	z	11,17
Shoulder	x	6,87
Shoulder	y	6,72
Shoulder	z	8,53
Shoulder gridle	x	2,90
Shoulder gridle	y	51,64
Shoulder gridle	z	3,98

Table 2 CV inter trial variability within session

left side		CV in %
Wrist	x	47,82
Wrist	y	34,45
Wrist	z	40,78
Elbow	x	54,74
Elbow	y	6,52
Elbow	z	36,07
Shoulder	x	11,64
Shoulder	y	10,42
Shoulder	z	13,34
Shoulder gridle	x	12,85
Shoulder gridle	y	83,71
Shoulder gridle	z	10,22

Table 3 CV for day to day variability

Left side		CV in %
Wrist	x	112,12
Wrist	y	18,64
Elbow	x	151,00
Elbow	y	6,48
Shoulder	x	18,08
Shoulder	y	18,93
Shoulder	z	25,15
Shoulder gridle	x	9,23
Shoulder gridle	y	75,90
Shoulder gridle	z	11,95

The day to day variability presented in Table 3.
The variability during one session is presented in Table 2. It consist of 3 trials with 3 cycles of each trial.

5. DISCUSSION

For this analysis it was necessary to make some assumptions. The simplification for the shoulder girdle that it is used as a fixed bodysegment is a limitation. If we consider that the forward stoke is a symmetric movement so it is admissible. The determination of the centers of rotation by palpating landmark provided an uncertainty factor (The determination of the center of rotation descriped by SOMMER HJ., et al. (1980); YOUM Y., et al. (1980), ANDREWS JG., et al. (1979 for the wrist joint, YOUM Y., et al. (1979) CHAO E.Y., et al. (1978) for the elbow joint and HÖGFORS C., et al. (1987). POPPEN N.K., et al. (1976) for the glenohumeral joint).

But as shown with the elbow and wrist joint in the CV factor, (interday variability) the determination of jointcenter was repeatable. If we look at the CV of the elbow in ab-adduction in Table 3, it shows a very high CV. The range of motion with 6 grad is very low, but the standard deviation shown in Fig. is minimal. This shows a low day to day variability. All variabilities show a low value and a good reproducibility in the range of motion. To analyse different patterns with low differences of about 1-3 degree we have to consider the system inherent inaccuracies (KOLLMITZER J. (1994)).

The simplification for the wrist with one center of rotation result in a increase of CV. Nevertheless the accuracy is high enough to see differences. The determination of the phases corresponds with the results from LOGAN SM., et al. (1985) and PLAGENHOFF S. (1979).

6. CONCLUSION

With this evaluated model it is possible to describe joint angles for the shoulder girdle, shoulder, elbow and wrist in all three planes during forward paddling. Differences in the pattern or paddle styles are detectable. The model will be used for further investigations studying different shaft types. The results could be contribute to a better understanding of the biomechanics of the forward paddle stroke.

REFERENCES

ANDREWS JG., YOUM Y. (1979). A biomechanical Investigation of Wrist Kinematics. Journal of Biomechanics. Vol. 12: 83-93.
BERGHOF R., ZWICK E.B.(1995) Gaitbase - A relational database for the managment of gait laboratory data. ESMAC Annual Scientific Meeting. p. 49 Enschede. Roessingh Research and Development bv.
CAMPAGNA P.D., BRIEN D., HOLT L.E., ALEXANDER A.B., GREENBERGER H. (1982).

A biomechanical comparison of olympic flatwater kayaking and a dry-land kayak ergometer. Can. J. of appl. Sport Sci. Vol. 7(4): 242

CHAO E.Y., MORREY B.F. (1978). Three-dimensional rotation of the elbow. Journal of Biomechanics. Vol. 11: 57-73.

HÖGFORS C., SIGHOLM G., HERBERTS P. (1987). Biomechanical model of the human shoulder- I. Elements. Journal of Biomechanics. Vol. 20(2): 157-166.

KOLLMITZER J.(1994) Errors and practical tests for 3-D gaitanalysis system setups with videometry and forceplates. Third international symposium on 3-d analysis of human movement. pp. 49-52. Stockholm, Sweden.

LENZ J. (1994) Leistungs- und Trainingslehre im Kanusport. Leibzig. Kanuverband Sachsen-Anhalt.

LOGAN SM., HOLT LE. (1985). The flatwater kayak stroke. National Strength Conditioning Association Journal. Vol. 7(5): 4-11.

MANN RV., KEARNEY JT. (1980). A biomechanical analysis of the olympic-style flatwater kayak stroke. Medicine and Science in Sports and Exercise. Vol. 12: 183-188.

NOLAN GN., BATES BT. (1982). A biomechanical analysis of the effects of two paddle types on performance in north american canoe racing. Research Quarterly for Exercise and Sport. Vol. 53(1): 50-77.

NOVAKOVA H., SUKOP J. (1979). Kinematicka analyza techniky padlovani vrychlostni kanoistice. (Kinematic analysis of paddling technique in speed canoeing.). Teor Praxe tel Vych. Vol. 27(4): 234-238.

PLAGENHOFF S. (1979). Biomechanic analysis of olympic flatwater kayaking and canoeing. Research Quarterly. Vol. 50(3): 443-459.

POPPEN N.K., WALKER P.S. (1976). Normal and abnormal motion of the shoulder. J. Bone Jt. Surg. Vol. 58a: 195-201.

SOMMER HJ., MILLER NR. (1980). A technique for kinematic modeling of anatomical joints. Journal of Biomechanical Engineering. Vol. 102: 311-317.

WINTER D.A. (1984). Kinematic and kinetic patters in humen gait: Variability and compensatin effects. Human Movement Science. Vol. 3: 51-76.

YOUM Y. (1987). Instantaneous center of rotation by least square method. Journal of Bioengineering. Vol. 2: 129-137.

YOUM Y., DRYER RF., THAMBYRAJAH K., FLATT AE., SPRAGUE BL. (1979). Biomechanical analyses of forearm pronation-supination and elbow flexion-extension. Journal of Biomechanics. Vol. 12: 245-255.

YOUM Y., ADRIAN E., FLATT E. (1980). Kinematics of the wrist. Clinical Orthopaedics and Related Research. Vol. 149: 21-32.

ZSIDEGH M. (1981). Survey of the physiological and biomechanical investigations made into kayaking, canoeing and rowing. Hung. Rev. Sports Med. Vol. 22(2): 97-115.

The Engineering of Sport, Haake (ed.)© 1996 Balkema, Rotterdam. ISBN 90 5410 822 3

The accuracy of kinematic data collected from underwater three-dimensional analysis

M. A. Lauder, P. Dabnichki, R. M. Bartlett & S. Aritan
Biomechanics Research Group, The Manchester Metropolitan University, Crewe & Alsager Faculty, Stoke-on-Trent, UK

ABSTRACT: A scaled down underwater environment was used to investigate the effect of changes to the experimental set-up on the accuracy of two-dimensional and three-dimensional reconstructions from video analysis. It was found that the position of the camera and the calibration object relative to the glass interface influenced the level of error in the reconstruction of known lengths. The results provide evidence that there is a need for an augmented calibration procedure in order to minimise the influence of unaccounted random errors in underwater video analysis that uses underwater windows/housings

1 INTRODUCTION

The use of underwater video recording is perhaps the most accessible method of technique analysis in swimming and certainly the use of video in sports biomechanics is widespread. Theoretically all underwater recordings taken from underwater housings, underwater windows or periscope systems are subject to the effects of refraction. In two and three-dimensional reconstructions that use the Direct Linear Transformation algorithm (Abdel-Aziz and Karara, 1971) this effect would cause a violation of the colinearity rule. Walton (1988) accounted physically for the effect of refraction when filming above and below water simultaneously by placing a domed glass barrier in front of the camera lens effectively incorporating the barrier into the lens system.

There has been only one study that has reported that the effect of refraction in a method of underwater video analysis. Yanai *et al.* (1996) reported that the effect of refraction on the image when using a panning periscope system was to enlarge the pincushion distortion of the lens at the interface between the two media. However it would seem that a full analysis of the effect of this distortion on the level of error was not undertaken and as such the effect of refraction on reconstruction accuracy remains unaddressed in the literature.

It was the aim of this study to establish the influence of experimental set-up on the level and distribution of error in two-dimensional kinematic data and the effect of employing different reconstruction protocols on the level and distribution of error in three-dimensional kinematic data derived from underwater video analysis

2 METHODS

A series of experiments were designed that would establish the effect of experimental set-up on the level of error in two-dimensional and three-dimensional reconstructions from video data collection using underwater windows.

2.1 *Experimental setting*

In order to achieve adequate control of the parameters that were to be examined a glass tank (1.0 m x 1.0 m x 0.75 m) was constructed from 19 mm glass of good optical quality. Based on the dimensions of the tank a scaling factor of 1:6 was used to establish the dimensions of the scaled down calibration object and the step increments between trials examining each parameter. This scaling factor allowed the full range of step increments for each of the variable distance experiments in a previous study (Dabnichki *et al.*, 1996; *these proceedings*) to be replicated and for realistic calibration object to camera distances to be investigated. The position of the camera and calibration object relative to the glass were controlled using an overhead runway with carriages for the mounting of the camera and calibration object (Figure 1). The tank and calibration frame were illuminated by two 200 Watt lamps carefully positioned to minimise reflections and to ensure all points on the frame could be identified.

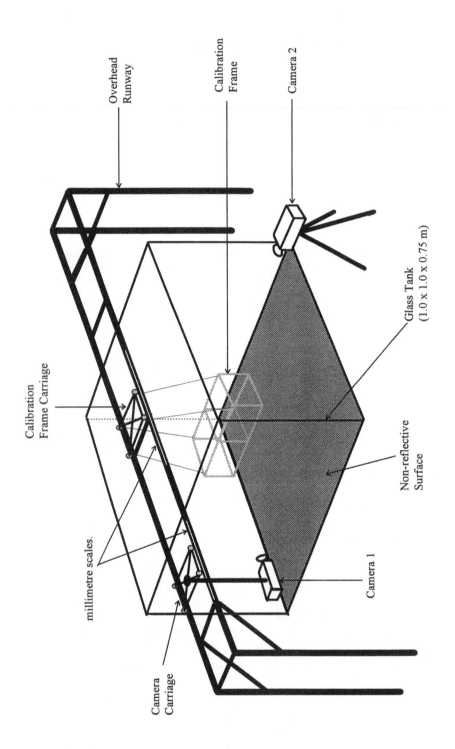

Figure 1. Experimental set-up for underwater experiments.

Overhead Runway

Calibration Frame

Camera 2

Calibration Frame Carriage

millimetre scales.

Camera Carriage

Glass Tank (1.0 x 1.0 x 0.75 m)

Non-reflective Surface

Camera 1

304

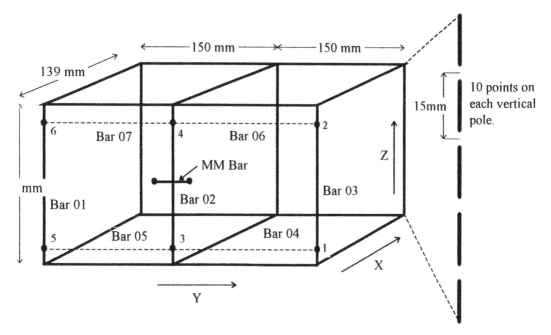

Figure 2. The calibration structure and bars defined by the six points used for reconstruction.

2.2 Camera position and calibration frame

Two Panasonic VHS video cameras (WV-F15E) fitted with x12 zoom lenses (WV-LZ15/12E) with macro setting options and connected to Panasonic VHS recorders (NV-180) were used to record the underwater images. One camera was mounted beneath the carriage on the overhead runway and used to film the front-on view of the calibration object in the two-dimensional and three-dimensional experiments. The second camera was mounted on a tripod (Monferrotto) and recorded the side-on view of the calibration object for the three-dimensional experiments. Both cameras were aligned vertically and horizontally and checks were made using engineering squares to ensure that the focal axis of each camera was perpendicular to the glass (a factor of prime importance in underwater filming and photography). For the two-dimensional experiments the focal axis of camera 1 was aligned with the central point on the middle pole of the frame (slightly offset for the three-dimensional experiments so all points were visible). The focal axis of camera two was aligned with the centre of the calibration frame.

The calibration frame was constructed from 5 mm square aluminium tubing (Figure 2). The dimensions of the frame were 139 mm x 300 mm x 150 mm in the X, Y and Z axes respectively. For brevity all dimensions are reported in millimetres due to the scaling of the object. The scaled down

calibration frame dimensions equated effectively to a life sized calibration frame of dimensions 0.8 m x 1.8 m x 0.9 m, similar to the dimensions of the calibration frame reported in Payton and Lauder (1995). The squareness of the calibration frame was checked using a surface plate and angle blocks. The frame was stained and etched to create 60 control points with known coordinates (10 control points on each of the 6 vertical poles). The position of the control points were determined using a Vernier Height Gauge (Bensons Ltd) with an accuracy of ±0.01 mm. The calibration frame was suspended from the second carriage (Figure 2) and carefully aligned to ensure that it was both vertical and horizontal. This was achieved with the use of a spirit level and the level of the water such that the calibration frame was suspended just below the water surface.

3 EXPERIMENTAL DESIGN

Four independent groups of experiments were conducted. Three of the groups were designed to examine the effect of experimental set-up on two-dimensional reconstruction accuracy and were based on the protocol of the experiments conducted by Dabnichki et al. (1996). Each experimental group had a variable number of identical trials depending on the step increment and the limits of the parameter of interest. The final experimental group was designed to examine variables associated with three-dimensional reconstruction accuracy. For each trial the image was recorded for a period of ten seconds. The design of each group of experiments is described briefly in the following sections.

3.1 *Variable distance between the camera and the glass (MC experiments)*

For the first group of experiments the distance between the focal point of the camera and the closest edge of the glass tank was varied by increments of 40 ± 0.5 mm between trials. A total of thirteen trials were undertaken with the camera to glass distance ranging from 103 mm to 583 mm ± 0.5 mm. This range was used as it incorporated the limits of the lens that was used. The distance between the glass and the calibration plane (defined as the closest side of the calibration frame to the camera) was kept constant at 490 mm for all trials. This distance represented the mid-point of the calibration plane to glass range used in the following experimental group.

3.2 *Variable distance between the calibration plane and the glass (MP experiments)*

This group of experiments examined the effect of varying the distance between the calibration plane and the glass on reconstruction accuracy. A total of nine trials were undertaken with the distance between the calibration plane and the glass varying from 330 mm and 650 mm by increments of 40 ± 0.5 mm. The camera to glass distance was kept constant at 343 mm (the mid-point of the camera to glass range).

3.3 *Variable distance of the object out of the calibration plane (MM experiments)*

A total of eleven experimental trials were conducted that examined the effect of varying the distance between the calibration frame and the reconstructed object. For all trials the camera was fixed at a distance of 343 mm from the glass and the calibration plane was fixed at 490 mm from the glass. The distance between the calibration plane and the object was then varied by increments of 8 ± 0.5 mm to give five trials progressively closer to the camera/glass and five trials progressively further away from the camera/glass. This range simulated the possible extrapolation that can occur when filming the underwater stroke of swimmers as the ability to capture a complete stroke within the calibration volume can prove to be difficult.

3.4 *Three-dimensional reconstruction accuracy*

This experimental group sought to examine the effect of varying the number of calibration points used in the three-dimensional DLT reconstruction algorithm on the reconstruction errors in each axis for two experimental set-ups. The error in each axis (X, Y and Z) was determined for the reconstruction for the points used in the calibration of the volume and the subsequent reconstruction of control points using the calibrated volume.

For both experimental set-ups the distance between the cameras and the glass was constant (343 ± 0.5 mm). In the first experiment the distance between the glass and the nearest edge of the calibration frame for camera 1 was 490 ± 0.5 mm and for camera 2 the distance was 354 ± 0.5 mm. The distance between the nearest edges of the object and the glass for both cameras was the same (354 ± 0.5 mm) for the second experimental set-up. The rationale for two experimental set-ups was that often in underwater analysis the distance between the object and the glass for each camera is different due mainly to limitations with the positioning of underwater windows.

3.5 *Digitising procedure*

All digitising was undertaken manually using an M-image video capture board and an Archemedes 440 computer with associated software (Bartlett, 1990).

3.5.1 *MC and MP experiments*

Each trial within these two groups was calibrated using a vertical and horizontal calibration object in the centre of the calibration plane. A six point model was defined as shown in Figure 2 and used to digitise fifty frames of each trial. Each digitised trial was recorded and the raw coordinate data exported for further processing.

3.5.2 *MM experiments*

For these experiments only one calibration was performed as the objective was to determine the level of error when the object was moved outside the calibration plane. For this purpose the calibration trial selected was in the middle of the range of trials. Fifty frames were digitised for each trial using a two point model to reconstruct a single known length (15.22 ± 0.01 mm). The length was positioned in the centre of and parallel to the calibration plane. The coordinate data for each trial were recorded and exported for further processing.

3.5.3 *Three-dimensional experiments*

Three-dimensional reconstructions were performed using 12, 34 and 54 calibration points. The calibration points were selected such that the whole calibration volume was defined (Challis and Kerwin, 1992). For each calibration condition the absolute mean error and the error of each calibration point in each axis were recorded. The calibration files were then used to reconstruct the same number (12, 34 54) of different (where possible) control points of known coordinates. The digitised coordinates of the control points were recorded and then exported. In total 12 data sets pertaining to - 2 experimental set-ups x 2 reconstructions x 3 numbers of calibration points were recorded.

3.6 Data processing and reduction

For the MC and MP experiments the raw coordinate data of the six points were used to compute seven known lengths or bars (Figure 2). Bars 01 to 03 were vertically aligned (105 ± 0.01 mm), whilst bars 04 to 07 were horizontally aligned (all 150 ± 0.01 mm). The position of the bars allowed an indication of any variance in measurements across the field of view.

From this data the RMS error (expressed as a percentage of the known length) between the known lengths and the calculated lengths for each bar was determined. Due to different calculations of RMS error being reported in the literature the RMS errors for this study were calculated according to the equation:

$$\text{RMS error (\% length)} = \left\{ \sqrt{\frac{\sum (x_R - x_i)^2}{N}} \right\} \cdot \frac{100}{x_R} \quad (1)$$

where x_R is the reference value, x_i is the measured value and N is the number of observations.

For the MM experiments the RMS error between the known length and the reconstructed lengths was calculated for each trial according to the equation given above.

The raw coordinate data for the twelve three-dimensional experimental conditions were used to calculate the RMS error in each axis between the known coordinates of the points and the reconstructed coordinates of the points for each condition.

3.7 Statistical analysis

The Kolmogorov-Smirnov test was used to assess the normality of the distribution of error for the reconstruction of each bar length in each trial of the experimental groups MC, MP and MM. For the MC and MP experiments the absolute error was used to perform separate two-way MANOVA's in order to assess for differences between trials within each group and between the seven reconstructed bar lengths. A one-way MANOVA was used to assess differences between trials in the MM experimental group.

For the three-dimensional data a three-way MANOVA for repeated measures was used that assessed for differences between errors in experimental set-up (base ratio for 12, 34, 54 points), axis (X, Y & Z) and reconstruction of calibration or control points. The Kolmogorov-Smirnov test was used to assess the normality of the distribution of error in each axis for each of the twelve experimental conditions defined above.

The level of significance was set at $p<0.05$ for all statistical analyses.

4 RESULTS.

Table 1 RMS error expressed as percentage length of each bar for the Moving Camera (MC) experimental group.

Experimental Trial[†]	Mean RMS error Vertical Bars (% length)	Mean RMS error Horizontal Bars (% length)
MC103	0.96	1.37
MC143	0.62	2.19
MC183	0.82	2.01
MC223	1.26	1.40
MC263	1.18	1.19
MC303	0.86	2.18
MC343	0.84	1.82
MC383	1.23	1.36
MC423	1.40	1.63
MC463	1.14	1.86
MC503	1.37	1.00
MC543	1.05	1.71
MC583	1.48	0.82

† MC indicates Moving Camera. The numerical values indicate the distance between the camera and the glass in mm.

Table 2 RMS error expressed as percentage length of each bar for the Moving Calibration Plane (MP) experimental group.

Experimental Trial[†]	Mean RMS error Vertical Bars (% length)	Mean RMS error Horizontal Bars (% length)
MP330	1.26	2.17
MP370	1.00	1.80
MP410	1.96	1.00
MP450	0.92	2.08
MP490	1.07	2.19
MP530	1.12	2.17
MP570	1.09	1.60
MP610	1.07	1.88
MP650	1.49	0.90

† MP indicates Moving Calibration Plane. The numerical values indicate the distance between the calibration plane and the glass in mm.

Table 3 RMS error expressed as percentage length for the Moving Object out of the Calibration Plane (MM) experimental group.

Experimental Trial†	Mean RMS error (% length)*
MM40+	13.0 (-3.8)
MM32+	14.4 (-1.9)
MM24+	16.3 (-0.5)
MM16+	13.6 (-3.2)
MM08+	15.3 (-1.5)
MM00	16.8 (0.0)
MM08-	17.0 (+0.2)
MM16-	17.3 (+0.5)
MM24-	18.3 (+1.5)
MM32-	15.8 (+1.0)
MM40-	19.2 (+2.4)

† MM indicates Moving out of the calibration plane. The numerical values indicate the distance between the calibration plane and the object in mm. The signs (+) and (-) indicate towards and away from the camera/glass.
* results also expressed as RMS error difference from calibration error (MM00).

Table 4. RMS error (mm) in each axis for the three-dimensional experimental conditions.

Condi tion†	RMS X_1	RMS Y_1	RMS Z_1	RMS X_2	RMS Y_2	RMS Z_2
A_12	1.81	7.44	4.41	2.26	7.32	7.76
A_34	2.40	5.35	3.79	2.82	4.53	4.07
A_54	2.09	6.43	3.42	1.86	5.83	3.51
B_12	2.76	6.96	4.29	2.43	6.56	4.33
B_34	2.55	6.54	3.96	2.33	6.30	4.06
B_54	2.49	6.51	4.52	2.34	5.80	4.08

† A and B represent the two camera/base set-ups. 12, 34 & 54 represent the number of calibration points used in the reconstruction. X_1, Y_1 & Z_1 represent the RMS error in reconstruction of calibration points and X_2, Y_2 & Z_2 represent the RMS error in reconstruction of control points.

5 ANALYSIS OF RESULTS

5.1 *Two-dimensional experiments*

The range of RMS errors presented in Table 1 (Moving Camera experiments) was found to be from 0.80% to 2.6%. The maximum was greater than the results presented in Dabnichki *et al.* (1996) for the same experimental set-up (Maximum RMS error 1.98%) but overall the values were not excessive and reflected well on the use of the scaled down experimental setting. The statistical analysis showed significant main effects between experiments, bars and interaction of experiments and bars (p<0.001). Post hoc analysis revealed that at the greater distances (> 383 mm) there were less differences observed between experimental set-ups (increments) than were observed between experimental set-ups at the shorter distances. Differences were found between the vertical bars with the RMS error in bar 03 significantly greater (mean 1.82%) than the error in bars 01 (mean 0.70%) and 02 (mean 0.76%). For the horizontal bars, bar 07 (mean 1.79%) was found to differ significantly from the other horizontal bars (means: bar 04, 1.33%; bar 05, 1.61%; bar 06, 1.59%).

Post hoc analysis of the interaction effects showed that for the vertical bars the distance of the camera from the glass did not significantly influence the level of error. For the horizontal bars it was revealed that at the longer camera to glass distances the number of significant differences between bars increased. However the error at the shorter camera to glass distances was on average greater than at the greater distances (103 mm to 303 mm, mean 1.72%; 383 mm to 583 mm, mean 1.40%). Kolmogorov-Smirnov tests showed the distribution of the error for all bar lengths across the experimental group to differ significantly from normal.

The results for the variable distance between the object and the glass (Table 2) showed the range of RMS error to be 0.70% to 2.52%. This was in line with range from the previous moving camera experiments. Significant main effects were again found between experiments, bars and the interaction of experiments and bars (p<0.001). Post hoc analysis for differences between experiments showed not differences between experiments grouped around the mid-point of the object to glass range (450 mm to 530 mm). Significant differences were found between experiments at the longer object to glass distances and the shorter camera to glass distances. Post hoc analysis between vertical bars showed that the mean RMS error in bar 02 (1.11%) was noticeably less than the mean RMS error in bar 01 (1.19%) and bar 03 (1.36%). For the horizontal bars again bar 07 (mean 2.03%) was shown to be noticeably different from the other horizontal bars (means: bar 04, 1.57%; bar 05, 1.70%; bar 06, 1.73%).

Interaction effects showed that for the vertical bars the distance of the object to the glass did not significantly influence the level of error. For the horizontal bars no significant differences were observed between bars for experiments around the mid range of the object to camera distances. At the longer object to camera distances the overall error was less than at the shorter distances (330 mm to 450 mm, 1.76%; 530 mm to 650 mm, 1.64%) and there were more significant differences between horizontal bar lengths across experiments at the longer object to camera distances. The Kolmogorov-Smirnov tests revealed that for this group of experiments bars 06 and 07 had error distributions that were not significantly different from normal.

The RMS error for the object out of the calibration plane experiments ranged between 13.0% and 19.2% (Table 3). These values were considerably larger than the previous experiments but can be accounted for by the size of the object and the resolution of the digitising system. A significant main effect was found for bar lengths between experiments ($p<0.004$) and the post hoc analysis showed significant differences between the experiments where the object was furthest from the calibration plane (MM32+ and MM40+ with MM24- and MM40-). The results presented in Table 3 showed a trend for the RMS error to increase as the object was moved further out of the calibration plane away from and towards the camera/glass but in opposite directions. For this group of experiments the Kolmogorov-Smirnov test showed four of the eleven experiments to have distributions of error that did not differ significantly from normal.

5.2 Three-dimensional experiments

Table 4 shows the RMS error for all axis across all experiments. The range of RMS error was 1.81 mm to 7.76 mm and was highest for the Y-axis (3.51 mm to 7.44 mm). Again these values support the view that the methods used were appropriate.

MANOVA results showed significant main effects between axes ($p<0.001$), significant interaction effects between the experimental set-up (number of control points) and calibration/reconstruction ($p<0.001$), the axis and calibration/reconstruction ($p<0.001$) and the interaction of axis, experimental set-up (number of points) and calibration/reconstruction ($p<0.004$).

Post hoc analysis showed the RMS error in X, Y and Z axis to differ significantly from each other. Overall the RMS error in the Y-axis was greater than the RMS error in the Z-axis which in turn was greater than the error in the X-axis. No significant differences were observed between axes when the calibration was performed using 12, 34 or 54 calibration points ($p<0.16$).

The interaction of experimental set-up (number of control points) with errors in the reconstruction of calibration and control points showed many interesting differences. No differences between experimental set-ups were found when using 12 calibration points however differences between

calibration and reconstruction errors were found. When the number of calibration points was increased (34 and 54) no differences between calibration and control point reconstruction errors were observed however differences between experimental set-ups were highlighted. Further analysis showed that the overall error in experimental set-up two was greater than the overall error in experimental set-up one. The RMS error when using 12 calibration points was significantly greater for both calibration and reconstruction conditions when compared to the RMS error when using 34 and 54 calibration points.

The results of the post hoc analysis for the interaction of axis and reconstruction showed no differences in the RMS errors for each axis when calibration and control point reconstructions were compared. except when the difference between the RMS error for the axis was high for the experimental condition.

The Kolmogorov-Smirnov tests revealed that the distribution of error in each axis across all experimental conditions did not differ significantly from normal ($p<0.05$).

6 DISCUSSION

6.1 Two-dimensional experiments

The results from the study suggest that the methods employed were suitable for the assessment of accuracy in underwater video analysis. The RMS errors (reported here as a percentage of the length of the object) when expressed as absolute RMS errors were within ranges reported elsewhere in the literature for DLT reconstructions (Shapiro, 1978; Angulo and Dapena, 1992; Challis and Kerwin, 1992; Chen et al., 1994), for close range techniques (Fraser, 1982) and underwater photogrammetric reconstructions (Hohle, 1971).

The distribution of the error however was a cause for concern as for the two-dimensional experiments very few of the distributions were shown to be representative of a normal distribution. This could be explained by the relatively few frames or data points that are often associated with biomechanical analyses. A previous study (Dabnichki et al., 1996) randomly sampled 200 frames from a total 1500 and found that no experiments had a distribution of error significantly different from normal. The distribution of error in each axis for the three-dimensional experiments conducted in the present study was found not to differ significantly from normal indicating a greater variance in the error with three-dimensional analyses. Some smoothing and filtering procedures require of the error that it is normally distributed whilst others require prior knowledge of the level of error and its distribution (Craven and Wahba, 1979). With a limited number of frames (data points) as shown with this study it cannot be assumed that the distribution of error is normal and therefore certain smoothing/filtering techniques cannot reliably be applied to the data. The implications of this are beyond the aims of this study however it would seem appropriate that research into the effect of error distribution on the derivatives and smoothing in kinematic data be undertaken.

The results presented in Table 1 reflect the effect of different camera to glass distances on the level of error. The statistical analysis showed that the number of significant differences between experimental trials at the longer distances were less than the number of differences between experimental trials at the shorter distances. Along with this it was found that the mean error at the greater camera to glass distances was less than the error at the shorter distances. The finding was the opposite of the results presented in Dabnichki *et al.* (1996) where an increased distance between the camera and the object resulted in an increased level of error. At the closer camera to glass distances the diffraction of light due to the glass would be picked up by the camera where as at the longer distances this effect would be minimised. This effect could have introduced additional distortion of the image at the closer distances therefore resulting in poorer image clarity. The finding from the present study indicates that the results from experimental accuracy studies that are conducted on land may' not be directly transferable to underwater settings.

The results for reconstructions of the vertical and horizontal bars showed that bar 03 (vertical) and bar 07 (horizontal) differed significantly from their counterparts. There exist two possible explanations for this. The first may be that the level of distortion caused by the glass barrier was larger in that particular region of the field of view. The second may be that subjective errors were introduced by the operator in locating point 6. The former explanation could be rejected as the experimental setting sought to provide the best possible image for reconstruction and the glass was of good optical quality. The second explanation may be more plausible as it was noted that although the image at the time of filming was ideal, the image at digitising in the region of the waters surface was less than ideal. This observation has previously been described by the term 'grey' area (McIntyre and Hay, 1975).

Another interesting finding was that for the vertical bars no main effect was found for the level of error across the range of camera to glass distances. For the horizontal bars however the error was overall greater at the shorter distances. It could be inferred from this finding that the effect of refraction in video analysis was greater for horizontal object reconstructions when the distance between

the camera and the glass was small. A physical explanation for this is that the deformation of the glass due to the water pressure was greater in the horizontal direction than in the vertical direction therefore resulting in higher errors in the reconstruction of the horizontal bars across the field of view. The obvious recommendation therefore would be to set the camera far back from the glass as possible due to the error at the longer camera to glass distances being less and differences due to camera positioning being smaller. This would seem to go against the natural choice of camera position which would be to place the camera as close to the glass as possible. Walton (1988) presented a correction to refraction by placing a domed glass front on an underwater housing and moving the camera forwards and backwards until a striped image was aligned. It may be appropriate to employ a similar procedure for cameras that are placed behind plane glass windows where the camera is moved back from and towards the window until a certain image is achieved. This warrants further investigation although it should be highlighted that limitations such as field width and the pool construction may restrict the ideal positioning of cameras.

The range of RMS errors for the experiments where the distance between the object and the glass was changed were very similar to that achieved in the moving camera experiments. The findings highlighted by the statistical analysis however were not. In these experimental trials no significant differences were found between experimental trials at the mid point of the object to glass range (450 mm to 530 mm). For the moving camera experiments there were always significant differences highlighted between experiments with less differences between experiments observed at the longer distances. The differences between experiments in this group were between the longer and the shorter distances with the overall error at the shorter distances greater than the error at the longer distances. The latter finding being consistent with the results from the moving camera experiments.

Errors in the vertical and horizontal bars again showed that the RMS error in bar 07 was significantly greater than the error in the other horizontal bars. For the vertical bars in the experimental group bar 02 was found to have a significantly smaller error than bars 01 and 03. The level of distortion introduced by the lens could explain this finding as generally the distortion is greater at the extremes of the field of view. This observation was also reported in Yanai *et al.* (1996) for periscope systems and it has been suggested that refraction magnifies this effect. The findings of this study suggest that this may also be the case for underwater video analysis through plane glass windows.

For the vertical bars the effect of the distance

between the object and the glass did not seem to influence the error between bars. However for the horizontal bars there was an increase in the number of differences between bars at the greater distances suggesting again that field width may an important consideration in the experimental set-up.

To summarise the findings of the MP experiments the results indicate that for underwater experimental set-ups errors were less as the object to glass distance was longer and that at the mid point of the range no differences between experimental set-ups were found. In terms of real life distances the object to camera distance based on these findings would be recommended to be about 3.00 ± 0.25 m. At the shorter distances the error was increased and although the error was less at the longer distances, the further the object is from the camera in the water the poorer the image due to the absorption of light (Townsend, 1978).

The results for the experiments where the object was moved out of the calibration frame showed large RMS errors when expressed as a percentage of length. These findings were due to the resolution of the screen and the size of the object used for reconstruction. As the object was moved out of the calibration plane towards the camera the RMS error increased as the bar length was overestimated. When the object was moved in the other direction the RMS error in bar length decreased with distance from the calibration plane. These findings were consistent with the findings from Dabnichki *et al.* (1996) and when the experimental results were applied to the theoretical optical error model of Nigg and Cole (1995) a linear correlation between theoretical model and experimental results of $R^2 = 0.68$ was found. When the object was moved out of the plane towards the camera the increase in the RMS error at the greatest increment (-3.8%) was greater than the increase in RMS error at the greatest increment when moved in the other direction (+2.4%). This was consistent with the findings of the previous experiments where the error was greater as the distances between the camera and the glass and the object and the glass were decreased. This could again be accounted for by the increased distortion due to the effect of refraction at shorter distances between the refractive surfaces.

6.2 Three-dimensional experiments

For the three-dimensional experiments the ranges of RMS error in the axes were consistent with those reported in the literature. The RMS error in the X axis was shown to be significantly less than the error in the Y and Z axis. Shapiro (1978) reported the error in the Y axis to be greatest (X, 1.28 mm; Y, 2.18 mm and Z, 1.58 mm) and other researchers have reported similar findings re-

porting errors in the long axis of the calibration frame to be consistently higher (Challis and Kerwin, 1992; Chen *et al.*, 1994). These findings again suggest that field width may effect the level of error.

When comparing the number of calibration points used in the DLT reconstruction several significant findings were observed. Firstly the RMS error using 12 calibration points was greater than using 34 or 54 calibration points for both calibration errors and control point reconstruction errors. Challis and Kerwin (1992) reported that for calibration errors the fewer the control points used the more 'flattering' the results (i.e. less reported RMS error in the axes). This was not the case and could be explained by the effects previously discussed relating to the two-dimensional experiments in that the level of error was found to be dependant on the position of the control points (calibration plane) relative to the glass interface. The size and position of the calibration volume may therefore be important if implementing an augmented calibration procedure for three-dimensional underwater filming.

With 12 calibration points differences between experimental set-ups were not found whereas with 34 and 54 calibration points differences were found. It was also found that with 34 and 54 calibration points significant differences between calibration reconstruction errors and errors in the subsequent reconstruction of control points were not found. With 12 points differences were observed and support the findings of Challis and Kerwin (1992) who reported increased error in subsequent reconstructions of control points when using a small number of calibration points.

Overall experimental set-up two was found to give a greater RMS error than experimental set-up one. This again may be explained by the effect on the level of error of the position of the control points relative to the glass. The side on camera view for experimental set-up one had an longer distance between the calibration object and the glass. It was observed from the previous experiments that at the longer object-glass distances the level of error was less. Given that the calibration points used defined the whole volume and therefore covered a depth of 300 mm of the field of view then similarly based on the previous findings the range of error in the points would also be less. This would theoretically result in a more accurate solution to the DLT equation by least squares and therefore less RMS error in the reconstruction. There may however be a greater error distribution of error across different areas of the calibration volume as some areas would be further from or closer to the glass/perspex barrier. In Dabnichki *et al.* (1996) differences were found in the reconstruction error for different areas of a calibration plane and the error associated with the position of the object in the calibration volume should be investigated.

7 CONCLUSIONS

The findings of the two-dimensional experiments suggest that the position of the camera and the position of the object relative to the glass influences the level of error associated with the reconstruction of known lengths. The results of the three-dimensional experiments suggest that no major differences between land three-dimensional reconstructions and underwater three-dimensional reconstructions exist. However there is some evidence that the positioning of the cameras and calibration object relative to the glass/perspex barrier may influence the error in areas of the calibration volume. Further investigation is required to establish the effect of additional factors such as turbulence and position of the object within the calibration volume before a procedure can be developed that will minimise the influence of unaccounted random errors. The study demonstrates that the scaled down environment was suitable for the assessment of error in underwater video analysis.

8 ACKNOWLEDGEMENTS

The authors would like to gratefully acknowledge the technical assistance provided for data collection by Mr Gerald Wright.

9 REFERENCES

Abdel-Aziz, Y.I. and Karara, H.M. (1971) Direct linear transformation from comparator coordinates into object space coordinates in close range photogrammetry. In: *Proceedings of the ASP Symposium on Close Range Photogrammetry:* 1-18. Falls, Church. American Society of Photogrammetry.

Angulo, R.M. and Dapena, J. (1992) Comparison of film and video techniques for estimating three-dimensional coordinates within a large field. *International Journal of Sport Biomechanics.* 8: 145-151.

Bartlett,R.M.(1990) The Definition, design, implementation and use of a comprehensive sports biomechanics software package for the Acorn Archemedes 440 Microcomputer. *Communication to the XIII International Symposium of Biomechanics in Sports.* Prague.

Challis, J.H. and Kerwin, D.G. (1992) Accuracy assessment and control point configuration when using the DLT for photogrammetry. *J. Biomechanics.* 25: 1053-1058.

Chen, L., Armstrong, C.W. and Raftopoulos, D.D. (1994) An investigation on the accuracy of three-dimensional space reconstruction using the direct linear transformation technique. *J. Biomechanics.* 27: 493-500.

Craven, P. and Wahba, G. (1979) Smoothing noisy data with spline functions. *Numer. Mathem.* 31: 377-403.

Dabnichki, P., Aritan, S., Lauder, M.A., and Tsirakos, D. (1996) Accuracy of the kinematic data collection, filtering and numerical differentiation. In: Haake, S. (ed) *Proceedings of the International Conference on The Engineering of Sport.* Rotterdam: A.A. Balkema.

Fraser, C.S. (1982) On the use of Nonmetric cameras in analytical close-range photogrammetry. *The Canadian Surveyor.* 36(3): 259-279.

Hohle, J. (1971) reconstruction of the underwater object. *Photogrammetric Engineering.* 37: 948-954.

McIntyre, D.R. and Hay, J.G. (1975) Dual Media Cinematography. In: Lewillie, L. and Clarys, J.P. (Eds.) *Swimming II. Proceedings of the second International Symposium on Biomechanics in Swimming:* 51-57. Baltimore: University Park Press.

Nigg, B.M. and Cole, G.K. (1994) Optical Methods. In: Nigg, B. and Herzog, W. (Eds.) *Biomechanics of the musculo-skeletal system:* 254-286. Chichester: John Wiley & Sons.

Shapiro, R. (1978) Direct linear transformation method for three-dimensional cinematography. *Research Quarterly for Sport and Exercise.* 49: 197-205.

Townsend, D. (1978) *Underwater photography, movies and still.* London: George Allen & Unwin Ltd.

Walton, J.S. (1988) Underwater traking in three-dimensions using the Direct Linear Transformation and a video based motion analysis system. *Proceedings of the 32nd International Technical Symposium on Optical and Optoelectronic Applied Science and Engineering.* Vol. 980: Underwater Imaging. San Diego, California, August.

Yanai, T., Hay, J.G., and Gerot, J.T. (1996) Three-dimensional videography of swimming with panning periscopes. *J. Biomechanics.* 29(5): 673-678.

9 Vibration analysis

The Engineering of Sport, Haake (ed.) © 1996 Balkema, Rotterdam. ISBN 90 5410 822 3

Vibrations on the golf course

Erik A. Ekstrom

Mechanical Engineering Department, Lehigh University, Bethlehem, Pa., USA

ABSTRACT

Avid golfers and senior golfers complain of sore hands, elbows, and shoulders due to impacts from playing golf. Lately, the golf world has targeted this as an area that needs attention and vibration dampers, shock absorbers and other products intended to alleviate this problem have been developed and are being marketed. This study focuses on the frequencies of vibrations that are transmitted through a golfer's hands to the skeleton and the range of frequencies that commercial golf-shaft dampers are able to absorb.

Tests on a low-handicap, young male golfer using a simulated golf shaft and grip driven by an electromagnetic shaker show a bandwidth of vibration transmission from the shaft to the distal end of the left radius of about 50 Hz; other subjects have similar transmissibility characteristics. Typical shaft grips have no effect on this transmissibility.

Properly tuned vibration absorbers fitted inside the butt of golf shafts can reduce dramatically the amplitude of free vibration of steel or composite shafts with driver or iron heads, which may not be within the passband of the golfer's hands, particularly in the case of steel shafts. Identical absorbers that are not properly matched to the shaft and club head do almost nothing. Frequency matching is just as important as vibration damping to the golfer's skeletal response. A golfer's hands in a normal grip make the most significant contribution to the vibration isolation of the skeleton from impacts generated while playing golf.

During a "typical" golf shot, steel clubs and graphite clubs act in different ways. The "typical" shot excites different natural frequencies modes in the steel shaft than in the graphite shaft. It is also noted that the magnitudes of the amplitudes at the natural frequencies of the graphite shafts are much lower than the amplitudes of the steel shaft. The type of head that is applied has a significant impact as well.

This brings us to the question of which of these vibrations are harmful to the golfer. This paper demonstrates that the average human has a band pass of about 50 Hz at his/her wrist. If the frequency is any higher than 50 Hz it will not travel up a person's arm further than the wrist.

In an effort to curb dangerous vibrations, Vendor1* designed several shock absorbers. This paper demonstrates that these dampers were quite effective, supporting the manufacturer's claim that the shock absorbers reduce vibrations up to 70 percent. They were especially effective at dampening out vibrations in the frequency range of 130 Hz to 180 Hz. Tests were done on both steel shaft and graphite dampers.

These three areas will enable future manufacturers of protective equipment to know what frequency ranges are important.

1 INTRODUCTION

When the head of a golf club impacts the ground or a golf ball, impulsive forces are generated and transmitted to the golfer's hands. In an effort to reduce the vibrations delivered to the golfer, several manufacturers have developed dampers that are inserted into the butt ends of golf club shafts.

Vendor1 boasts that its damper can reduce shaft vibrations by as much as 70 percent. The first part of this paper studies the properties and the effectiveness of Vendor1's three shock absorbers (Damper-a, Damper-b, Damper-c) and the Vendor2* shock absorber, which was submitted to the U.S.G.A..

As more appliances such as these are

designed, manufactured, and marketed for the alleviation of shock- and vibration-induced discomfort and injury, it becomes essential for designers and consumers to know the frequency ranges of skeletal responses. Take golf as an example: devices intended to dampen the higher harmonics of golf-club transverse vibrations are suppressing frequencies that the hands and wrists cannot transmit to the bones of the forearms. Protective devices should supplement and extend the natural vibration suppression of the hands and wrists rather than duplicate it. A golf club is taken to be representative of a piece of sports/recreation equipment that is used to strike a ball, shuttlecock, or opponent's foil and has a compliant handle, cover, or grip by which it is hand-held. The effects of the resulting impacts are transmitted through the hands to the user's skeleton. Effective prevention of injury/discomfort requires knowledge of the frequencies of excitation, transmission, and susceptibility.

The second part of this paper reports on the measurement of vibration transmission of a particular configuration of hand excitation and forearm response. Finger and wrist responses have a much higher bandwidth, but these frequencies are not transmitted to the primary skeleton. Nevertheless, Southmayd and Hoffman[2] describe elbow injuries among golfers as equal to the sum of hand plus wrist injuries. The experiment configuration is not a simulation of the motions experienced by a golfer's hands. It is closer to wood chopping, but the purpose of the work is to obtain a bandwidth or passband and a qualitative picture of possible resonance and associated damping factors for a generic excitation that is within with the capabilities of an existing electromechanical shaker.

2 APPARATUS AND INSTRUMENTATION

2.1 *Apparatus used to Analyze Dampers*

A test fixture was constructed to hold a clear lucite tube, with an inner diameter of .579 in.(see Figure 1). The Vendor1 and Vendor2 dampers, which were made for steel shafts, were inserted into the lucite tube. A vertical axis MB Electronics electromechanical shaker with a 10 KVA power amplifier was used to drive the fixture with the lucite in a horizontal position in sinusoidal motion in a vertical plane. This was used to make a qualitative determination of the resonant frequencies of the dampers.

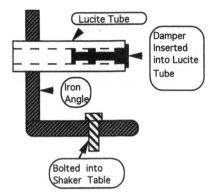

Figure 1. Shaker table fixture #1.

A True Temper Dynamic Gold S-300 steel shaft and a UST Tour Weight graphite shaft were tested; then a MetaPro 590 Series 1 Driver was placed on the steel shaft and a Ben Hogan Apex 6 iron was placed on the graphite shaft and the two shafts were again tested. The dampers being tested were submitted by Vendor1 and Vendor2. A pendulum apparatus was used to produce a transverse impulse (see Figure 2) and a one-gram piezoelectric-crystal accelerometer (PCB model 309a) was used to obtain the response data. This data was collected and analyzed using an Ono Sokki Mini FFT analyzer.

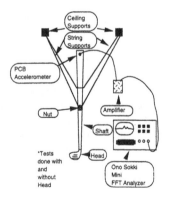

Figure 2. Pendulum apparatus.

The Vendor1 dampers are made out of a viscoelastic material and have two "stems" and three "heads". The stem to head ratio has to be kept between 1:1 and 5:1 in order for it to work.[3] The dimensions for the Vendor1 dampers, A & B, can be seen in Figure 3.

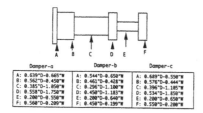

Figure 3. Damper dimensions.

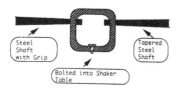

Figure 4. Shaker table fixture #2.

2.2 *Apparatus used to Analyze the Wrist Bandwidth*

A second test fixture was constructed from which two solid-steel shafts extended horizontally (see Figure 4). A typical, compliant, commercially available golf grip was installed on the 5/8 in. diameter uniform shaft. The other, a tapered shaft, was machined to be a solid steel replica of the straight shaft with grip. A vertical axis MB Electronics electromechanical shaker with a 10 KVA power amplifier was used to drive the fixture with its horizontal shafts in sinusoidal motion in a vertical plane.

The motions of the outer ends of the shafts were measured by one-gram piezoelectric-crystal accelerometers (PCB model 309a), and the same accelerometers were used to measure the vertical motion of the distal process of the radius of the right forearm of a young adult test subject. The subject was left-handed and used a conventional interlocking golf grip. An Ono-Sokki CF-920 dynamic analyzer was used to acquire the shaft-motion frequency response data and the radius frequency response data and to compute and display the transfer-function magnitude and phase-angle plots.

3 EXPERIMENTS

3.1 *Testing of Dampers*

Areas Tested:
1. Observed frequency response of each damper using a transparent fixture on the shaker table.

2. Measured the natural frequencies of the steel shaft and the graphite shaft.

3. Repeated test 2 to see what happened to these natural frequencies quantitatively when the dampers were inserted.

4. Repeated test 2 to measure the frequencies of the shafts when the heads are attached to the shafts.

5. Repeated test 3 to determine the effects the dampers have on the natural frequencies.

Methods of Testing:

1. The natural frequencies of each damper are determined using a transparent fixture attached to a shaker table (see Figure 1). From preliminary testing, the dampers appeared to be excited within a frequency range of 5 Hz to 200 Hz. Therefore, the shaker table will be set up to sweep through this range of frequencies. The damper will be watched, with the aid of a strobe light, through the transparent lucite shaft of the fixture, and when the damper has the greatest amplitude, we will assume this to be the resonant frequency(ies). The shaker table fixture is assumed to be rigid enough to vibrate at the same rate as the shaker table. It has been speculated that the dampers are most effective at their resonant frequencies and this will be determined.

2. The natural frequencies of the shafts will be tested using the Ono Sokki Mini FFT analyzer, a one gram PCB accelerometer, an amplifier, and a bolt(see Figure 2). Both a steel shaft and a graphite shaft will be tested. The bolt will strike the hanging shaft and the accelerometer will pick up the signal. The signal will then be sent through an amplifier an intó the Ono Sokki Mini FFT analyzer, where the resonant frequencies will be read off the screen. The same tests will be repeated with different dampers inserted into the shafts.

3. Natural frequencies of the steel shaft with a head will also be determined in the same manner as step #2. The same procedure will be used for the graphite shaft, except the graphite shaft will have an iron head attached to it.
*Note: No grips were used during experimentation.

3.2 *Testing of Bandwidth*

The horizontal simulated golf shafts were 37 inches above the floor. The subject assumed a normal golf stance and gripped the shaft as if addressing a ball, to the extent that the shaker and fixture permitted. With the accelerometer on the shaft (the grip material was cut away to gain access to the butt end of the straight shaft), the shaker was driven smoothly through a range of frequencies and the magnitude and phase angle of the response were recorded. Then the accelerometer was moved to the skin of the subject, over the distal process of the right radius, and held firmly in place by a Velcro-closed strap. The subject assumed the same stance and attempted to grip the shaft in the same way while the shaker was driven through the same frequency range. The analyzer computed, displayed, and plotted the resulting transfer function relating radius-bone motion to steel shaft motion. Earlier work by Kim, Voloshin, Johnson, and Simkin [3] on a cadaver showed that one can expect a bandwidth from such an arrangement of about 150 Hz.

Opinions varied widely as to the bandwidth one could expect from the club-golfer interface, so the first series of tests was run over a frequency range of 10 Hz to 1500 Hz, using the shaft with the compliant grip in anticipation of smoother results with fewer and more heavily damped resonances. The shaker is run open loop; the motion profile is roughly constant displacement amplitude up to 150 Hz and constant acceleration after that. Three low shaker-table power levels were used. At the middle setting, without any load on the shaker, the amplitude of the shaker-piston motion at 10 Hz is about 0.01 inches and declines with frequency to about 0.001 inches at 100 Hz while the acceleration is increasing. The acceleration is constant at about 3 g's above 150 Hz. With the apparatus in place and the subject gripping a shaft, the three power settings produced 3.2 g's, 2.0 g's and 1.2 g's above 100 Hz.

Three overlaid transfer-function frequency-response plots for the subject holding the shaft with compliant grip are shown in Figure 5. The frequency sweep rate was 1.6 decades/minute. These runs were duplicated using the solid-steel shaft and the resulting frequency responses are shown in Figure 6. The solid-steel experiments were repeated at a slower sweep rate of 0.55 decades/minute and a frequency range of 2.5 Hz to 250 Hz. The results are shown in Figure 7.

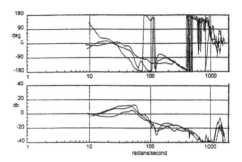

Figure 5. Transfer function obtained from frequency sweeps at three different power settings on the electromechanical-shaker power amplifier. The subject assumed a golfer's stance and gripped the solid-steel shaft with a representative, compliant grip installed.

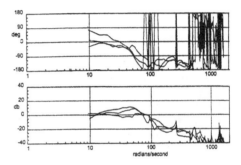

Figure 6. Transfer function obtained from frequency sweeps at three different power settings on the electromechanical-shaker power amplifier. The subject assumed a golfer's stance and gripped the solid-steel shaft that was shaped like a golf shaft with a grip.

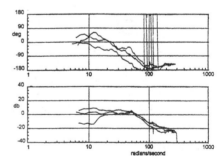

Figure 7. Transfer function obtained from frequency sweeps at three different power settings on the electromechanical-shaker power amplifier. The subject assumed a golfer's stance and gripped the solid-steel shaft that was shaped like a golf shaft with a grip. The frequency range is narrower and the sweep rate slower than Figure 6

3.3 *Analytical Model*

Figure 8 shows the mid-power frequency responses from each series of tests and the frequency response of a linear, constant-coefficient transfer function that is judged to be a reasonably good good fit to all of the data. That transfer function is given by:

$$G(s) = \frac{\left(\frac{s}{884}\right)^2 + \frac{2(.4)s}{884} + 1}{\left(\frac{s}{1325} + 1\right)\left(\frac{s}{2820} + 1\right)\left(\left(\frac{s}{283}\right)^2 + \frac{2(.4)s}{283} + 1\right)}$$

Where G(s) is the ratio of shaft displacement, velocity or acceleration to radius displacement, velocity or acceleration respectively. The transfer function is equivalent to a set of first-order, linear, constant-coefficient ordinary differential equations

$$\underline{\dot{x}} = \underline{A}\underline{x} + \underline{b}u(t)$$

where

$$\underline{A} = \begin{vmatrix} 0 & 1 & 0 & 0 \\ 0 & 0 & 1 & 0 \\ 0 & 0 & 0 & 1 \\ -.299252\times10^{12} & -.117791\times10^{10} & -.475502\times10^{7} & -.437140\times10^{4} \end{vmatrix}$$

$$\underline{b} = [0 \quad .382942\times10^{6} \quad -.167399\times10^{10} \quad .579659\times10^{13}]^{T}$$

The first element of the state vector, x_1, is the response of the radius and u(t) is the shaft motion. If u(t) is the shaft displacement, then x_1 is the radius displacement, or both can be velocities or accelerations.

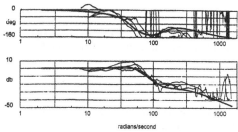

radians/second

Figure 8. The midpower results from figures 5, 6 and 7 are extracted and replotted over the Bode plot of an approximating, linear, constant coefficient, analytical transfer function. Many variations could be argued to be just as valid as the transfer function shown.

4 DISCUSSION

4.1 *Discussion of Damper Results*

Only the steel shaft damper inserts were able to be tested using the transparent fixture attached to the shaker table because the graphite insert requires a smaller fixture inside.

The frequency ranges in which the Damper-a and Damper-c vibrated the most varied slightly depending on whether the frequencies were swept up or down. Damper-a touched the bottom of the transparent tube at 140 Hz and stopped touching the bottom at 180 Hz, when the frequencies were swept in an increasing manner. It touched the bottom at 170 Hz and stopped touching the bottom at 140 Hz, when swept in a decreasing fashion. Damper-c started touching the bottom of the transparent tube at 130 Hz and stopped touching at 180 Hz, when swept up. When swept down it started touching at 170 Hz and stopped touching between 130 Hz and 120 Hz. As you can see Damper-c appears to have a natural frequency range that starts about 10 Hz to 15 Hz lower than Damper-a, but ends at about the same frequency (180 Hz).

Spectrum analysis on the Ono Sokki Mini FFT analyzer was used to produce plots of amplitude vs. Hz for the freely supported steel shaft and the freely supported graphite shaft. Spectrum analysis was also used to generate the graphs for each of the dampers inserted into the respective shafts. The X and Y coordinates were read off of these plots.

The Vendor1 Damper-c appeared to be the most effective damper for the steel shafts. For the steel shaft alone, it reduced the first mode's (51.0 Hz) amplitude by 0.605 mV (83%) and it reduced the second mode (149.5 Hz) by 3.655 mV (98.9%). When the head was added, it reduced the amplitude of the first mode (31.5 Hz) by 0.9 mV (46 %) and the second mode (102.00 Hz) by 6.29 mV (86.6%)

The Vendor1 Damper-a appeared to produce the second best results by reducing the amplitude of the first frequency mode (51.5 Hz) of the steel shaft by 0.251 mV (34.6%) and the second mode (147.5 Hz) by 3.48 mV (94.2%). When the head was added, it reduced the first mode (31.50 Hz) amplitude by 1.13 mV (57.95%) and the second mode (102.50 Hz) by 5.9 mV (81.27%).

The Vendor2 damper appeared to do very little. In fact, it produced very similar results to simply stuffing a paper towel into the shaft. The Vendor2 damper reduced the amplitude of the first mode (53.0 Hz) of the steel shaft by 0.432 mV (59.67%), while the paper towel reduced it by 0.442 mV (61.05%). The Vendor2 damper reduced the amplitude of the second mode (147.5) by 2.435 mV (65.91%) and the paper towel reduced it by 1.404 mV (38.0%). When the head was added, the Vendor2 damper appeared to increase the amplitude of the first mode (33.0 Hz) by 0.64 mV (32.82%), while the paper towel increased it by 0.29 mV (14.87%). The Vendor2 damper reduced the amplitude of the second mode (103.50 Hz) by 2.16 mV (29.75%) and the paper towel reduced it by 2.38 mV (32.78%).

When no head was placed on the graphite shaft, only one frequency mode was observed between 0 Hz and 200 Hz, but when the iron head was attached, two frequency modes were observed in the same region.

When the Vendor1 Damper-b was inserted into the graphite shaft it reduced the amplitude of the first frequency mode (72.0 Hz) by 3.194 mV (97.57%), whereas the paper towel didn't appear to reduce the amplitude. After the iron head was attached, the Vendor1 Damper-b reduced the amplitude of the first frequency mode (45.0 Hz) by 4.9 mV (92.98%) and the amplitude of the second mode (147.5 Hz) by 9.79 mV (96.45%). Repeating the same tests using a paper towel yielded results of 1.81 mV (34.35%) and 2.46 mV (24.24%).

The results to the proceeding paragraphs can be seen in Figures 9-12 pictured below.

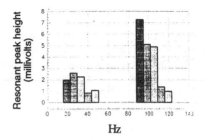

Figure 10. Steel shaft with a driver head. From left to right: no damper, Vendor 2, paper towel, Damper-a, Damper-c

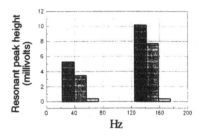

Figure 11. Composite shaft with an iron head. From left to right: no damper, paper towel, Damper-b

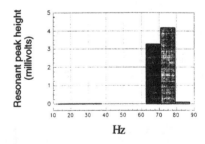

Figure 12. Composite shaft without a club head. From left to right: no damper, paper towel, Damper b

4.2 Discussion of Wrist Bandwidth Results

Fitting transfer functions to frequency-response data is something of an art form. In this situation it is complicated by the frequency-dependent forcing function, which is the response of the fixture to the shaker excitation. For example, the dips in the transfer functions in Figure 5 are neglected as nonlinear responses to resonances of the fixture. The resonances at about 250 Hz in Figure 6 are also neglected despite the fact that the underlying forcing functions are flat in the 100 to 500 Hz range. The slower sweep rate experiments for Figure 7 do not show resonances

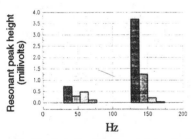

Figure 9. Steel shaft without a club head. From left to right: no damper, Vendor 2, Damper-a, Damper-c

at or below 250 Hz and the magnitude responses do not change slope by 40 db/decade. The peaks are down 20 db from unity. The choice of damping factor is almost arbitrary. The modest peaks at about 50 Hz in Figure 7 may be due to nonlinear responses to dips in the forcing function at that frequency or may be genuine under damped behavior. The computed response is insensitive to the particular value of damping factor in the vicinity of critical damping and 0.4 was chosen to indicate mild under damping. The choice for numerator damping factor is similarly uncertain.

To test the bandwidth of our apparatus, tests were run where the Ulna was driven directly from the shaker table. This resulted in a bandwidth of 200 Hz. This indicates that our testing apparatus has a sufficient bandwidth for this type of test.

5 CONCLUSIONS

5.1 *Damper Conclusions*

Our data supports our earlier proposal that the dampers will be most effective if they operate at their natural frequencies. The most effective damper was the Vendor1 Damper-c. Its natural frequencies were the closest to the frequencies the steel shaft produces. The Vendor1 natural frequency range is between 120 Hz and 180 Hz. The lowest two natural frequencies of the golf club are 51.0 Hz and 150 Hz without the driver head and 32 Hz and 102 Hz with the driver head. Our data indicates that Vendor1 can get better damping results if it moves the natural frequencies of its dampers to match the natural frequencies of the club.

The data also illustrated the fact that the Vendor2 damper was about as effective as a paper towel stuffed into the butt end of a golf shaft.

5.2 *Wrist Bandwidth Conclusions*

Tests in which a young adult golfer grips a simulated golf shaft driven by an electromagnetic shaker show that the bandwidth of transmission of transverse vibration into the distal end of the forearm radius is about 50 Hz and the response is slightly under damped. Four other test subjects were tested: a young adult female, a young adult male, and a middle-aged male. All four response plots yielded similar low- and high-frequency asymptotes and bandwidth (see Figure 9).

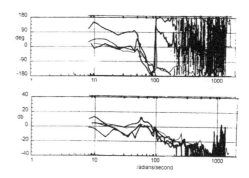

Figure 13. The four test subjects, consisting of three males and one female, two left handed and two right handed, yielded low- and high-frequency asymptotes and bandwidth.

Two conclusions seem warranted. One is that, despite the inherent variability of in-vivo testing, the results are definitely repeatable and the bandwidth is defined. Magnitude effects are weak and a linear transfer function is a good representation of the measured responses. The second conclusion is that the compliant grip seems not to affect the frequency responses in the ranges tested. Figures 5 and 6 are not sufficiently different to warrant separate transfer functions. This is not to say that the grip has no effect. Presumably, at the higher frequencies where the subject felt his fingers responding, the grip would contribute to comfort, but it does not seem to have any influence at the frequencies that are transmitted into the skeleton.

The final conclusion of this experiment is that the steel shaft's frequency mode at 31.5 Hz (this occurs when the driver head is attached) is the most important frequency mode to dampen when the safety of a persons wrist is in question. An important note is that hand may have a higher bandwidth, in which higher frequency modes would also be important.

* The names of the manufacturers have been changed for the purposes of this paper.

ACKNOWLEDGMENTS

The Lehigh University affiliated authors wish to thank Mr. Frank Thomas, technical director of the United State Golf Association, for his help and encouragement and Dr. Douglas Winfield, manager of applied testing and research,

USGA, and Denise Doak, Lehigh student, for being willing test subjects.

Also a special thanks to Prof. Stanley Johnson, Ph.D., Professor of Mechanical Engineering and Mechanics at Lehigh University, who not only is my advisor, but he also was the mastermind behind the entire operation.

REFERENCES

[1]   United States Patent Number: 5,362,046. Issued November 8, 1994.

[2]   Southmayd, W., and M. Hoffman, <u>Sports Health: The Complete Book of Athletic Injuries</u>, Putnam Publishing, New York, 1981.

[3]   Kim, W., A. Voloshin, S.H. Johnson, and A. Simkin, "Measurement of the impulsive bone motion by skin-mounted accelerometers." J. of Biomechanical Engineering, Vol. 115, pg. 47-52.

The Engineering of Sport, Haake (ed.) © 1996 Balkema, Rotterdam. ISBN 90 5410 822 3

The validation and updating of dynamic models of golf clubs

M. I. Friswell, M. G. Smart & S. M. Hamblyn
University of Wales Swansea, UK

G. Horwood
TI Apollo Ltd, Oldbury, West Midlands, UK

ABSTRACT: This paper compares the results of the dynamic modelling of a typical golf club and measurements made on the same club. The shaft was modelled using beam elements and the head is represented as a rigid body. The natural frequencies of the club are measured and compared to those predicted from the analytical model. Mass and inertia properties of the head are estimated from the measured data to produce an improved analytical model.

1 INTRODUCTION

The quality of the dynamic response of a golf club can have a considerable impact on the 'feel' of the club, and the quality of the contact with the ball. Matching the shaft to a player's swing may increase the impact velocity, although in practice this effect is probably small. The size and shape of the 'sweet spot' is related to the natural modes of the club: hitting the ball within the 'sweet-spot' will excite few modes of the club, and leave the golfer more satisfied with his shot. Horwood (1994) discussed the dynamics of the shaft in a qualitative manner. The dynamics of the club relate directly to the shaft, and the mass and inertia properties of the club head. The low frequency range is of most interest and within this range the head may be assumed rigid. The modelling of the 'static' club, that is when the grip is stationary, is considered as a first step to modelling the dynamics of the club through the swing.

One difficulty in the analysis and testing golf clubs is the boundary condition at the grip. If the analysis is required to simulate the club during the swing then the interface between the grip and the golfer must be considered. Mather (1996) and Swider *et al.* (1994) tried to replicate this boundary condition in static tests of the clubs. The approach here is somewhat different, since the purpose of the current work is to validate the model of the club. Thus it is preferable to fix the

grip end of the club, so that the boundary condition is less uncertain. The grip is removed so that the shaft is clamped directly. Once the finite element model of the clamped club has been validated, this model may be used in further studies by modelling the golfer's grip over a range of stiffnesses.

Finite element modelling and modal testing have been performed on a club with a TI Apollo Acculite steel shaft with a Bioedge series No. 1 wood head. The wood was swing weighted to D1 using lead tape.

2 FINITE ELEMENT MODELLING

2.1 *A simple beam model*

One of the first questions to be answered when modelling golf clubs is the type of element that should be used. Swider *et al.* (1994) used shell elements to model a composite shaft and head. Whereas this makes the incorporation of geometry changes relatively easy, it should not be thought that this method will necessarily produce an accurate model. First, the frequency range of interest for the clubs is very low, and is determined by the frequencies that have some effect on the club dynamics during the swing and on impact with the ball. Within this frequency range the modes will involve the bending of the shaft and so plate representations of the shaft are not really required. Similarly, within the frequency range of interest the head will act as a rigid body: the flexible modes of the head will be at very high frequencies. The accuracy of the shell and brick models rely on the accurate measurement or estimation of the shaft thickness and the geometry of the head. Iwata *et al.* (1990) used 3D brick elements to model the impact between the ball and club. This approach may be necessary during impact, although the deformation of the ball is far larger than the deformation of the head.

The approach taken in this paper is to produce a simple model of the shaft using beam elements for the shaft and a rigid body (i.e. just use the mass and inertia properties) for the head. Brylawski (1994) used a similar approach, but modelled the shaft as a continuous beam using partial differential equations, rather than the finite element modelling approach adopted here. The uncertain parameters in this model will be identified from measured data.

The element matrices will involve 2 nodes and 6 degrees of freedom per node. The element matrices are assembled from the standard bending elements in two planes, the shaft torsion element and the axial extension element (see for example Dawe, 1984). Within the element it is assumed that there is no interaction between these 4 vibration mechanisms. Shear effects are neglected in the bending elements.

2.2 *Estimating the inertia matrix*

The main difficulty with modelling the head as a rigid body, is the estimation of its inertia matrix, with respect to axes fixed at the end of the shaft. The approach adopted by Johnson (1994) was to measure the inertia matrix. Measuring the inertia should produce reasonable estimates, but is time consuming, it requires an isolated club head and will inevitably still contain errors. Since the objective in this paper is to estimate the inertia matrix, it will suffice to produce a reasonable estimate of the inertia. This is readily available from an FE model of the club head (Iwata *et al.*, 1990, Swider, *et al.*, 1994) or from a CAD model (Mitchell, *et al.*, 1994). Here we use an initial estimate of the inertia matrix based on all the mass of the head being located at a single point. This point will be slightly further away from the end of the shaft than the centre of gravity. If the position of the head mass from the end of the

shaft is (x_m, y_m, z_m) where the x direction is along the shaft, then the inertia matrix is

$$m \begin{bmatrix} y_m^2 + z_m^2 & x_m y_m & x_m z_m \\ x_m y_m & x_m^2 + z_m^2 & y_m z_m \\ x_m z_m & y_m z_m & x_m^2 + y_m^2 \end{bmatrix} \quad (2.1)$$

where m is the mass of the head (see for example, Bedford and Fowler, 1995). If the z direction is chosen to be along the one of the principal axes of the club head, then $y_m = 0$ and only one pair of the off diagonal elements in the inertia matrix are non-zero. Of course, this is difficult to achieve exactly in an experiment, but it should be possible to get this approximately right.

2.3 *The Finite Element Model of a Golf Club*

The golf club to be modelled has a TI Apollo Acculite steel shaft with a Bioedge series No. 1 wood head. The dimensions of the shaft are given in Table 1, where the shaft has been split into segments of constant diameter. The thickness values were obtained by cutting open a shaft specimen. Note that the first shaft segment does not include the length that is clamped. Similarly the last shaft segment does not include the shaft incorporated into the head. The estimated mass of the manufactured shaft is computed from these properties, with a first segment length of 279 mm and a combined penultimate and last segment length of 413 mm, is 87 g. This compares to a specified manufactured mass of 101±2.5 g. As a first approximation the head has a mass of 270 g, located at a position $(x_m, y_m, z_m) = (40, 0, 20)$ mm. This gives an inertia matrix of

$$\begin{bmatrix} 1.08 & 0 & 2.16 \\ 0 & 5.40 & 0 \\ 2.16 & 0 & 4.32 \end{bmatrix} \times 10^{-4} \text{ kg m}^2.$$

The shaft was modelled using beam elements based on a tube of constant diameter and thickness for each element. Apart from the first and penultimate segments given in Table 1, each segment was modelled using a single element. The first shaft segment was modelled using 2 elements and the penultimate segment using 6, making a total of 26 elements and 156 degrees of

Table 1 Properties of the TI Apollo Acculite Shaft (Note, the bending stiffness of the last segment is increased to allow for the plastic sleeve)

Shaft Segment Dimensions		
length (mm)	diameter (mm)	thickness (mm)
Grip End		
114.3	15.24	0.25
12.7	15.7	0.20
76.2	16.3	0.18
57.2	15.7	0.19
25.4	15.2	0.20
25.4	14.7	0.22
25.4	14.2	0.20
25.4	13.7	0.20
12.7	13.2	0.20
12.7	12.7	0.20
25.4	12.2	0.20
25.4	11.7	0.20
25.4	11.2	0.26
12.7	10.7	0.28
12.7	10.2	0.29
25.4	9.7	0.30
25.4	9.1	0.30
25.4	8.8	0.31
289.6	8.5	0.37
22.9 (see above)	8.5	0.37
Head End		

Table 2 Initial, Experimental and Updated Natural Frequencies (Hz)

plane	Experi-mental	Initial Model	Updated Model
z	4.45	4.20	4.47
y/ϕ_x	4.50	4.21	4.48
y/ϕ_x	49.0	46.5	49.0
z	49.5	50.6	49.5
y/ϕ_x	66.0	88.3	66.0
z	132	126	125
y/ϕ_x	156	229	156
z	267	283	282
y/ϕ_x	299	450	306
axial	467	439	467

freedom. The last shaft segment was covered by a plastic sleeve that was modelled as a metal shaft element whose bending stiffness only was multiplied by 3. Typical material properties for steel were used, namely a Young's Modulus of 210 GN/m², a density of 7800 kg/m³ and a shear modulus of 80 GN/m². The natural frequencies are given in Table 2, and mode shapes shown in Figures 1 and 2, for this model.

As expected, since the z direction has been assumed to lie along a principal inertia axis of the head, there are modes that just involve bending in the x-z plane. The first bending mode in the x-y plane does not involve much coupling with the torsional modes, but the higher modes show great coupling between bending in the x-y plane and torsion.

2.4 Errors in the Finite Element Model

The finite element model is used to calculate the measured outputs and also the sensitivity of those outputs to changes in the unknown parameters. The errors in the model of the club can arise from three main areas which will now be described briefly.

Model structure errors: These errors occur when the governing physical equations or principles are uncertain or complex. For example the model may be assumed to be linear although the actual structure behaves in a non-linear way. For the golf club, it may be that the beam and rigid mass/inertia model is not sufficiently accurate and that shell and brick elements would be better. The end of the shaft, where the grip would be, is assumed fixed but some flexibility may be present in the experiment. Damping is very difficult to model and in many numerical models is ignored completely.

Model parameter errors: Even if the underlying structure of the model was correct some of the parameters may be uncertain. For example, if the boundary is assumed flexible it is often difficult to theoretically estimate the stiffness of the connection. In the case of the golf club, the inertia of the club head is difficult to estimate accurately. These errors are most amenable to correction by model updating.

Discretisation errors: Most numerical models of structures, including finite element analysis, approximate the motion of the continuous structure by a discrete system. If the level of discretisation, that is the number of degrees of freedom, is insufficient then the model order will be too small to accurately model the dynamics of

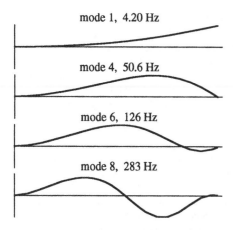

Figure 1 Vertical modes from the initial finite element model of the club (only the *x-z* plane is shown)

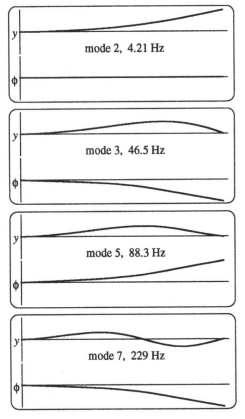

Figure 2 Combined horizontal and torsional modes from the initial finite element model of the club - *y* is the horizontal deflection and φ is the torsion angle (the modes are scaled so that the *x* and *y* directions have the same scaling and φ is between ±30°)

the structure. The requirements of model updating are considerably more stringent that straight-forward design analysis. The natural frequencies should be fully converged, that is the difference between the predicted frequencies and those for a model with a very large number of degrees of freedom should be much smaller than the difference between the predicted and measured frequencies. In this paper we have ensured that enough elements are used so that the discretisation errors are small.

3 EXPERIMENTAL MODAL ANALYSIS

Experimental Modal Analysis (EMA) is now an established technique for many industries, for example automobiles and aerospace applications. The idea is to apply a force to a structure, and from measurements of the force applied and the response, to estimate the natural frequencies, damping ratios and mode shapes of the structure. The purpose here is not to review EMA in detail, but to highlight the special features in testing and updating golf clubs. Ewins (1984) gave more detail on vibration testing and modal analysis.

3.1 *Errors in experimental data*

The quality of measured time series data has improved considerably with the arrival of computerised data acquisition systems. In the hands of an experienced operator, the algorithms available to estimate the frequency response functions and then the natural frequencies, mode shapes and damping ratios are very accurate. Even with modern sophisticated systems errors may still occur.

The data obtained from the structure under test will be used to update the parameters of an analytical model. It is therefore vital to predict and if possible, eliminate the likely errors in the measurements. Errors may be either random or systematic. Random errors may be reduced by careful experimental technique, the choice of excitation method and by averaging the data. Impact excitation such as hammer excitation puts very little energy into a structure and can produce noisy data. Identification, including modal extraction, work satisfactorily providing the noise is random with zero mean and a large quantity of data is used to identify a relatively small number of parameters.

Systematic errors are difficult to remove from the data and are a serious problem in model updating. These errors arise from many sources, including inadequate modelling of the mounting of the club, mass loading due to the accelerometers, a poorly designed stinger and leakage. Two major problems have to be considered in the case of a golf club. The first is that the mass loading of the accelerometer will be significant when the accelerometer is placed on the shaft. Although the accelerometer used only weighs 3.5 g, this compares to a weight per unit length of 0.56 g/cm at the lightest part of the shaft. This mass loading due to the accelerometer makes the accurate measurement of mode shapes on the club very difficult. Using a shaker to apply the excitation force can add stiffness to the structure. Only the axial force from the shaker is measured and the assumption is that no other force is transmitted to the club. A stinger is placed between the shaker and the club, with the properties that stinger is relatively stiff axially, but flexible in bending. Since the shaft is very flexible, and many modes involve significant bending, the stinger requirements are difficult to apply. An alternative is to excite the club using an instrumented hammer. This works well at the head, although the flexibility within the shaft makes triggering the analyser very difficult on the shaft.

Errors in the measured data can never be eliminated. The estimates of the natural frequencies are usually very good, whereas mode shape and damping estimates usually contain significant noise. Although the individual elements of the mode shape vector contain relatively high levels of noise, the general shape of the mode will be quite accurate. The object is to reduce the effect of these errors by good experimental technique. There is no substitute for high quality measured data.

3.2 Measurements from a golf club

The grip was removed from the club, which was then clamped using a purpose made block. This block consisted of a hole only slightly larger than the shaft diameter and a slot to allow the club to be rigidly clamped. The club was excited using a hammer excitation at the head and by measuring the response at the head. The impact and response measurements were taken in the same direction during each test and the tests were repeated in two orthogonal directions. The club was arranged so that the vertical, or z, direction was approximately parallel to one of the principal axes of inertia of the head. This would decouple the vertical modes from the horizontal / torsion

modes. Of course the principal axis may only be approximated, and the alignment error will cause excitation of the vertical modes when exciting the club horizontally, and vice-versa. In practice the out of plane modes were visible as low level peaks in the frequency response functions and are easily identified. Frequency bandwidths of 10 Hz, 100 Hz and 500 Hz were used; the lower frequency range was required to accurately estimate the first two modes at around 4.5 Hz. The natural frequencies were easily estimated directly from the frequency response functions since the club is very lightly damped. In each direction the modes are well separated and so the experimental and analytical modes may be paired. More formal methods of pairing modes, such as the modal assurance criterion (MAC, Allemang and Brown, 1982) could not be employed because of the difficulty in obtaining mode shapes due to the mass loading of the accelerometer. Table 2 shows the first 10 natural frequencies and a comparison with the finite element estimates.

4 FINITE ELEMENT MODEL UPDATING

As demonstrated in the previous sections, the dynamic response of the club is not identical to the predicted response. Ideally, if the measurements are accurate, the uncertain parameters of the numerical model should be changed to more closely reflect the properties of the physical structure. This is model updating. Mottershead and Friswell (1993) gave an extensive survey of the field and Friswell and Mottershead (1995) review most the popular methods in detail.

4.1 The uncertain parameters and the reduced model

It should be possible to model the shaft reasonably accurately, given its diameter and thickness as they change along the shaft. The primary unknowns will be the inertia properties of the head, and to a lesser degree the mass of the head. Thus the stiffness matrix of the club will remain unchanged as the unknown parameters are varied. Further the mass matrix may be written in the following form

$$\mathbf{M} = \mathbf{M}_0 + \delta m \mathbf{M}_m + \delta I_{xx} \mathbf{M}_{xx} +$$

$$\delta I_{yy} \mathbf{M}_{yy} + \delta I_{zz} \mathbf{M}_{zz} + \qquad (4.1)$$

$$\delta I_{xy} \mathbf{M}_{xy} + \delta I_{xz} \mathbf{M}_{xz} + \delta I_{yz} \mathbf{M}_{yz}$$

where δm is the change in head mass, $\mathbf{M}_0$ is the mass matrix with the initial estimate of the head mass and inertia included,

$$\begin{bmatrix} \delta I_{xx} & \delta I_{xy} & \delta I_{xz} \\ \delta I_{xy} & \delta I_{yy} & \delta I_{yz} \\ \delta I_{xz} & \delta I_{yz} & \delta I_{zz} \end{bmatrix}$$

is the change in the inertia matrix of the head and the other matrices are given by, for example,

$$\mathbf{M}_{xx} = \frac{\partial \mathbf{M}}{\partial I_{xx}} \qquad (4.2)$$

and include either one (for the moments of inertia), two (for the cross products of inertia) or three (for the mass) unity elements, the other elements being zero. The unknown parameters are written as the parameter vector

$$\theta = \begin{Bmatrix} \theta_1 \\ \theta_2 \\ \theta_3 \\ \theta_4 \\ \theta_5 \\ \theta_6 \\ \theta_7 \end{Bmatrix} = \begin{Bmatrix} \delta m \\ I_{xx} \\ I_{yy} \\ I_{zz} \\ I_{xy} \\ I_{xz} \\ I_{yz} \end{Bmatrix} \qquad (4.3)$$

The mass matrix may then be written in the standard form

$$\mathbf{M} = \mathbf{M}_0 + \sum_{i=1}^{7} \theta_i \mathbf{M}_i \qquad (4.4)$$

where the matrices $\mathbf{M}_i$ may be obtained from equations (4.1) and (4.3).

Once the mass matrix is in the form of equation (4.4) we can reduce the size of the mass and stiffness matrices. This is vital to ensure the efficient application of iterative updating methods. For example, if we apply the transformation matrix $\mathbf{T}$, which is generally a rectangular matrix, and often consists of the lower modes of the golf club, then the reduced mass and stiffness matrices are

$$\mathbf{K}_R = \mathbf{T}^T \mathbf{K} \mathbf{T}$$

$$\qquad (4.5)$$

$$\mathbf{M}_R = \mathbf{T}^T \mathbf{M}_0 \mathbf{T} + \sum_{i=1}^{7} \theta_i \mathbf{T}^T \mathbf{M}_i \mathbf{T}$$

$$= \mathbf{M}_{R0} + \sum_{i=1}^{7} \theta_i \mathbf{M}_{Ri}$$

The full length modes shapes for a non-zero parameter vector may be recovered from the modes of the reduced model using the transformation matrix.

4.2 A sensitivity analysis

If measured eigenvalues (natural frequency squared) only are used for updating then the sensitivity matrix is easily calculated using the eigenvalue derivatives (Fox and Kapoor, 1968). The derivative of the i th eigenvalue, λ_i, with respect to the j th parameter, θ_j, is

$$\frac{\partial \lambda_i}{\partial \theta_j} = \phi_{Ri}^T \left[\frac{\partial \mathbf{K}_R}{\partial \theta_j} - \lambda_i \frac{\partial \mathbf{M}_R}{\partial \theta_j} \right] \phi_{Ri}$$

$$\qquad (4.6)$$

$$= \phi_{Ri}^T \left[\mathbf{K}_{Rj} - \lambda_i \mathbf{M}_{Rj} \right] \phi_{Ri}$$

where ϕ_{Ri} is the i th eigenvector of the reduced model. The derivatives of the natural frequencies in Hertz may be obtained from equation (4.6) as

$$\frac{\partial \omega_i}{\partial \theta_j} = \frac{1}{8\pi^2 \omega_i} \frac{\partial \lambda_i}{\partial \theta_j}. \qquad (4.7)$$

For updating the eigenvalues will be used rather than the natural frequencies.

Table 3 shows the sensitivity matrix for the first 8 natural frequencies to the 7 uncertain parameters, based on the initial finite element model, and calculated using equation (4.7). Of particular note are the zero elements in this matrix. The natural frequencies are insensitive to the inertia terms I_{xy} and I_{yz} because of the symmetry imposed by the z axis being coincident

Table 3 Sensitivities of the Natural Frequencies to the Mass and Inertia of the Head

Mode Number	vibration plane	Sensitivity with respect to						
		m	I_{xx}	I_{yy}	I_{zz}	I_{xy}	I_{xz}	I_{yz}
1	z	-7.4	0	-3.5×10^1	0	0	0	0
2	y/ϕ_x	-7.4	-1.9×10^{-3}	0	-3.5×10^1	0	-5.2×10^{-1}	0
3	y/ϕ_x	-4.3×10^{-5}	-2.7×10^4	0	-1.3×10^4	0	-3.8×10^4	0
4	z	-7.6×10^{-2}	0	-2.5×10^4	0	0	0	0
5	y/ϕ_x	-3.5	-2.4×10^5	0	-8.3×10^3	0	8.9×10^4	0
6	z	-2.7	0	-3.9×10^4	0	0	0	0
7	y/ϕ_x	-3.6	-1.3×10^6	0	-2.7×10^5	0	1.2×10^6	0
8	z	-4.5	0	-1.7×10^4	0	0	0	0

with a principal axis of inertia of the head. Furthermore, we will assume that we have the z axis aligned correctly so that these parameters are forced to remain zero. The first bending modes in both directions are relatively insensitive to the inertia terms since these modes do not involve much rotation at the club head. The natural frequencies of the vertical (z direction) modes are only sensitive to I_{yy} and the horizontal / torsional modes are only sensitive to I_{xx}, I_{zz} and I_{xz}. Thus the dynamics of the 2 planes decouple, although remember that this occurs because of the alignment of the z axis.

4.3 The weighted least squares method

Friswell and Mottershead (1995) described a large number of updating techniques. The weighted least squares method will be used in this work. The objective of the method is to improve the correlation between the measured data and the analytical model, given by a penalty function involving the modal data. For example, if the natural frequencies alone are used then this may be the sum of squares of the difference between the measured and estimated eigenvalues. Because of the nature of this penalty function, the solution requires the problem to be linearised and thus optimised iteratively. The method allows a wide choice of parameters to update and both the measured data and the initial analytical parameter estimates may be weighted. This ability to weight the different data sets gives the method its power and versatility but requires engineering insight to provide the correct weights. Often the weighting matrices are chosen as the inverse of the corresponding estimated variances.

There are problems relating the measured data and the corresponding analytical estimates. The natural frequencies and mode shapes in the experimental and theoretical data must relate to the same mode, that is they must be paired correctly. Arranging the natural frequencies in ascending order of magnitude is not sufficient, especially when two modes are close together in frequency as they will be with a golf club. The experiment has highlighted how the measured natural frequencies were matched up to the mode shapes. Generally mode shape data contains far more errors than the natural frequencies, and so the information lost by not measuring mode shapes is relatively small. Also the mode shapes are not very sensitive to changes in the updating parameters. For example, the first mode of a golf club will be, approximately, the standard first bending mode of a beam, for a large range of shaft cross sections and head properties.

The weighted least squares algorithm can explicitly weight the parameter changes in a penalty function as well as the errors in the measurements. The method minimises the penalty function

$$J(\delta\theta) = \varepsilon^T \mathbf{W}_{\varepsilon\varepsilon} \varepsilon + \delta\theta^T \mathbf{W}_{\theta\theta} \delta\theta \qquad (4.8)$$

where $\varepsilon = \delta\mathbf{z} - \mathbf{S}\delta\theta$ is the error in the first order Taylor series of the measurements, $\delta\mathbf{z}$ is the error in the measurements, $\mathbf{S}$ is the sensitivity matrix and $\delta\theta$ is the change in the parameter estimate from the current iteration to the next. Minimising this penalty function with respect to $\delta\theta$ gives the change in parameters as

$$\delta\theta = \left[\mathbf{S}^T\mathbf{W}_{\varepsilon\varepsilon}\mathbf{S} + \mathbf{W}_{\theta\theta}\right]^{-1} \mathbf{S}^T \mathbf{W}_{\varepsilon\varepsilon} \delta\mathbf{z}. \quad (4.9)$$

The penalty function, equation (4.8), limits the change in $\delta\theta$ at every iteration, but does not limit the overall change in the parameters. Equation (4.8) may be easily changed so that the second term weights the difference between the parameter at the next iteration and the original parameter estimate. If the number of measurements is greater than the number of parameters, then an over-determined problem may be obtained by not weighting the parameters at all, i.e. $\mathbf{W}_{\theta\theta} = \mathbf{0}$.

4.4 *Updating a golf club model*

Only the measured eigenvalues (natural frequency squared) will be used to update the golf club model by equation (4.8). The sensitivity matrix is easily calculated using the eigenvalue derivatives, equation (4.6), with a model reduced using the first 25 modes of the initial model. The updating task will be separated into three parts. First the mass will be updated using the first bending modes in the vertical and horizontal directions. Next the inertia term I_{yy} will be updated using the second vertical bending mode. Finally the inertia terms I_{xx}, I_{zz} and I_{xz} will be updated using the second, third and fourth horizontal / torsional modes. In the second and third cases there are the same number of parameters as measurements. Assuming that the natural frequencies are sensitive to the parameters, and the parameters have been chosen with this in mind, the updating algorithm should be able to reproduce the measurements for some choice of parameters. In this case, if the parameter weighting matrix is zero, $\mathbf{W}_{\theta\theta} = \mathbf{0}$, then the value of the measurement weighting matrix, $\mathbf{W}_{\varepsilon\varepsilon}$, is irrelevant. Our estimates of the inertia matrix are so uncertain that the zero parameter weighting matrix is a sensible choice, in this case. The measurement weighting matrix for the updating of the head mass is the identity matrix so that both natural frequencies are weighted equally.

The updated value of the mass of the club head is 237 g, and the updated inertia matrix is

$$
\begin{bmatrix}
1.86 & 0 & 1.21 \\
0 & 5.87 & 0 \\
1.21 & 0 & 3.20
\end{bmatrix} \times 10^{-4} \text{ kg m}^2.
$$

The updated natural frequencies are given in Table 2. Notice that the first two natural frequencies are improved, but are reproduced exactly. This is because a single parameter, the head mass, has been updated using two frequencies. The resulting updated natural frequencies are an average of the measured frequencies. The natural frequency of the second vertical bending mode has been reproduced exactly, although the frequency of the third and fourth modes have not been improved. The error in these frequencies indicate that there are some errors that have not been parameterised in the model. The fifth horizontal / torsional mode has improved considerably after updating, although some error is still present. The natural frequency of the first axial mode has improved considerably.

5 CONCLUSION

The golf club is a fascinating structure. The shaft is symmetrical, which would produce repeated natural frequencies if tested in isolation. The addition of the asymmetrical head causes the natural frequencies to separate, and to couple vibration in bending and torsion. By choosing a frame of reference so that one axis is aligned with the principal axis of inertia of the head the bending vibration in one plane is decoupled from the torsion. This decoupling is a vital aid to inferring the measured mode shapes. The usual modal analysis techniques using a roving accelerometer or roving hammer excitation, are impractical on the golf club. Strain gauges may be suitable and this is the subject of further work.

The natural frequencies from the initial finite element model showed considerable errors when compared to measured frequencies. By updating the mass and inertia properties of the head the agreement between the measurements and the analytical model is considerably improved.

REFERENCES

Allemang, R.J. and Brown, D.L., 1982, A Correlation Coefficient for Modal Vector Analysis, *1st International Modal Analysis Conference,* Orlando, Florida, 110-116.

Bedford, A. and Fowler, W., 1995, *Engineering Mechanics: Dynamics,* Addison-Wesley Publishing Company.

Brylawski, A.M., 1994, An Investigation of Three Dimensional Deformation of a Golf Club During Downswing, in *Science and Golf II: Proceedings of the World Scientific Congress of Golf,* eds. A.J. Cochran and M.R. Farrally, 265-270.

Dawe, D.J., 1984, *Matrix and Finite Element Displacement Analysis of Structures,* Oxford University Press.

Ewins, D.J., 1984, *Modal Testing: Theory and Practice,* Research Studies Press Ltd., John Wiley.

Fox, R.L. and Kapoor, M.P., 1968, Rates of Change of Eigenvalues and Eigenvectors, *AIAA Journal,* **6**(12), 2426-2429.

Friswell, M.I. and Mottershead, J.E., 1995, *Finite Element Model Updating in Structural Dynamics,* Kluwer Academic Publishers.

Horwood, G.P., 1994, Golf Shafts - A Technical Perspective, in *Science and Golf II: Proceedings of the World Scientific Congress of Golf,* eds. A.J. Cochran and M.R. Farrally, 246-258.

Iwata, M., Okuto, N. and Satoh, F., 1990, Designing of Golf Club Heads by Finite Element Method (FEM) Analysis, in *Science and Golf: Proceedings of the World Scientific Congress of Golf,* ed. A.J. Cochran, 274-279.

Johnson, S.H., 1994, Experimental Determination of Inertia Ellipsoids, in *Science and Golf II: Proceedings of the World Scientific Congress of Golf,* eds. A.J. Cochran and M.R. Farrally, 290-295.

Mather, J.S.B, 1996, The Role of Club Response in the Design of Current Golf Clubs, *14th International Modal Analysis Conference,* Dearborn, Michigan, 397-403.

Mitchell, S. Newman, S.T., Hinde, C.J. and Jones, R., 1994, A Design System of Iron Golf Clubs, in *Science and Golf II: Proceedings of the World Scientific Congress of Golf,* eds. A.J. Cochran and M.R. Farrally, 390-395.

Mottershead, J.E. and Friswell, M.I., 1993, "Model Updating in Structural Dynamics: A Survey," *Journal of Sound and Vibration,* **167**(2), 347-375.

Swider, P., Ferraris, G. and Vincent, B, 1994, Theoretical and Experimental Dynamic Behavior of a Golf Club Made of Composite-Material, *Modal Analysis: The International Journal Of Analytical And Experimental Modal Analysis,* 9(1), 57-69

The Engineering of Sport, Haake (ed.) © 1996 Balkema, Rotterdam. ISBN 90 5410 822 3

Engineering 'feel' in the design of golf clubs

Alan Hocknell, Roy Jones & Steve Rothberg
Loughborough University, Leicestershire, UK

ABSTRACT: For the competent golfer, 'feel' in the golf shot is an important but ill-defined concept. A survey of competent players' perceptions of the golf shot indicated that the trajectory of the ball in flight, the sensation in the hands and, most importantly, the sound of the impact are the three main components of 'feel'. The input to the golfer's hands and ears is due to vibrations set-up in the club and in the air during the club-ball impact. The vibration amplitude and frequency content of the signals reaching the hands and ears were measured for a variety of club-ball combinations in tests carried out using a golf robot. A group of 8 low handicap golfers were asked to use the same equipment and to state their perception of the 'feel' in each case. Whilst there was some obvious subjectivity in their replies, comparison with the results from the robot tests demonstrated that the vibration measurements could be used to provide a quantitative description of two important aspects of 'feel'. Of particular interest was the poor 'feel' associated with modern, hollow metal drivers. Using the experimental data, the features of the club design responsible for poor 'feel' could be identified. With this knowledge, the club designer can attempt to alter the 'feel' characteristics of the club within the constraints of other club performance criteria.

1 INTRODUCTION

'Feel' is an extensively used term which is taken to encompass the physical and psychological feedback experienced by a golfer in playing a golf shot. Physical feedback to the hands, eyes and ears is influenced by the biomechanics of the swing in terms of rhythm, tempo and grip pressure, and by the equipment used, which varies mostly in terms of the materials employed in its construction. The golf impact event, however, has a duration of between 450μs and 500μs and is thus too short for the golfer to react to this feedback in a single swing. 'Feel' therefore has a psychological element, whereby the golfer becomes increasingly programmed to expect certain physical feedback. Further psychological influences on 'feel' include the golfer's levels of confidence and concentration, both of which are external to the golf swing itself.

This experimental study is concerned primarily with the capture and analysis of data relevant to the physical feedback experienced by a golfer during and immediately following the impact between a golf club and ball. The approach taken is to identify and obtain measurements of the physical

quantities which most influence the 'feel' of a golf impact. These measurements, obtained for two golf clubs with different club head constructions, are then compared against competent golfers' perceptions of 'feel' when using the same equipment in order to identify measurements corresponding to good and bad 'feel'. Finally, the experimental data is used to identify the features of the two club designs which influence the production of good and bad 'feel' characteristics. The results show quantitatively how the club designer can potentially modify designs to improve 'feel', other club performance criteria permitting.

2 MEASUREMENT OF 'FEEL'

The analysis of 'feel' is complicated by its subjective nature and by the vague language which to date has been the main descriptor of physical input to the human nervous system. However, a survey of low handicap and professional golfers, conducted as part of this study, identified three main components of 'feel'. These are the trajectory of the ball in flight, the sensation in the hands and, most importantly, the sound of the impact. In this study, the physical feedback from the equipment to the golfer's hands and ears is of primary interest. This input is in the form of mechanical vibrations excited in the club and in the air during the club-ball impact. Measurements of these vibrations are used to quantify the sensation in the golfer's hands and the sound of the impact in terms of the amplitude and frequency content of the vibration.

Vibration excited in the golf club head during impact is transmitted along the shaft to the golfer's hands. This vibration can be detected with a piezo-electric accelerometer attached to the shaft immediately below the grip. The accelerometer is aligned in the direction of the impact and is thus predominantly sensitive to accelerations occurring as a result of lateral and torsional vibration of the shaft. Acceleration was chosen as the most convenient descriptor of the shaft motion to facilitate investigation of disturbances on top of a fairly large steady velocity. Sound generated by surfaces of the club and ball vibrating as a result of the impact can be detected using a sound meter positioned close to the impact site. Output from both instruments was stored on a digital oscilloscope at a high sample rate.

3 EXPERIMENTAL METHOD

The two golf clubs investigated in this study are a modern hollow metal driver and a traditional persimmon wood. These clubs, which will be referred to as types 'M' and 'W' respectively, were chosen as examples of club head constructions having different 'feel' characteristics. Both were fitted with the same regular steel shaft and grip. Three golf ball types, also known to have different 'feel' characteristics, were tested with each club. The ball types differ in materials and construction and were chosen to represent those in common usage. Type 'A' has a two-piece construction comprising a solid polymer core and a tough ionomer cover. Type 'B' has a three-piece construction comprising a smaller diameter solid polymer core surrounded by a synthetic rubber thread winding and an ionomer cover. Type 'C' has a liquid filled core, similar winding to type 'B' and a soft balata rubber cover.

A hydraulic robot golfer was used to swing the two golf clubs. The accuracy and repeatability of the robot swing was essential in obtaining accelerometer and sound meter output for each club-ball combination under equivalent conditions, thus affording reliable comparison of equipment in terms of signal amplitude and frequency content. The same equipment was used by a group of eight low handicap golfers. Accelerometer and sound meter outputs were again recorded, but the relative amplitudes of signals recorded from different club-ball combinations could not be compared as the impact velocity generated by even the best golfers varies significantly between shots. However, the frequency content of the signals was used to verify results from the robot. The golfers were asked to describe the perception of physical feedback to the hands and ears for each shot. Recognising the subjective nature of the eight golfers' responses, a wider survey of golfers' perceptions of 'feel' was carried out by means of a questionnaire completed by a total of thirty low handicap golfers and club professionals prior to the experiments. An average opinion of club-ball combinations considered to produce good and bad 'feel' was thus constructed and compared against data from the accelerometer and sound meter.

4 COMPARISON OF PERCEIVED 'FEEL' AND MEASURED VIBRATION

The results of the study of golfers' perceptions of 'feel' are summarised in Figure 1. The figure shows the average opinion of the equipment tested in terms of the relative hardness of the impact on the golfers' hands and the relative pitch of the sound made by the impact. The letters 'WA' etc. in the figure denote the equipment combination of club type 'W' and ball type 'A'.

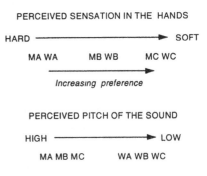

Figure 1. Summary of results from the 'feel' perception study.

The sensation in the golfer's hands resulting from the impact is perceived to be softer for the club-ball combinations MC and WC than for other combinations. The study also suggests that the softest combination, WC, also provides the most feedback and is preferred in general. The sound generated by the impact is perceived to be of significantly higher pitch for impacts involving club type 'M', with MA being the highest pitch. The lower pitch of sounds from club type 'W' are generally preferred, but a significant percentage of the participants found the higher pitched sounds from club type 'M' acceptable. Ball type 'B' was found to have properties intermediate between types 'A' and 'C' and will be omitted from further discussions to improve clarity.

Measurements of shaft vibration resulting from the impact are shown for club-ball combinations MA, MC, WA and WC in Figure 2. It is clear from the figure that the vibration amplitude measured immediately below the grip is marginally greater for combinations involving ball type 'C' than for ball type 'A'. The figure also suggests that vibration amplitudes are very similar for both club types.

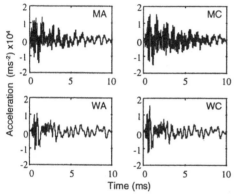

Figure 2. Accelerometer measurements of post-impact shaft vibration for each equipment combination.

The vibration signals measured on club type 'M' contain high frequency information in the range 5-8kHz which will be discussed later. However, vibration of the shaft in the frequency range 0-2.5kHz is believed to be of greater importance to the sensation in the golfer's hands. Power spectral density estimates for each club-ball combination are presented in Figure 3. These show the extent to which vibrations in the frequency range 0-2.5kHz are excited in the shaft as a result of the club-ball impact. The frequencies at which the peaks occur are in agreement with the general form of previously reported frequency spectra taken from accelerometer measurements on the shaft of a single club-ball combination (Hedrick & Twigg, 1995) and correspond to natural modes of vibration of the shaft-club head system. Figure 3 shows that the WC combination excites the natural modes of vibration in the range 500Hz-2.5kHz more than other equipment.

In particular, the modes around 700Hz, 1kHz and 2kHz are excited most by WC and progressively less by the MC, WA and MA combinations. This progression from WC to MA is in agreement with the trend of reducing perceived softness of the sensation in the golfer's hands and the perception that the WC combination provides the most feedback.

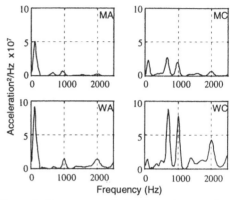

Figure 3. Power spectral density estimates of post-impact shaft vibration for each equipment combination.

Figure 3 additionally shows a lower mode of the shaft-club head system at around 100Hz which is excited strongly by the type 'A' ball and less by type 'C'. This is believed to contribute to the harder sensation in the hands perceived by golfers using the type 'A' ball. It is therefore suggested that a desirable sensation in the hands can be achieved by exciting modes of vibration of the shaft-club head system in the frequency range 500Hz-2.5kHz more strongly than modes in the region of 100Hz.

Power spectral density estimates of the sound produced by the impact are shown for the club-ball combinations MA, MC, WA and WC in Figure 4. An important feature of the sound spectrum produced by impacts involving club type 'M' is the presence of frequencies in the range 5kHz-11kHz. These high frequencies are absent from the corresponding spectra of club type 'W' and it is therefore suggested that they are responsible for the less preferable higher pitch sound perceived by golfers using club type 'M'. Figure 4 also shows that, for each club, there is little difference in the sound spectra produced by impacts involving either ball type. This is in agreement with golfers' perceptions that the club type is most important in determining the sound characteristics of the impact. It is additionally noted that frequencies present in all of the sound spectra in the range 0-2.5kHz appear unrelated to the modes of vibration of the shaft shown in Figure 3

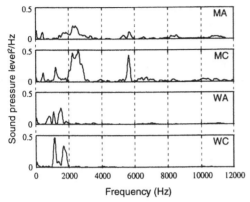

Figure 4. Power spectral density estimates of the sound produced by impacts involving each equipment combination.

The agreement between perceived 'feel' and measurements of vibration shows that data from an accelerometer mounted on the shaft immediately below the grip and a sound meter positioned close to the impact site can be used to provide a quantitative description of two important aspects of 'feel'.

5 RELATIONSHIP TO GOLF EQUIPMENT DESIGN

Measurements which describe 'feel' quantitatively can be further analysed to determine features of the equipment design responsible for the production of good and bad 'feel' characteristics.

The frequency spectra of the vibration measured on the shaft of both club types, shown in Figure 3, contain peaks at similar frequencies in the range 500Hz-2.5kHz. This is to be expected, as the shaft-club head systems have similar masses and stiffness and these frequencies correspond to the cantilever modes of vibration of each system. There are three possible reasons for the different level of excitation of these modes measured for each club-ball combination. Firstly, ball type 'C' is known to be in contact with the club face for approximately 500μs as opposed to the shorter contact of 450μs for ball type 'A'. The impact force therefore exists for a longer time with ball type 'C' and this larger impulse may give rise to greater shaft vibration in the frequency range 500Hz-2.5kHz than occurs with ball type 'A'. Secondly, the solid wooden club head structure may be stiffer than the thin walled hollow metal club head, in which case greater transmission of vibration from the club head to the shaft could occur with club type 'W'. Finally, the vibration

energy given to the type 'W' club head during the impact is contained within the frequency range 0-3.5kHz. In the type 'M' club, vibration of the hollow club head in the frequency range 5-11kHz indicates that a percentage of the total vibration energy is contained at higher frequencies, potentially reducing transmission to the shaft in the frequency range 500Hz-2.5kHz. The higher frequency vibration of the hollow club head excites vibrations at distinctly similar frequencies in the shaft of club type 'M'. These higher frequency shaft vibrations are apparent upon comparison of the frequency spectra measured on the shafts of club-ball combinations MA and WA, as in Figure 5. The extent to which high frequency vibrations in the shaft contribute to golfers' perceptions of 'feel' is currently undetermined.

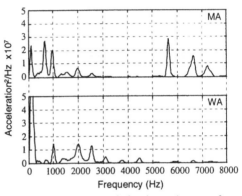

Figure 5. Power spectral density estimates of post-impact shaft vibration for two equipment combinations.

As mentioned earlier, there is little agreement between the frequencies of vibration of the shaft and frequencies present in the sound spectra produced by the impact. Additionally, the sound spectra in Figure 4 show that impacts involving the solid club head, type 'W', do not produce sound frequencies above 3.5kHz and that there is a gap in the sound spectrum between 3.5kHz and 5kHz for impacts involving the type 'M' club. Thus, it is believed that components of the sound spectra in the frequency range 0-3.5kHz are generated mostly by vibration of the ball.

Confirmation that frequency components in the range 5-11kHz, produced by impacts involving club type 'M', are due to vibration of the hollow club head was obtained by a further experiment in which a vibration excitation was applied to a stationary 'M' type club. The vibration excitation was provided by a piezo-ceramic tile attached to the club head and the response of the club head was measured using two optical transducers. Single frequency excitation of the club head over the range 0-20kHz allowed the natural frequencies and corresponding modeshapes of vibration of the hollow head to be observed using a technique known as Electronic Speckle Pattern Interferometry. Broadband excitation of the club head facilitated measurement of the frequency response functions of points on the surface of the hollow club head using a laser Doppler vibrometer. The frequency response function of a point on the face of the hollow club head is shown in Figure 6, where it can be seen that the first distinct mode of vibration appears at around 5.7kHz. This provides verification of the argument that sound frequency components below 5.7kHz are not due to vibration of the club head.

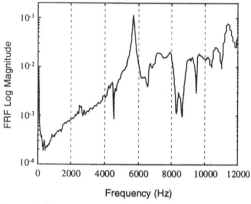

Figure 6. Frequency response function measured on the face of a type 'M' club head.

By comparison with Figure 4, it can be seen that the first mode of vibration of the head contributes significantly to the frequency spectrum of the sound produced by impacts involving club type 'M'. Other, higher modes of club head vibration are apparent in the sound frequency spectrum to a lesser extent. Thus it is concluded that vibration of the hollow club head is responsible for the production of sound frequencies greater than 5kHz. Frequency response functions measured elsewhere on the club head have similar magnitude to that shown in Figure 6, suggesting that vibrations of the face, top surface and sole of the hollow club head all contribute to the sound of the impact.

6 CONCLUSION

This study has shown that measurements taken with an accelerometer attached to the shaft of a golf club immediately below the grip and a sound meter positioned close to the impact site can be used to provide a quantitative description of two important aspects of 'feel'. This quantitative description is in agreement with the 'feel' perceived by a group of low handicap and professional golfers.

Analysis of the measurements shows that the sensation in the hands is dominated by the natural frequencies of vibration of the shaft-club head system in the range 0-2.5kHz and that the club head and ball combination influences the excitation of these natural frequencies. The sound of the impact is divided into two distinct frequency ranges. Sound in the frequency range 0-3.5kHz is generated mostly by the ball and is apparent in impacts involving all club head and ball combinations. Sound in the frequency range 5-11kHz is generated by vibration of hollow metal club heads and is thus absent from impacts involving solid club heads.

Whilst it is recognised that the measurements taken in this study are relevant to only one part of the total 'feel' concept, it is believed that quantitative information of the type presented in this paper can assist equipment designers in engineering 'feel' in the design of golf clubs.

REFERENCE

Hedrick,M; Twigg,M (1995) : The feel of a golf shot: Can we measure it? In: Golf the Scientific Way. (Ed: Cochran,AJ), Aston Publishing Group, UK, 131-133.

This study was carried out as part of an Engineering and Physical Sciences Research Council grant investigating the design analysis of sculptured surface products. The authors would like to acknowledge the assistance of Dunlop Slazenger International Limited.

The Engineering of Sport, Haake (ed.) © 1996 Balkema, Rotterdam. ISBN 90 5410 822 3

Cricket bat design and analysis through impact vibration modelling

S. Knowles, J. S. B. Mather & R. Brooks
Department of Mechanical Engineering, University of Nottingham, UK

ABSTRACT: Modern sports equipment, for games such as baseball and golf is the subject of increasing amounts of research. Such research, together with the use of stiff, light-weight composite materials has spawned many novel features and designs. One sport which has seen little such development is cricket.

Modal analysis has proved a particularly useful tool in measuring the performance of sports equipment and can be applied to cricket bats to give the response of the bat to a given input. During impact, however, the vibrational response has a major effect upon the force and duration of contact. A computer model has been developed to simulate the case of a ball impacting upon a cricket bat. The use of composite materials increases the scope of the designer and makes the impact and vibrational analysis all the more important. By altering the design and material properties the vibrational response of the bat can be 'tuned' to achieve increased performance.

Impact tests performed on two different cricket bats are used to verify the computer predictions. These two sources of data can be used as an effective tool in the design and analysis of a high performance composite bat.

NOMENCLATURE

A	beam cross-sectional area
α	approach of bat and ball ($= w_2 - w_1$)
E	Young's modulus
e	coefficient of restitution (= $-v_{out}/v_{in}$)
F	contact force
I	second moment of area
k_2	Hertz contact constant
L	beam length
λ	eigenvalue where $\lambda^4 = \rho A \omega^2 / EI$
m	mass
ρ	beam density
Q	factor for evaluating ω_i for beams with nonclassical boundary conditions
t	time
T	torsional stiffness
T^*	TL/EI
$\Delta\tau$	time interval used in numerical calculations
$\bar{\tau}$	variable of integration
v_o	initial velocity
w	deflection
ω	undamped natural frequency
x	distance along length of the beam
X	mode shape function of the beam

Subscripts

1	bat
2	ball
i	representing the ith mode
n	representing the nth time interval
SF	representing a simple/free beam
CF	representing a clamped/free beam

INTRODUCTION

The increasing demand among consumers for the latest high-performance sports equipment is fuelling scientific and engineering research by sports equipment manufacturers. Such research, together with the use of stiff, light-weight composite materials has spawned many novel features and designs, particularly for golf clubs and baseball bats. One sport which has seen little such development is cricket. The techniques and results outlined in this paper could also be applied to other types of sports, particularly baseball bats.

In order to understand the factors affecting bat performance an understanding of the mechanics of impact are required. In the simplified case a cricket bat and ball can be modelled as a sphere impacting upon a beam. This analysis uses a combination of contact and vibration theory [1], however the problem cannot be accurately defined using classical boundary conditions.

The batsman's hands can be regarded as a clamp and the bat can be modelled as a beam attached to a torsion spring representing a relatively flexible handle opposite a free end [2].

Modifying the impact theory [1] to take account of the non-classical boundary conditions [2] gives a model for the impact of a cricket ball on a bat which can be programmed into a computer. This model can then be used to assess the performance of various bat designs.

In order to validate the model, experimental tests have been performed on two different bat designs and deflection measurements have been taken at various impact points.

Modal analysis as used on baseball bats to assess the size and position of the "sweet spot" [3], can also be applied to a cricket bat. This paper also uses modal analysis to assess the accuracy of the impact response theory and to validate the experimental results.

THEORY

By combining the general equation for forced vibration of a continuous beam [4] with the Hertz contact theory, a generalised equation can be derived. This links the forcing function to the beam's vibrational response [1] for impact at some point x = c.

$$\alpha = \left[\frac{F}{k_2}\right]^{\frac{2}{3}} = v_o t - \frac{1}{m_2} \int_o^t dt \int_o^t F \, dt$$

$$- \frac{1}{\rho A} \sum_{j=1}^{\infty} \frac{[X_i <c>]^2}{\omega_i \int_o^t X^2 dx} \int_o^t F<\bar{\tau}> \sin \omega_i \, (t-\bar{\tau}) \, d\bar{\tau} \quad (1)$$

The above equation, containing the force function on both sides, does not have an analytical solution. A numerical solution, however may be found by using the small increment method, in which the contact force is regarded as constant over any time increment $\Delta\tau$.

For the nth time interval t = $n\Delta\tau$, equation (1) can be written as

$$\alpha_n = \left[\frac{F_n}{k_2}\right]^{\frac{2}{3}} = v_o n \Delta\tau - \frac{(\Delta\tau)^2}{m_2} \sum_{j=1}^{n} D_{n-j+1} \, F_j$$

$$- \frac{1}{\rho A} \sum_{j=1}^{\infty} \frac{[X_i <c>]^2}{\omega^2 \int_o^L X^2_i \, dx} \sum_{j=1}^{n} \quad (2)$$

$$F_j \, [\cos \omega_i \, (n-j) \, \Delta\tau - \cos \omega_i \, (n-j+1) \, \Delta\tau]$$

where the term $(\Delta\tau)^2 \sum_{j=1}^{n} D_{n-j+1} \, F_j$ is the

numerical evaluation of $\int_o^t dt \int_o^t F \, dt$

The mode shape function X_i and the natural frequencies ω_i are dependent on the beam 'properties and the boundary conditions. For a beam with a torsion spring at one end and free at the other the mode shape function is given by the following equation [2].

$$X(x) = - (\sin \lambda X + \sin \lambda x)$$

$$+ \left(\frac{\sin \lambda L + \sin \lambda L}{\cos \lambda L + \cos \lambda L}\right) (\cos \lambda x + \cosh \lambda x) \quad (3)$$

The natural frequency ω_i and eigen values λ_i are calculated from the equation

$$T^* = \frac{\lambda L \, (\sin \lambda L \cosh \lambda L - \sin \lambda L \cos \lambda L)}{1 + \cos \lambda L \cosh \lambda L}$$

and lie between the corresponding values for a simple/free beam and a clamped/free beam depending on the torsional stiffness (T = 0 ω = ω_{SF}; T = ∞ ω = ω_{CF}). This natural frequency can thus be expressed as;

$$\omega_j = \omega_{iSF} + Q_i \, (\omega_{iCF} - \omega_{iSF})$$

where Q varies with the torsional stiffness T*. Tables of Q values for a range of values of T* [2] greatly simplify the evaluation of ω_i and λ_i.

These values can then be used in equation (2) to evaluate how the ball and beam deflections, contact force and ball velocity vary with time.

Table 1 - Physical Properties of the Two Test Bats

Properties	Traditional Bat	GRP Composite bat
Overall mass (kg)	0.974	1.625
Overall Blade Density $\rho(kg/m^3)$	440	560
Blade Cross Sectional/Area A (m^2)	3.4×10^{-3}	4×10^{-3}
Blade Flexural Rigidity EI (N/m^2)	3300	4000
Torsional Stiffness of Handle (Nm/rad)	1550	1200

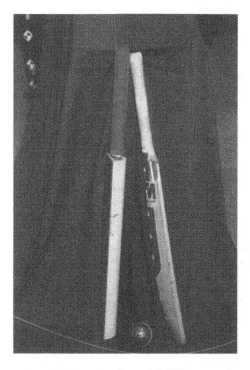

Figure 1. The two test bats - left GRP composite bat, right traditional bat

TESTS

Cricket bat behaviour has been investigated using three different tests:
1. Modal analysis
2. Theoretical computer model
3. Cricket ball impact tests

Two different bats have been studied; a traditional style lightweight willow cricket bat and a rather more heavy glass reinforced plastic (GRP) composite test bat. A photograph of the two bats is shown in Figure 1. Table 1 shows the key properties of the two bats.

The values of EI for the traditional bat are based on an average cross-section along the length of the blade. For the GRP bat with a constant cross-section of several component materials, the overall flexural rigidity is calculated from $(EI)_{bat} = (EI)_1 + (EI)_2 + ...$ The torsional stiffnesses were measured in static deflection tests.

Modal Analysis

For the modal analysis of the two bats a standard impact excitation test was used. The accelerometer was positioned at the bat tip and five hammer impacts were made at each of twenty-three impact points along the blade. For both bats the test was performed with a clamped handle boundary conditions in order that the results could be compared with the other methods.

Theoretical Computer Model

A computer program has been written to solve equation (2) numerically. Results have been obtained for various impact points along the length of the blade at various impact velocities. For each case the results were output in graphical form showing impact point deflection versus time. The summation included the first ten modes although modes four and above make a negligible contribution.

Cricket Ball Impact Test

This test is used to measure the deflections caused by impact of a cricket ball upon the bat. The experimental set up is shown in Figure 2. An accelerometer is attached to the underside of the bat at the impact point for each of the nine positions along the length of the bat. The signal from the accelerometer is fed into a PC via a charge amp and an ADC. The data acquisition is triggered using an infra-red beam and detector. As the ball breaks the beam the data sampling commences. The time between the first reading and the impact is used to determine the impact velocity.

The acceleration/time data is imported into a Microsoft Excel spreadsheet. A numerical integration can then be performed to give first the

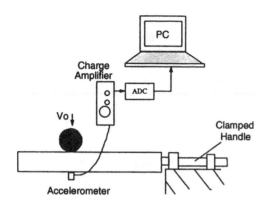

Figure 2. Impact test layout

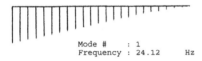

```
Mode #    : 1
Frequency : 24.12    Hz
```

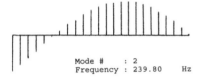

```
Mode #    : 2
Frequency : 239.80   Hz
```

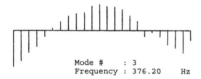

```
Mode #    : 3
Frequency : 376.20   Hz
```

Figure 3. Deflected mode shapes - traditional bat

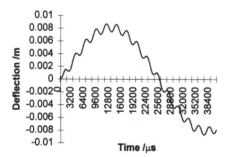

Figure 4. Computer model - impact point deflection, traditional bat

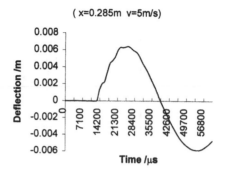

Figure 5. Impact test - impact point deflection, traditional bat

Table 2 - Resonant Frequencies of the Traditional Bat

	Frequency Hz		
	Modal Analysis	Theoretical Model	Impact Test
Mode 1	24.1	20.3	19.6
Mode 2	239.8	379	247
Mode 3	376.2	1100	-

velocity and then a deflection / time curve, showing the magnitude and frequency of the bat oscillations. These measurements are confirmed by a simple pen plot taken at the bat tip. For each impact point several sets of impact data are acquired. The peak bat deflection is also found for each test and an average for each impact point is calculated.

RESULTS

Frequency

Figure 3 shows the deflected mode shapes and the resonant frequencies for the traditional willow cricket bat.

Frequency data was also obtained from the computer model, Figure 4, and the impact test, Figure 5. A comparison of the results from these three methods is given Table 2, overleaf.

It can be seen that the theoretical model is not very accurate at predicting the frequencies of the higher modes. This is due to the non-uniform

Table 3. Resonant Frequencies of the GRP Composite Bat

	Frequency Hz		
	Modal Analysis	Theoretical Model	Impact Test
Mode 1	15.47	14.7	16.5
Mode 2	311	339	326
Mode 3	713	1087	-

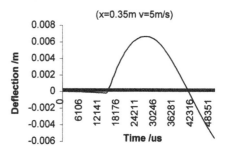

Figure 8. Impact test - impact point deflection, GRP composite bat

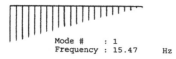

Mode #   : 1
Frequency : 15.47   Hz

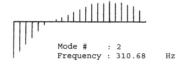

Mode #   : 2
Frequency : 310.68   Hz

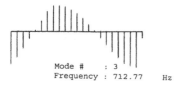

Mode #   : 3
Frequency : 712.77   Hz

Figure 6. Deflected mode shapes - GRP composite bat

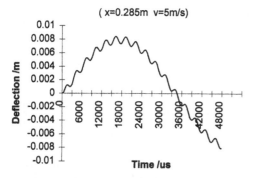

Figure 7. Computer model - impact point deflection, GRP composite bat

cross-section along the length of the willow blade, an estimated average value for the flexural rigidity and cross sectional area being used.

The composite GRP bat does have a constant cross-section and therefore the match between the modal analysis data, the theoretical computer model and the impact test (Figures 6, 7 and 8 respectively) is much closer as can be seen in Table 3.

It should, however, be pointed out that for the third mode the frequency found by the theoretical model differs quite significantly from the experimental value.

Deflection

From the experimental results (Figures 5 & 8) and the theoretical computer model (Figures 4 & 7) for each of the bats the peak deflection at the impact point can be determined. Data was acquired at nine evenly spaced points along the length of the bat (at 65 mm intervals, from 25 mm from the handle to 545 mm). The values of peak deflection are shown in Figures 9 & 10.

It should be noted that these graphs do not represent mode shape as they encompass the contributions of all modes and represent the response of each point to an impact at that particular position.

DISCUSSION

The results of the frequency analysis ie. the comparison between the modal analysis, the theoretical model and the impact tests show very good agreement for the first mode of both bats. It is this mode which dominates the response of the bat, particularly for impact in the end two-thirds of the blade which is the main hitting area.

The second mode of the composite GRP bat is also predicted by the theory with reasonable

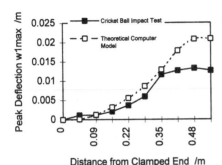

Figure 9. Peak impact - point deflection, traditional bat

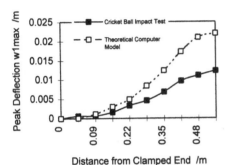

Figure 10. Peak impact - point deflection, GRP composite bat

accuracy with significant errors occurring only at the third mode. The higher modes of the traditional bat are not very accurately predicted due to the non-uniform cross-section of the blade. The theoretical model assumes constant properties along the length of the blade and consequently for other than the first mode where the bending of the handle is dominant, the experimentally recorded frequencies vary significantly from those predicted by the theory.

From Figure 10 it is evident that the theoretical model is consistently over estimating the bat deflection by a factor of around 1.8 although the shape of the two curves matching very closely. A similar over estimate (1.65) exists for the impact at the tip of the traditional bat. For intermediate points (0.155 to 0.35), however, the theoretical model and the test results are in good agreement. Again this variation is likely to be caused by the assumption of a uniform cross-section beam in the theory.

The fact that the theoretical model over estimates the impact point deflection implies that in practice some of the energy which the theory predicts as going into the transverse vibration of the bat, is being lost.

The theory based around equation (2) assumes all the kinetic energy initially in the ball is transferred into the bat vibrations and the outward velocity of the ball. This neglects the effects of vibrations and energy absorbtion within the ball. In practice some of this energy is lost as heat and noise during the impact resulting in less energy being transferred to the bat and ball after impact. A cricket ball is certainly not a purely elastic body and would not, therefore, be expected to undergo impact without some energy loss. In order to assess the effect of these losses a brief set of impact tests were performed using a much softer 'bouncier' ball.

Effects of Ball Elasticity

A very soft, highly elastic, relatively heavy ball - refered to as a power ball due to its very bouncy nature - was used to impact the GRP composite cricket bat in the same way as for the cricket ball impact tests.

Prior to the impact test, a simple bounce test was performed on the two balls to compare their relative elasticities. When a purely elastic body collides with a rigid, immoveable solid it should rebound at the same speed as it approached ie. the coefficient of restitution e ($= -v_{OUT}/v_{IN}$) equals unity. Table 4 shows the properties of the two balls and their e values when dropped on an immoveable steel block.

Table 4. Physical Properties of Two Test Balls.

Properties	Cricket Ball	Power Ball
Young's Modulus (N/m²)	90×10^6	0.85×10^6
Poisson's Ratio v	0.33	0.5
Mass (kg)	0.156	0.102
Diameter (m)	0.074	0.060
Coefficient of Restitution e	0.59	0.81

From the power ball impact test, the bat deflections were compared with the corresponding values predicted using the power ball properties input into the computer model.

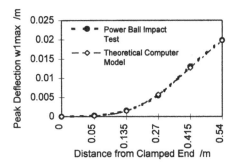

Figure 11. Power ball impact tests - peak impact point, deflection, GRP composite bat

Figure 11 shows how the peak impact point deflections compare in theory and in practice along the length of the blade of the GRP composite bat. Clearly the agreement between the experimental and theoretical results is far better than obtained for the cricket ball, the two curves lying almost on top of each other.

From this investigation into the effect of the inelastic behaviour of the ball, it is evident that the discrepancy between the theoretical model and the real life situation could be accounted for by taking this effect into account. To do this an appreciation of inelastic impact mechanics is required and is, therefore, a current area of investigation.

Applications of the Model

From the comparison with the modal analysis and the cricket ball impact tests, the theoretical model has been shown to be a useful tool in analysing the behaviour of cricket bat structures during impact with a cricket ball. The model can predict, with reasonable accuracy, the first vibrational mode of a bat and for the uniform blade structure the second mode is also identified. It is the first mode of vibration which is dominant in the evaluation of the bat response.

The results of the power ball impact test show that for the impact point deflection curve, the constant factor between the theory and experiment is likely to be caused by the inelastic behaviour of the ball. Because almost all the local impact point deflection takes place within the ball for almost all cricket bat materials, the effect of the ball's inelasticity will be almost the same for all potential cricket bat designs. This allows the theory to be used with confidence, in a qualitative manner to compare alternative bat designs.

The theory, as well as calculating the bat deflection with time, also evaluates the ball's displacement and can consequently be used to predict the output velocity of the ball for a given set of impact conditions. This is particularly useful when assessing the performance of a particular bat design. The following section uses the model to assess the relative merits of some different bat designs.

DESIGN ANALYSIS

In order to assess the contribution of various design parameters on the overall bat performance, the model of the GRP composite bat was used with changes in the following parameters:
- flexural rigidity   EI
- handle stiffness   T
- mass per unit length (of the blade) ρA

For the analysis of each of these factors the values of the other properties are those originally used, as given in Table 1. In each case an impact of 5 m/s, 6cm from the free end was used.

The flexural rigidity was varied over a range of EI = 1000-8000 Nm^2 in stages of 1000 Nm^2 (original value EI=4000 Nm^2) Figure 11 shows the effect of these changes on the coefficient of restitution e (= v_{OUT}/v_{IN}). The shape of this curve indicates that the stiffer the bat (the higher EI) the greater the rebound velocity of the ball. Large increases in the value of EI, however, only give a small increase in e.

A range of handle stiffnesses from T = 500-3000 Nm/rad in 500 Nm/rad steps were studied (original value T=1200 Nm/rad). Figure 12 illustrates how the coefficient of restitution is affected by these changes. As can be seen there is very little effect.

Figure 13 shows the effect of changing values of ρA from 0.8 kg/m to 3.0 kg/m in 0.2 kg/m steps (Original value ρA=2.24kg/m). Changing the value of ρA which, for a bat of a given length, is equivalent to increasing its mass, has a dramatic effect on the hitting power of the bat. A very lightweight bat gives a very low coefficient of restitution. As the mass of the bat is increased so does e. For heavyweight bats, increasing the mass still further has less of an effect than for lighter bats.

The obvious conclusion from a purely analytical viewpoint is that a very stiff, very heavy bat would be capable of hitting the ball the furthest. In practice the performance of a batsman also depends on his ability to swing the bat

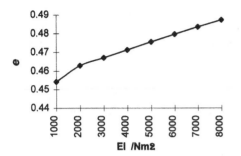

Figure 12. Effect of varying the flexural rigidity EI on the coefficient of restitution

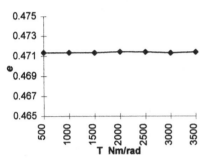

Figure 13. Effect of varying the torsional stiffness T of the handle on the coefficient of restitution e

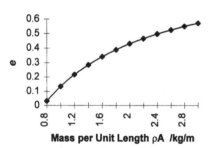

Figure 14. Effect of varying the mass per unit length ρA on the coefficient of restitution e

effectively. For this reason these results can be seen as the first step in the analysis and prediction of cricket bat performance.

CONCLUSIONS & FURTHER WORK

As identified in the discussion the torsion spring/free beam model does not take any variations in cross-section into account. Modification to the mode shape function (2) and the frequency equation can be made to accommodate this. It is also envisaged that the other vibrational modes of the bat will be investigated in order to assess their effect on the measured transverse deflections and the inclusion of the effect of an inelastic collisions should improve the overall accuracy of the model.

The hands of a batsman cannot provide a rigid clamp around the bat handle. An analysis of various boundary conditions and how closely they can simulate a hand-held bat will enable a better understanding of the factors affecting bat performance.

Notwithstanding these desirable modifications, it has been shown that the model, in its present form, provides a good basis for gaining a greater insight into the performance of cricket bats. Furthermore, this model has been validated for the dominant first mode of vibration using two independent experimental techniques. As a design tool, the model is already giving a good indication of the parameters controlling performance and thus will allow novel design solutions involving composite materials to be investigated with confidence.

ACKNOWLEDGEMENTS

The authors wish to acknowledge the financial support and encouragement of CadCam Technology Ltd. and the EPSRC (UK) for the award of a CASE studentship.

REFERENCES

[1]   Goldsmith, W. (1960). Impact, the Theory and Physical Behaviour of Colliding Solids, *Ed. Arnold Ltd*, London, pp. 108-111.
[2]   Gorman, D.J. (1975). Free Vibration Analysis of Beams and Shafts, *John Wiley & Sons*, pp 7-16.
[3]   Tognavellir, K & Dunbar, E. (1994). How Sweet   it is!! - Can your Basebll Bat Measure Up?, *Sound & Vibration*, January 94 pp 7-14.
[4]   Timoshenko, S.P. (1937). Vibration problems in Engineering, 2nd Edition, *Van Nostrand Company Inc.*

The Engineering of Sport, Haake (ed.) © 1996 Balkema, Rotterdam. ISBN 90 5410 822 3

Author index

Alaways, L.W. 289
Allyn, D. 43
Angue, J.C. 137
Aritan, S. 119, 125, 303
Ashby, M.F. 175
Ashcroft, M.W. 21

Barbier, F. 137
Barker, A.J. 187
Barker, M.B. 153
Bartlett, R.M. 303
Blackford, J.R. 161
Brody, H. 79
Brooks, R. 339
Burgess, S.C. 83

Chang, Y.-W. 141
Chisholm, S.J. 21
Conlan, T.M. 289
Cooke, A.J. 91
Crompton, R.H. 71

Dabnichki, P. 119, 125, 257, 303
Davidson, A.M. 153
DeVaney, T.T.J. 63, 131

Ekstrom, E.A. 315

Fairbairn, D.R. 51
Franks, I.M. 263
Friswell, M.I. 323
Fuller, R. 43

Gleeson, N. 37
Gobush, W. 193
Grant, C. 169, 245
Grove, S.M. 97
Gunther, M.M. 71

Habermann, W. 131
Hamblyn, S.M. 323

Hanna, R.K. 3
Hicks, N. 21
Hocknell, A. 333
Horwood, G. 323
Hose, D.R. 11
Hossein Alizadeh, M. 31
Hubbard, M. 195, 289
Huffman, R.K. 195

Immohr, J. 221
Iwnicki, S. 125

Johnson, S.H. 251
Jones, R. 333
Joseph, S.H. 205

Kakihara, T. 211
Kanehiro, H. 211
Katsanis, D. 97
Khan, M.A. 263
Kidd, M.D. 217
Knowles, S. 339
Kobenz, I. 131
Kollmitzer, J. 297
Kranzl, A. 297
Kruger, M. 43

Lammers, M. 43
Lauder, M.A. 119, 257, 303
Li, Y. 71
Lieberman, B.B. 251
Lo, K.-C. 141
Loch, N.E. 217

Madabhushi, S.P.G. 51
Mather, J.S.B. 221, 339
McGarry, T. 263
Miles, J.A. 289
Morgan, J.E. 103
Müller, W. 63, 131

Nixon, S.A. 169, 245

Pavier, M. 271
Penrose, J.M.T. 11
Preston, S.B. 281
Pudlo, P. 137

Rajalakshmi, M. 51
Rakowski, S. 37
Rees, D. 37
Regener, D. 153
Reuben, R.L. 217
Rothberg, S. 333

Samastur,M. 131
Satoh, K. 211
Smart, M.G. 323
Smith, R.A. 229
Standring, J. 31
Stewart, S. 205
Stoner, L.J. 43
Su, F.-C. 141
Subic, A.J. 281
Sugimoto, T. 111

Takeda, S. 211
Trowbridge, E.A. 11
Tsirakos, D. 119

Usmani, A.S. 51

Walker, C.A. 239
Wang, L.-H. 141
Wang, W. 71
Waśkiewicz, Z. 147
Wearing, J.L. 187
Wegst, U.G.K. 175
Wu, H.-W. 141

Zdravkovich, M.M. 21
Zwick, E.B. 297

Printed and bound by CPI Group (UK) Ltd, Croydon, CR0 4YY

23/10/2024

01777686-0009